International Political Economy Series

Series Editor

Timothy M. Shaw, University of Massachusetts Boston, Boston, USA;
Emeritus Professor, University of London, London, UK

The global political economy is in flux as a series of cumulative crises impacts its organization and governance. The IPE series has tracked its development in both analysis and structure over the last three decades. It has always had a concentration on the global South. Now the South increasingly challenges the North as the centre of development, also reflected in a growing number of submissions and publications on indebted Eurozone economies in Southern Europe. An indispensable resource for scholars and researchers, the series examines a variety of capitalisms and connections by focusing on emerging economies, companies and sectors, debates and policies. It informs diverse policy communities as the established trans-Atlantic North declines and 'the rest', especially the BRICS, rise.

NOW INDEXED ON SCOPUS!

More information about this series at
https://link.springer.com/bookseries/13996

Abel Polese
Editor

Informality, Labour Mobility and Precariousness

Supplementing the State for the Invisible and the Vulnerable

palgrave
macmillan

Editor
Abel Polese
Dublin City University
Dublin, Ireland

Tallinn University
Tallinn, Estonia

This book has been made possible thanks to funding from the European Commission [H2020-MSCA-RISE-2018-Shadow, grant no: 778118 and H2020-MSCA-RISE2019-New Markets, grant no: 824027]. The book was finalized during a research stay funded by the Japan Society for the Promotion of Science Visiting Fellowship Scheme (2020/2021).

ISSN 2662-2483 ISSN 2662-2491 (electronic)
International Political Economy Series
ISBN 978-3-030-82498-3 ISBN 978-3-030-82499-0 (eBook)
https://doi.org/10.1007/978-3-030-82499-0

Cover credit: Rob Friedman/iStockphoto.com

This Palgrave Macmillan imprint is published by the registered company Springer Nature Switzerland AG
The registered company address is: Gewerbestrasse 11, 6330 Cham, Switzerland

Preface: Life Lessons from the Informality Secession Project

The story behind any book is, in many respects, unique. But this book incarnates a sequence of behind-the-scene events that are worth filling a separate book by themselves. There is love, friendship, betrayal, academic politics, money. That would be enough to write a sort of academic thriller or even create a fictional character digging through academic politics to solve some kind of extremely intriguing academic mystery. That could be, for instance (and how exciting for non-academics), about how the third author of a paper managed to get upgrades to first author.

Some years ago, I joined an allegedly anonymous book project for the sake of working with people I liked and that were, until a certain point, my friends. I believe you should work only with people whose aura you enjoy and this project would give me the chance to meet more often with some friends, organise events together and eventually be an excuse to work with other like-minded people. However, being at different stages of our careers our priorities were not the same. Likewise, the reasons for engaging with this book project turned out to be radically different, a thing that made us part at some point. Yet, at least initially, my naive reasoning was that when you work with reasonable people, you can eventually come to an agreement on most things and keep on going. This assumption is, in theory, accurate. What I failed to take into account at that time was that life events may transform reasonable people into "unreasonable" ones for some periods and, no matter how short these spells are, these can actually compromise long consolidated relationships.

I can compromise on many things but quality is not one of these. I learned it the hard way: if you have ten good books and one easier to criticise, that last one will incidentally be reviewed more often than any other of your books. And these negative reviews of that very book will get on your nerves in a way that no positive review of your other books will manage to compensate.

A bad review stands like a coffee spot on a white shirt in your academic career. Besides, by contributing something not worth the paper is printed on, you undermine trust in science and your own trust in what you are doing. That's why, at this stage of my life, I am willing to work more than initially planned, I can go the extra mile and work shoulder to shoulder with authors to improve the paper, version after version. But I will not sign off anything that I believe is of poor quality.

So when you see a co-editor pushing for getting a book published with papers that are of dubious quality and the other co-editor makes it clear to be willing to sign off anything for the sake of devoting to the project as little work as possible, you understand that you have a problem. When all this matches betrayal on other issues that you discover being, day by day, deeper than you knew the day before, you see problems turning into troubles and years of collaborations crumbling into a painful academic divorce.

A real divorce is, in an average, much more enduring and painful. But untying a large web of collaborations that you had entrusted someone over the years is a real pain in the back and eventually may take much longer than a real divorce. One of these collaborations was this book and I would have gladly just walked off the project and leave everything to them. But many of the authors were my close collaborators with whom I had worked step by step on every new version of their manuscript. Besides, the series editor had entrusted the project to me (and the people I proposed as co-editors) so just leaving would have probably worsened my relationship with the series editor, someone I appreciated and respected a lot.

It took a lot of work and diplomacy (and nerves) to get the contract cancelled, to agree on an official version of the divorce that would be communicated to the authors, to then ask them whom they would like to continue working with. I renamed the folder in my computer "informality secession project" and started reworking the narrative of this book into something that I felt was more appropriate than its initial focus (then continued by my co-editors alone in their own project). I had never edited a book by myself but, given the advanced stage of the project, I decided not to take up any co-editors. The extra efforts required by working alone to the final phase of this book would eventually be compensated by the fact that no time would be spent on identifying someone, pulling them into the project, agree on details and eventually discuss issues that would inevitably be emerging. The preparation process was delayed several times and I saw my (ex) co-authors' book coming out many months before I was able to submit the final version of this manuscript. But Rome was not built in a day and I believe these papers altogether have something to say to informality debates. This book is of course just a drop in the ocean. There are possibly many things that could have been done "better" with a bit more time or energy at my disposal. For one thing, I had to reject a few chapters with a very interesting focus for lack of time and energy from my side. I believe anything can be made into a publication and it is just a matter of time. Some papers need a week's revision, some others a month and some others several months. The rejected authors (that I hope to include in a future publication) would just need too much extra time to put their chapter in shape and, given the initial delay and the fact that other authors had been waiting for too long, we simply could not afford that.

Looking back, I see with nostalgia how this project accompanied me through several stages of my life. The idea was initially discussed during long phone conversations about life and other things with one of the co-editors, who was very helpful and supportive in that particular period of my life. It was then developed through a very successful collaboration leading to the organisation of two conferences to which excellent speakers were invited. I had the chance to meet new colleagues, make new friends and visit my childhood dream city Samarkand thanks to the other co-editor. These collaborations were crucial to a fellowship that I was awarded and thanks to which I could spend two idyllic months in one of my favourite research institutes (and cities) in the world.

The secession phase was hard to digest at some moments but also pushed me to take up other ways of working, rely on new people and eventually consolidate new collaborations that are, as of today, still ongoing. There are several people that I need to thank for having contributed, in a way or another, to this project. In primis, I need to thank my ex co-editors since, without their input, I would have never started this book or gone through any of the amazing experiences I went through during our collaborations. I also want to thank the authors who patiently reworked their chapters multiple times and put up with my pedantry in preparing a text.

Support by the Palgrave editorial team, Anca Pusca and Arun Kumar Anbalagan, and in particular, encouragement from the series editor Tim Shaw was crucial. Never has in my life a simple dinner at a Japanese chain restaurant (in Boston) led to such an extensive collaboration across books (this is the third one with his series) and world regions. I also wish to thank the organisations and people who agreed to host me when I was working on the various phases of this manuscript: KIMEP University in Almaty, SIAR Research and Consulting in Bishkek, Kyiv International Institute of Sociology in Kyiv, Aleksanteri Institute in Helsinki, Kyungpook National University in Daegu, Moldova State University in Chisinau and Ritsumeikan University in Kyoto. This book was made possible also thanks to two generous grants from the European Commission H2020/MSCA RISE Scheme (grant No: 824027) and H2020-MSCA-RISE-2019-New Markets (grant no: 824027).

At various times I benefited from the support of colleagues and friends that I thank here in random order: Brian Harney, Rustamjon Urinboyev, Rano Turaeva, Nodira Kholmatova, Gian Marco Moisé, Nozima Akhrarkhodjaeva, Valentina Bodrug, Sayaka Ogawa, Shioya Akifumi, Zhamilya Utarbayeva, Olha Lysa, Donnacha Ó Beacháin, Aliya Tankibayeva, Anna-Liisa Heusala, Anna Korhonen, Akmaljon Abdullayev, Adrien Fauve, Mai Ngoc, Ainoura Sagynbaeva, Catherine Poujol, Jeremy Morris and may I be forgiven by the ones I forgot here.

Once again, this project confirms what I suggested some years ago in the foreword of my "the Scopus Diaries and the (il)logics of academic survival". No matter how hard you were hit, if you survive and manage to "dry out" from emotions in a situation, you can just take it as a package. Some memories will be sweeter than others but it was all worth it and, in spite of all hardship, it shaped you into what you are now.

Kyoto, Japan
August 2021

Abel Polese

Praise for *Informality, Labour Mobility and Precariousness*

"This is a timely and impressive intervention in the field of informality, state-society relations and economic precarity. Too often in social research the significant marginalised populations in societies rich and poor are overlooked. The chapters in the volume offer a unique and nuanced analysis of the livelihood strategies and 'nomadic' tendencies of those rendered invisible by economic transformations—particularly in Eurasia—where the majority of the case studies are situated. As usual, Abel Polese, together with contributors, offers enormous bang for buck in his editing—an in-depth original conceptual approach drawing on political science and sociology, and careful selection of cases, including overlooked places and topics such as migrants in Spain and the aftermath of Lebanon's sovereign debt crisis."

—Jeremy Morris, *Aarhus University*

"A must read for global citizens facing informality and conflicting moral orders. A journey towards setting up the boundary between corruption and solidarity. Learn from this book and act for a better world."

—Ibrahim Sidi Zakari, *Université Abdou Moumouni, Global Young Academy's member*

"Based on a range of solid case studies from different regions of the world, and enriched by the long-term commitment of each author to their fieldwork, this book has a lot to tell about the boundary between informality, solidarity and their embeddedness in localised cultural practices.

Its value is the capacity to go beyond traditional approaches to informality and governance to look for informality in places where it has not been searched before, from music production to the everyday construction of social relations. Because of this, it is a valuable piece of work for development workers, researchers and decision makers at all levels."

—Zeenath Solih, *The Maldives National University*

"This book offers a solid alternative approach, and excellent inspiration, to better understand precariousness and the main challenges faced by the weaker and the marginalised. Taken together, the case studies presented in the volume provide a timely contribution to some of the most urgent social theory debates. I expect this book to become a useful resource for anyone interested in how to address, both theoretically and in practice, issues related to migration, informal employment and development in from a critical perspective."

—Ahmad Harakan, *Universitas Muhammadiyah Makassar, Indonesia/Editor in Chief of* Journal of Contemporary Governance and Public Policy

"Labour Mobility and Precariousness is a fine collection of well-crafted chapters that offer original insights into the different meanings of informality in countries as diverse as Romania, Spain, Croatia, Russia, Armenia, Georgia, Azerbaijan, Uzbekistan, Kyrgyzstan, Moldova and Lebanon. The world needs a better understanding of informality, now more than ever, when even in long-standing democracies states rapidly lose credibility and trust of their own citizens and a growing number of people start to engage in informal practices in order to survive in the absence of essential public services or to act against policies perceived as harmful. This volume provides an important step into the next stage of thinking about informality. Abel Polese has successfully assembled this great collection of chapters that explore the intersection of informality and the state from a primarily ethnographic perspective. Where are the boundaries between the criminal and the informal? What level of informal activity is tolerated by authorities? When does informality become socially acceptable or even a community or society-wide norm? While addressing these questions, the authors discuss a wide range of informal activities including communication, economic, and labour practices in numerous sectors from kiosks to restaurants to hospitals. This collection provides a unique insight into the life of vulnerable and marginal actors, such as undocumented migrants, ethnic entrepreneurs or homeless people, who use informality to survive or integrate into a new and culturally different context. It also offers readers an exciting journey towards the world of informality based rich qualitative and mixed-method data."

—David Jancsics, *San Diego State University*

"Featuring studies on a broad variety of forms of informality, this long-awaited volume provides a much-needed reflection on informality by expanding its scope from economic to social and societal dynamics. Featuring contributions from Asia, Europe and the Middle East, the volume offers relevant food for thought to scholars interested in governance issues, no matter what their discipline."

—Marcello Mollica, *Associate Professor of Cultural Anthropology and Ethnology, University of Messina*

"There is a theoretical way things should be working and how things are working in practice. We pretend to live our lives as if everything is well defined and determined by sets of norms and rules to be observed by individuals, states and their citizens. However, alongside this reality lies another functioning reality, that of informality. Ubiquitous, operating in the shadows, informality is an essential component of everyday governance affecting the economy, the political and state-citizen relationships. This important volume sets out its theoretical and practical contours by focusing on cases ranging all the way from Spain to the east of Russia, from the north of Europe to the Mediterranean."

—Erhan Dogan, *Marmara University*

"This book offers a timely snapshot and extensive empirical insight into the multifaceted world of labour mobility and precariousness. The ongoing Covid-19 pandemic has, once again, exposed the limits and contradictions of privatisation, giving us numerous examples of how informal networks and practices can significantly mitigate the shortcomings of a weakened public sector, revamping debates whither the state in the coming decades. Labour mobility and precariousness sheds light on how alternative social and economic structures can help us rethink the meaning of solidarity and morality, something that will be crucial to answer some pressing questions that will be asked to our social systems in the years to come."

—Filippo Menga, *Associate Professor of Geography, University of Bergamo*

Contents

Staying

Competing

Notes on Contributors

Laura Adams is Freedom House's Senior Director for Strategy, Development and Learning. In this role, she coordinates organisational strategy and learning, and oversees a portfolio of evidence-based programmes that promote and protect human rights and fundamental freedom globally. Previously, Adams was a Democracy Fellow in the Learning Division of the Center of Excellence in Democracy, Human Rights, and Governance (DRG Center) at the U.S. Agency for International Development. Prior to that, Adams was in academia for fifteen years, most recently at Harvard University as Director of the Programme on Central Asia and the Caucasus, Academic Advisor to the MA programme in Russian, East European and Eurasian area studies, and a Lecturer in Sociology. Her areas of expertise include qualitative research and evaluation methods, authoritarianism, political transitions, civil society and social movements. She has published numerous articles on culture and politics in Central Asia, as well as an award-winning book, *The Spectacular State: Culture and National Identity in Uzbekistan*.

Mirela Alpeza is an Associate Professor at UNESCO Chair in Entrepreneurship at J.J. Strossmayer University of Osijek, Faculty of Economics in Osijek, Croatia. She received her M.Sc. in Entrepreneurship in 2002 and her Ph.D. in Entrepreneurship in 2009 at J.J. Strossmayer University of Osijek. From 2011 Mirela is director of CEPOR—SMEs & Entrepreneurship Policy Centre in Zagreb where she is involved in policy research and advocacy activities, project development and implementation, as well as management of wide network of CEPOR's external associates and partnering organisations. She is a member of the European Council for Small Business and Entrepreneurship, where she held the position of Country Vice-President for Croatia from 2013 to 2017; and TRANSEO–European Association for SME Transfer where she led international experts' working group Business Transfer Ecosystem in 2019–2020.

Gulzat Baialieva is a Ph.D. candidate at the Social Anthropology Department, University of Tuebingen, Germany. She holds an M.A. degree in Political Science from Central European University (Hungary) and a graduate diploma in European Civilizations from Bishkek Humanities University (Kyrgyzstan). Currently, she is teaching at the Social Anthropology Department at the University of Tuebingen. Her research interests include environmental anthropology, post-socialist transformations, digital ethnography and populism.

Karolina Bielenin-Lenczowska is a social anthropologist and linguist, Assistant Professor at the Institute of Ethnology and Cultural Anthropology, University of Warsaw. Currently (2019–2021) she is a visiting professor at the Department of Linguistics at Federal University of Santa Catarina, Florianopolis, Brazil. Her research interests include fieldwork methods, ethnographic and sociolinguistic aspects of migration and diaspora.

In 2008–2014, she conducted ethnographic research in the Republic of Macedonia and Italy, focusing on everyday transnational practices, multi-ethnic neighbour relations, Islam and gender. She is currently working on the cultural heritage of Brazilians of Polish origin living in southern Brazil, focusing on landscape, language ideologies and food practices.

Elena Druică is a Professor of Statistics for Business, Applied Econometrics, and Data Analysis with the University of Bucharest. She was formally trained in Mathematics, completing her Ph.D. in 2003. During her Ph.D., she studied genetic algorithms and optimisation processes, developing a new classification of objective functions that converge to the optimum in polynomial time. During her second doctoral work, she studied Economics with a particular focus on risk assessment. She was leading, or has been a member of 16 research, as well as consultancy-based projects, national and international. Her publications cover a broad range of statistical and econometric techniques, both in the area of unsupervised learning (like static and temporal clustering, fuzzy clustering, or regression and classification trees), as well as supervised learning techniques. She is proficient in various regression techniques, from simple linear methods to advanced generalised methods. She is also able to work with panel data and apply spatial modeling like geographically weighted regression, gravity models, or spatial models.

Anil Duman is currently an Associate Professor at Central European University, Budapest. She has received her M.A. and Ph.D. in Economics from University of Massachusetts, Amherst. Her broad research interests include political economy, industrial relations, welfare state policies, and redistribution. In her recent research, she has been specialising in the interrelations between labor market status and socio-economic inequalities. Her previous research focuses on analysis of skill formation, skill distribution and their relation to individual policy preferences across countries and over time. She is also involved in research projects examining the transformation of social protection regimes in several transition countries.

Ignacio Fradejas-García is a post-doctoral researcher at the University of Iceland (School of Social Sciences). He holds a Ph.D. from the Autonomous University of Barcelona researching the interplay between mobility and informality. With extensive fieldwork experience in The Gambia, Chile, Morocco, Haiti, RD Congo, Turkey, Romania and Spain, he does research in migration, transnationalism, (im)mobilities, humanitarianism and informality. He is the co-chair of the SIEF working group on Migration and Mobilities, and he has published peer-reviewed articles in *Mobilities*, *Social Anthropology* and *Migration Letters*.

Joseph P. Helou is an Adjunct Assistant Professor of political science and international affairs in the department of social sciences at the Lebanese American University in Beirut. He holds a Ph.D. in Politics from the University of Exeter, an M.A. in International Relations from the University of Toronto and a B.A. in Economics from the American University of Beirut. His research focuses on questions of international political economy, conflict and security, statehood (sovereignty, hybridity, areas of limited statehood and enforcement) and group politics. His current research agenda includes the exploration of informality during sovereign debt crises and other severe monetary and fiscal dislocations which adopts an inter-disciplinary approach that cuts through conceptual boundaries to illuminate the multifaceted dimension of informality.

Rodica Ianole-Călin is an Associate Professor at the Faculty of Business and Administration, University of Bucharest, Romania. She teaches Behavioral and Experimental Economics, Data Analysis, Microeconomics and Game Theory, while her research is focused on applied work in the areas of consumer behavior, financial decision-making, informality and health economics.

Aksana Ismailbekova completed her dissertation at the Max Planck Institute for Social Anthropology, Halle, Germany. Based on her Ph.D. dissertation, she wrote her monograph *Blood Ties and the Native Son: Poetics of Patronage in Kyrgyzstan*, which was published by Indiana University Press in 2017. Ismailbekova is a research fellow at Leibniz-Zentrum-Moderner Orient (ZMO). Her research work focuses on kinship, ethnicity, patronage, conflict and gender in Kyrgyzstan. She has been a collaborator of the Basel Institute on Governance on two research projects: "Informal Governance and Corruption" funded by the British Academy Anti-Corruption Evidence Programme (2016–2018) and "Harnessing Informality" (2018–2020) funded by the Global Integrity Anti-Corruption Evidence Programme.

Daniel Kashnitsky is a Junior Researcher at the Center for Qualitative Studies, Higher School of Economics in Moscow, Russia. His research interests are focused in the area of migration and health, social exclusion, racism and social determinants of health in the post-Soviet region. Daniel is a co-founder of the Regional Expert Group on Migration and Health in East Europe and Central Asia where he strives to use research data for stronger advocacy of the right to health of people on the move across the larger European region.

Professor Tanel Kerikmäe is Jean Monnet Professor in European Law and director of Tallinn Law School at Tallinn University of Technology; a board member of several high-ranked law journals and an author of more than 150 articles and publications, Tanel is active as an expert for public and private institutions, international organisations and has been the key expert of EU in Central Asia. He has been contributing to the research of European law and policies and created a research group of law & tech that received significant funding from EU Commission (H2020,

Jean Monnet Centre of Excellence), NATO, Folke Bernadotte Academy, USAID, Estonian Research Council, etc. He serves as a President of the Estonian branch of European Community Studies Association and has been teaching and supervising at several universities in and outside Europe, including the Baltic States. Currently, prof. Kerikmäe is responsible for several international projects and related to international research cooperation networks. Moreover, his expertise also includes the topics on EU legal policies, digital single market, automatisation, legal impediments to new technologies, and augmented and artificial intelligence in legal practice.

Miranda Jessica Lubbers is Associate Professor at the Department of Social and Cultural Anthropology of the Autonomous University of Barcelona (UAB), Spain. She is the Director of the research group GRAFO. Her current research projects address migration and transnationalism, poverty and livelihood strategies, social exclusion and social cohesion.

José Luis Molina is Professor at the Department of Social and Cultural Anthropology of the Universitat Autònoma de Barcelona (UAB), Spain. He is an Economic Anthropologist interested in the emergence of socio-economic structures, like migrant enclaves and transnational fields. Prof. Molina is interested in mixed-method approaches, combining ethnography and personal network analysis. He is specialised in Southeast Europe and Romania and is currently the president of the Research Ethics Committee of the UAB and co-editor of the REDES-Revista Hispana para el Análisis de Redes Sociales.

Petru Negură is a Humboldt fellow at Leibniz-Institut für Ost- und Südosteuropaforschung in Regensburg, Germany. Negura held his Ph.D. in sociology at the Ecole des Hautes Etudes en Sciences Sociales (EHESS) in Paris. He is an Associate Professor at Free University of Moldova and a researcher at the Institute of Legal, Political and Sociological Research in Chisinau. His academic interests deal with the sociology and history of intellectuals, education and social welfare in Eastern Europe and the ex-USSR. Negura's personal webpage: https://petrunegura.academia.edu/.

Johanna Paquin is a research and teaching fellow at the Chair for Political Economy and Economic Sociology at the Goethe University in Frankfurt, Germany. Her research interests are in the study of informality and institutions, and their effects on entrepreneurship, business practices, as well as economic development and transformation. Her doctoral research is concerned with the interplay between formal and informal institutions in the business environment of SMEs in Georgia. She received her B.A. in Political Science and Economics at the Technical University Darmstadt, Germany, as well as an M.A. in Development Studies at the University of Sussex, United Kingdom.

Helena Patzer is an Assistant Professor at the Institute of Ethnology and Cultural Anthropology, University of Warsaw. Her research interests include transnational migration, long-distance care, the critical anthropology of development, and urban studies. She has studied Filipino middle-class professionals in the United States and their involvement in development projects in the home country (PhD thesis: *Long-Distance Care. The Practice of Sustaining Transnational Ties by Filipino Immigrants in Boston*), as well as migrant communities in Poland, Denmark and Czech Republic. She was a recipient of a Kościuszko scholarship at Harvard University, she held a post-doctoral position at the Czech Academy of Sciences in Prague and was an Erasmus-Mundus post-doctoral fellow at Ateneo de Manila University. She is a co-editor of *Pretextual Ethnographies. Challenging the Phenomenological Level of Anthropological Knowledge-Making* (2018). She also directed "Money Tree", an ethnographic film based on her research on Filipino migrant families.

Abel Polese is a Researcher, Trainer, Writer, Manager and Fundraiser. His interest in theory and practice of development led him to work across disciplines, conduct research and design interventions in Africa, Asia, Europe, the Middle East and the Americas. An alumnus of the Global Young Academy, he is interested in mental health and support for early career researchers, scholars at risk and researchers from the Global South. He is the Author of *The SCOPUS Diaries and the (il)logics of Academic Survival : A Short Guide to Design Your Own Strategy and Survive Bibliometrics, Conferences, and Unreal Expectations in Academia*, a reflection on academic life, research careers and the choices and obstacles young scholars face at the beginning of their career. You can find him on Twitter at @Abiquitous and @scopusdiaries.

Ružica Šimić Banović is an Associate Professor in the Department of Economics, Faculty of Law, University of Zagreb. She holds a Ph.D.

degree from the School of Economics and Business, University of Ljubljana. As a Chevening scholar she received her M.A. in Public Policy from King's College London, University of London. She completed a fellowship in the Department of Economics at The George Washington University as a part of the Junior Faculty Development Programme funded by the U.S. Department of State. Before joining the academic community, Ružica was an adviser in the Croatian National Competitiveness Council and had worked in the private sector as well. Her research interests include institutional interaction in the postsocialist business environment (with an emphasis on the reform implementation and "economy of favours"), immigrant entrepreneurship and recent migration trends in Eastern Europe.

Vlatka Škokić is an Associate Professor of Entrepreneurship at University of Split, Faculty of Economics, Business and Tourism, Croatia. Prior joining the University of Split, she spent six years as an Assistant Professor (Lecturer) at the University of Surrey, School of Hospitality and Tourism Management, United Kingdom. Vlatka holds a Ph.D. from the University of Strathclyde, Scotland. Her research is focused on entrepreneurship, innovation and networks. Vlatka's research is published in the leading management and social science journals including *Annals of Tourism Research* and *European Management Journal*. Vlatka is experienced at teaching a range of subjects including entrepreneurship, family business and research methods (qualitative) at post-graduate and doctoral level. She is a Fellow of the Higher Education Academy (HEA).

Aneta Strzemżalska is an Associate Researcher at the Department of Anthropology of the European University at St. Petersburg, Russia. She has conducted research in republics of the South Caucasus since 2012. Her research interests include nationalism, ethnicity, political anthropology and informal practices. She is currently working on musical nationalism in Azerbaijan.

Mans Svensson is Associate Professor in Sociology of Law. He Svensson defended his thesis "Social Norms and Legal Compliance" in 2008 dealing with law's influence and dependency on underlying informal social control mechanisms. He is currently leading projects on the role of Social Norms in Fighting Corruption in Local Governments (Swedish International Centre for Local Democracy) and the Multilevel Orders of Corruption: Insights from a Post-Soviet Context (Swedish Research

Council). He successfully supervised seven Ph.D. students (two as main supervisor and five as secondary supervisor).

Rustamjon Urinboyev (Coordinator) is an Associate Professor at the Department of Sociology of Law. Rustam received his Ph.D. in sociology of law for a thesis entitled: "Living Law and Political Stability in Post-Soviet Central Asia" (2013). His thesis explored the interconnections between informal economy, community-based traditional governance institutions and political stability, and has been published as a monograph with Media-Tryck/Lund University (2013). Rustam has been a Visiting Scholar at the University of Cambridge, University of Helsinki, University of Copenhagen, Tallinn University of Technology and the University of World Economy and Diplomacy, Uzbekistan.

Aimar Ventsel is an Associate Professor at the Department of Ethnology of the University of Tartu. He received his Ph.D. at the Max Planck Institute for Social Anthropology and Martin Luther University (Halle). He has conducted research in Russian Far East since 2000 and has focused on informal practices, kinship, ethnicity and political and economic anthropology. He is currently working on subnational statehood of Russian ethnic regions and studies their economy.

Acronyms

CEDRE	Conférence économique pour le développement, par les réformes et avec les entreprises
CEE	Central and Eastern Europe
DJ	Disc Jokey
EBRD	European Bank for Reconstruction and Development
EU	European Union
GDP	Gross Domestic Production
HIV	Human Immunodeficiency Virus
ILO	International Labour Organisation
MENA	Middle East and North Africa
NGO	Non-Governmental Organisation
OECD	Organisation for Economic Cooperation and Development
SME	Small and Medium Enterprises
TSF	Transnational Social Field
USD	United States Dollars
VAT	Value Added Tax

List of Figures

Political vs. Everyday Forms of Governance in Uzbekistan: The Illegal, Immoral, and Illegitimate

Mixed Perceptions of State Responsibility Among Informal Sector Participants in MENA

State Collusion or Erosion During a Sovereign Debt Crisis: Market Dynamics Spawn Informal Practices in Lebanon

Perceived Pull and Push Factors of Healthcare Professionals Intention for Mobility: The Case of Romania

List of Tables

Mixed Perceptions of State Responsibility Among Informal Sector Participants in MENA

State Collusion or Erosion During a Sovereign Debt Crisis: Market Dynamics Spawn Informal Practices in Lebanon

Perceived Pull and Push Factors of Healthcare Professionals Intention for Mobility: The Case of Romania

Work, Subsistence and Distress of the Homeless in Moldova

Introduction

The (Im)Moralities of Informality: States, Their Citizens and Conflicting Moral Orders

Abel Polese

When a childhood friend came to visit me in Kyiv, and joined our weekly football gig, he was lucky enough to be hit in his eye at the very beginning of the game. My (former) father-in-law took him to the closest hospital where he was attended by the local staff and requested to come back and check his eye in the next days. When leaving the department, he put some money in the pocket of a nurse saying, "that's for some tea". The nurse did not check the pocket and tried to conceal her pleasant surprise by timidly saying "thank you" as we left.

To most, if not all, people active in the anti-corruption industry, this transaction that I witnessed has all the features of a bribe. The nurse could (or should) have refused the money since she had already a salary paid from the state and was not going to declare that extra income. But such a gesture has a different meaning in a hospital in London and Kyiv. In the latter context, doctors would work long hours for a salary that is a fraction

A. Polese (✉)
School of Law and Governance, Dublin City University, Dublin, Ireland
e-mail: abel.polese@dcu.ie

Institute of Social Studies, Tallinn University, Tallinn, Estonia

A. Polese (ed.), *Informality, Labour Mobility and Precariousness*, International Political Economy Series,
https://doi.org/10.1007/978-3-030-82499-0_1

of what they need to get to the end of the month. This is why I provocatively suggested that hospitals in Ukraine were de facto private (Polese 2006a, b). Other researchers have regarded such payments as recognition of personhood (Patico 2002) or basis for a mutual dependency relationship (Rivkin-Fish 2005). Trying to set the boundary between corruption and solidarity in a context where the state is absent, I was once told "if I receive it, it is a gift but if I demand it, then it is a bribe" (Polese 2009a). This statement could draw a line between the two but, yet, some argue that payments are so embedded in people's minds in a certain environment that patients feel some subliminal pressure to pay the doctors. The overall understanding of these payments is that patients could then expect to be served better or simply could think to be doing what it is expected from them (Gaal et al. 2006) would suggest that even "spontaneous payments" (when the doctor does nothing to solicit such payments) are not that spontaneous after all. At this point, one could say that whatever transaction does not pass through the hands of the state undermines credibility in institutions and eventually weaken a society.

The causality relationship is key here. Do such payments really weaken and undermine the state, or they are the result of a state that is weak and not trusted by its citizens? This question is on the basis of the book you have in your hands (or, nowadays, more likely the PDF you're glancing at on your screen). Practitioners may suggest either that people need to start respecting state rules for things to go in the right direction, or that state should work towards enhancing trust by its people so that they feel more prone to respect state rules. Both strategies are likely to work but it is rather a combination of the two that will save us. However, there seems to be a tendency to push citizens towards legality without really asking to what extent their expectations from their state are fulfilled. People complain. This is what has pushed the human race so far as the moon, looking for ways to solve issues in the technological world we know today. So a few people complaining that the state is not giving them as much as they want is more or less in the norm, that's what politics is about. Do something and some people will complain, do the opposite and other people will complain. But when a growing number of people start not only to complain but to act against the state, we are probably facing a problem of governance, not a standard deviation from societal norms that is present, to a limited but different extents, virtually everywhere.

Thanks to the framework that James Scott developed throughout decades of research, we know now that apparently insignificant actions, repeated thousands or millions of times by people unaware of one another, are a way to contribute to the construction of the political (Scott 2019). Eventually, a state is there to create standards, to harmonise, to propose attitudes and behaviours that can mediate individualism and egoism to find a way to make a society viable through common rules and standards that everyone is deemed to respect. However, not all attempts to do are successful (Scott 1998). The art of changing people's behaviour is a complex one. It can certainly be a matter of how many resources are invested into a given deed. After all, if I want everyone in my country to speak, say, Aramaic from today, I could change all the schoolteachers (assuming I can find some nowadays), force adults to enrol in language courses and adopt a law that anyone not speaking the new language could be executed on the spot. But in a modern and (allegedly) democratic state, there is a margin of agency left to the citizens. Besides, economic viability of a decision rests on the capacity to implement it with a limited number of resources thus leaving funds also for other interventions.

So, back to our nurse, the state has at least two ways to avoid such situations. One is to install cameras everywhere and strictly monitor hospital's staff behaviour. This may somehow curb some behaviours but not all of them. After all, no black market has ever been tamed from the supply side (Naylor 2004), meaning that it's probably the demand side that one needs to address. When Ukraine adopted a law forbidding policemen to have any money in their pockets (to ensure they would not pocket in bribes), policemen demanding bribes would not accept payments hand-to-hand. But they would gently ask people to put the agreed money into the side compartment of their cars. Technically, they were not handling cash and no bills could be found in their pockets. They were abiding by the law as diligent officers.

The other way is to get the state working properly or at least to have its people perceive it as a state. A state is not a state per se, it is a state if it also acts as a state (Polese 2009b). During the COVID-19 lockdown, several countries requested their restaurants to close down. Societal and cultural factors can certainly have influenced the way restaurant owners reacted and explain different behaviours across countries and regions. But inasmuch as you can complain or protest that your business is dying, there is a difference between a government that puts in place measures (subsidies to restaurants) and one that simply orders everything to shut down with no compensation foreseen. In the case of a total shutdown with no subsidies, some new forms of entrepreneurship emerged, some small businesses even managed to adapt, change their technology and eventually grow. But the majority of them suffered in silence, many had to close down. Newspapers sometimes reported some cases of SME managing to go online and boost their incomes. But little is known about restaurants that managed to stay open and closed at the same time. This was coming from a combination of flexible compliance. For instance, when in some countries restaurants with terraces were allowed to receive customers, many places with large windows just got them removed and pretended they had a terrace. They were not alone in their creative impulse and many more examples can be provided. These, and other forms of what Phelps calls inventiveness of informality (Phelps 2021) were coming as a way to supplement or replace the state that, in their views, was not sufficiently present. There is little doubt that, in the medium or long run, business will adapt, and new forms of entrepreneurship will emerge. But when a crisis happens only a few are able to respond properly. Business will close, entrepreneurship will suffocate, and damage is unevenly distributed. For instance, the hospitality and culture industries have suffered while much of the online business has significantly expanded. If the state claims to be there (at least nominally) but then people perceive it as absent or negligent, they will rely on whatever alternative structures are available at the time to make a living and, with the state formally occupying a position in the economic life of the country, most of these new structures will be (guess what) informal. This discourse can be visualised through the Fig. 1. Some cultural habits such as staying with your family for longer years (compared to cultures where at 18 you just go, find a job and live alone) or living with your elders (opposed to some cultures where your elders, health allowing, go and travel alone or mingle with their friends) had mostly developed as social protection mechanisms to replace market elasticity, unemployment subsidies or pensions that were deemed not sufficient for a living in some cases (Polese 2021).

There are two ways to look at the informal welfare gap. One is to say that "yes, this is the reality, but we are working to fill the gap and formalise governance as much as it is needed". But how much a transition (towards formalisation) should last to be considered "transitional" only? And how are people supposed to entertain themselves, make a living or take care of their health while their state is figuring out the best way to deliver such policies? The second way is to look at this situation as the current state of things that will not change until a suitable and viable alternative is identified, policies are conceived, tailored to the given

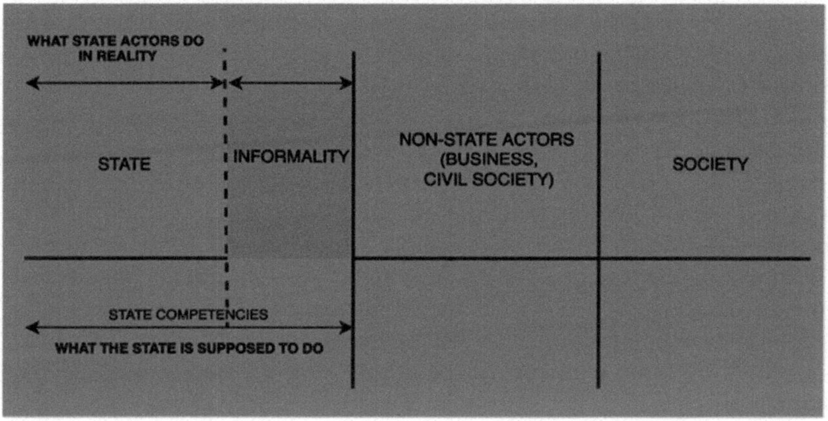

Fig. 1 Figure on informal welfare (*Source* Polese 2021)

context and then the institutions engage with behavioural change (since behaviours consolidated throughout the years need time and effort to be reversed). In such a context, informality comes to be part of what can be called informal governance. There is a theoretical way things should be working and how things are working in practice. This means that rules are there, but no one enforces them, that non-compliance is not punished, or that only a part of the citizens obeys the principles that the state is dictating in a given sphere of public governance. This may be because a policy is perceived as harmful or simply citizens do not see an advantage in respecting the law. Non-compliance is common to most, not to say all, systems. Even within a small family, there will be one member that is less prone to respect family traditions, follow hierarchies and do what they are expected to do, let alone in a complex multilayered society composed by millions of individuals. Now, if this non-compliance remains at marginal levels, it is usually treated as deviant behaviour and corrected or tolerated depending on its societal impact (there is a different attitude towards autistic people, serial killers or sociopaths, since these deviant behaviours are considered harmful and acceptable to different extents). But when the number of people engaging in this form of non-compliance with socially acceptable and political acceptable behaviour increases, we get to the question "how much is too much". Beyond that level, we do no longer have deviance but a societal tendency. There is no answer and the threshold between the acceptable and the problematic is arbitrary, or even subjective. But when corruption is said to be at 50%, that is one individual out of two paid a bribe in the past year to get to see a doctor or receive a document, this is no longer a deviation but a tendency, a norm or even an informal rule. It should not be addressed by punishing the perpetrators since it's too costly and probably dispersive. Efforts should be made to grasp what provokes this tendency and address it at the cultural, societal and/or policy level or the fight against windmills will continue forever. The figure below offers an overview of the possible modes of intervention for, or against, a newly formed informal practice. Until a practice becomes socially relevant, there will be no need to address it. Once it is performed by some segments of the population, the question of whether an intervention (and what kind of intervention) is needed will arise. That's still not the end of the story because intervention by the state, its institutions or any overarching entity can be aimed at institutionalising the practice, if they see social advantages in its use and understand how to do that or liquidate it. But that's not the end of the story since liquidation, once the state has decided to proceed that way, is not assured. Failed repression could thus lead to the practice staying or even citizens fighting back to keep it, which is why it was suggested that insurgency is a (rare) case of successful development of informal practices by channelling them into parallel institutions that take over from the state in some areas of governance (Polese and Kevlihan 2013) (Fig. 2).

This is the starting point of our book, which is an attempt to look at informality as a means to understand societal tendencies, gather feedback and better grasp societal needs not only through direct feedback (elections, referendum) but also indirect ones. In theory, direct feedback should be sufficient but there are several reasons why in practice it is not so. First, in politically repressive countries, political actions might be perceived as going against the state or the ruling elites, thus triggering a response that goes against the citizens. By force of this, citizens are more likely to abstain from such direct feedback. Second, people are not always aware of what they need and it is up to experts or overarching entities to conceive measures. This is the classical account of "I am hungry, give me a fish" against "he is hungry, let's teach him how to fish". Finally, not all actors have the energy and time to engage in feedback exercises and not all systems allow for proper feedback to be delivered on time and identifying clearly the issue. Not only looking at informality we can identify tendencies that are somehow invisible, untold and can be interpreted for

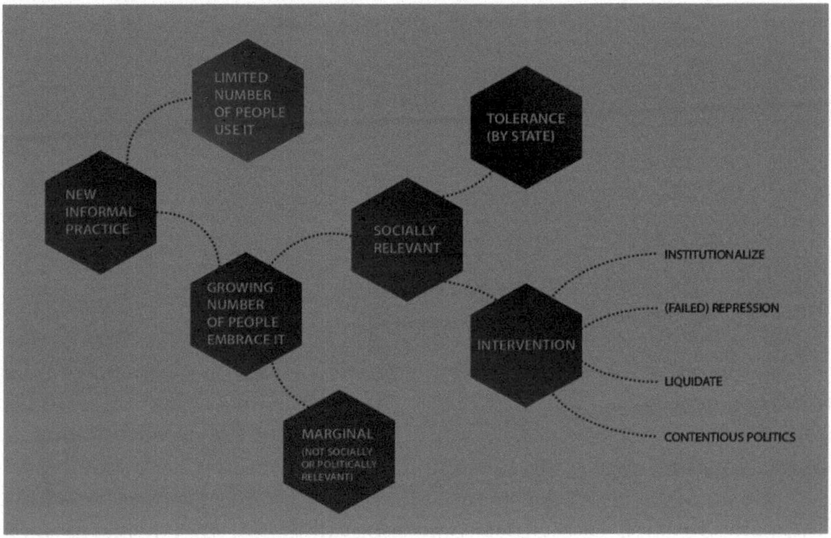

Fig. 2 The genesis of informality

a better understanding of the society. We can also gather new ideas about how to address emerging challenges.

Informality has been used to refer to phenomena from petty corruption to massive fiscal fraud and could potentially be stretched to include criminal activities (Polese et al. 2019b) so a question arises here. Where is the boundary between what can be tolerated, what should be corrected and what can be considered a policy response and what is merely a money-generating activity that falls into the criminal sphere? To answer this question, it becomes useful to turn to the distinction between the licit and the legal conceptualised by Van Schendel and Abraham (2005). In their matrix, one can clearly see the tension between what a state reputes illegal and what people may consider acceptable. This is of utmost importance when seeking to understand, or change, a given behaviour. When fighting an attitude that is despised by the society, coercion may suffice to draw the line. People socially accept the harmfulness and illicit nature of a given practice and most of the society will find it easier to accept that this is something they should not engage in, if they want to comply with societal values and if they do not want to risk being caught by state authorities (Table 1).

Table 1 The relationship between legality and licitness

	Legal	*Illegal*
Licit	State and society norms overlap	The society allows something that is forbidden by state institutions or codes
Illicit	The state does not punish actions that are stigmatised socially (by one or more communities)	State and society norms overlap

Adapted from Van Schendel and Abraham (2005)

The discourse becomes blurred, however, when a given practice is formally illegal but socially acceptable. National bans of foods and drinks may push some segments of the society towards the illegal but, as long as a given community accepts consumption of a given item, consumers will continue enjoying social acceptance and perceive their choice as somehow legitimate. In such cases, their only motivation not to engage in such behaviour is the fear of punishment by the state, which is a lesser detractor than punishment through social stigmatisation and state coercion. At a recent conference on criminal studies, a representative of the police force fiercely advocated for stronger state interventions to liquidate mafia-like structures. However, if these structures still enjoy the tacit support of the population, liquidating them may just create a void that will then be filled by the next criminal organisation. In areas where the state is only nominally present, where if I need food, a job, support I will turn to my local community rather than the state, I am more likely to endorse whatever structure is giving me food to eat, money to pay a doctor than an abstract entity officially looking after my welfare. In such situations, lives become entwined, cooperation becomes mutually beneficial, the state becomes superfluous eventually being replaced by trust networks that instead become essential for survival, or even decent living. When the police come to arrest a person charged with serious criminal activities, the neighbourhood rebels, throws objects at representatives of the state to defend that individual that the state deems a dangerous criminal but for them is the ultimate source of welfare that is not provided by the state.

The tension between the criminal and the informal is central to our discourse. Our authors here focus on activities that fall without the grace of the state, that civil servants and politicians may condemn, that international organisations will describe as unlawful and to address, but that eventually come to constitute the aggregate of activities that are part of the political. Public policy is not only the art of proposing laws, enacting rules but also the way officially illegal or extra-legal behaviours are tolerated, approached or even encouraged by failing or forgetting to be present in a given sphere of life. Taking the distance from criminal behaviours, we are keen to understand and theorise behaviours that states may nominally consider unlawful. But how to draw the line between what should be considered criminal and what can instead be studied through the lenses of informality to better understand how public policy works and how it can serve the interests of the citizens?

The framework used here takes into account direct and indirect harm. When an action harms directly fellow citizens, it may be regarded as both illegal and illicit (using van Schendel and Abrahams' classification) and difficult to justify from a community perspective. Trafficking, heroine selling, killing someone is usually an act that benefits an individual or a very restricted circle of individuals while harming people around (or from) that very circle. These are also activities that cannot be performed widely and regularly by a large number of individuals or else these will significantly harm the society at large. By contrast, smaller, more recurrent activities that benefit one or more individuals without directly harming the community around, are likely to deserve being deconstructed, analysed and interpreted from the lenses of informality.

Fiscal fraud and tax evasion are illegal and even criminal behaviours in some contexts, but they harm a society only indirectly, by affecting state capacity, limiting fiscal revenues and thus the amounts that a state can spend on welfare, protection or infrastructures. We are not talking here of the situation where one single company is hiding billions. But if under-reporting is a recurrent practice by a significant portion of the country's (or the area's) business actors, this should be considered beyond its economic or legal significance and lead to the question: why people persistently engage in such behaviours. Likewise, the flourishing of black market transactions to find certain goods like meat, alcohol, or even hard currency, should also lead to the question of why people do this. Now, it is undeniable that this will expose individuals to the agency of criminal organisations that might try to regulate the black market currency or even monopolise it. But the situation that led to this opportunity, the gap between what was needed by the society and what the state was offering directly or indirectly (by allowing firms to operate in that given sector of field), should be the subject of public policy discussion. Why members

Table 2 Direct (affecting fellow citizens), indirect (affecting a society) harmfulness and legality

	Direct harm (mostly illicit)	*Indirect harm (might be licit)*
Illegal	Murder, trafficking, heavy drug dealing, ethnic violence (might be licit in some cases)	Fiscal fraud, nepotism, ethnic or religious discrimination
Legal	Use legal action against unaware people to extort money or property; clauses written in a smaller font at the end of a contract	Laws that favour one (ethnic, religious) group over others (are licit for the favoured ones)

of a society decide massively to engage with a behaviour that is illegal, punishable and potentially dangerous? They do it because demand and supply do not match and, in a number of cases, offering a given service or product is considered acceptable by the local community. Starting from this definition, it has been proposed that the boundary between "political" and "criminal" informality can be drawn by referring to the kind of harm an action provokes on fellow citizens. Occupying and building a house on a territory that is not yours, or simply without the necessary permissions, is certainly something illegal as well as being caught paying bribes or hiding revenues but these activities are on a sort of grey territory and could still be considered a contribution to the construction of the political (Table 2).

Within the above boundaries, that is studying transactions that may go against the state but not directly against fellow citizens, this book will put forward the idea that informality is the way weaker and marginalised societal actors gain some voice in a state, renegotiate some policies and eventually contribute to the construction of the political and macro processes. In no way we claim that this is the only use of informality. But our evidence draws from a variety of situations where informality is used to contrast or survive an ineffective state where "ineffective" is not an absolute but relative term. No state is simply failed or successful. There are some areas where it performs better and some others where it manages to look after a significant majority of its citizens and others where it performs poorly and mostly lets them down. This happens all the time and a degree of dissatisfaction is inevitable. However, when this dissatisfaction or incapacity to look after its citizens is "contiguous", then there is some margin for improvement. Contiguous here refers to the fact that all, or the great majority of, the citizens suffering because of a given policy have a substantial number of things in common. They could be all from the same ethnic group, similar income range, a given region or administrative area (or even neighbourhood). When it is possible to "isolate" the group and it becomes clear that they tend to engage with informal activities in a given situation, the first thing to do is not to necessarily repress or punish them (unless they are causing direct harm to other citizens) but to look into the effect of given policies for that given group. To do this, the first step is possibly to define the boundaries of informality (or the kind of informality that we intend to study) and its relationship with shadow economies and other approaches.

Informality, the Informal Economies and the Shadow Economy

In the beginning, there was one informality. It was mostly evoked by economists and it generally referred to activities beyond the formal economy that started being noticed, or recorded, after WWII. In spite of the general belief that informality would decline, or even disappear, as a result of enhanced economic efficiency of a country, or a region (Lewis 1954), things seemed to take a different turn. By the seventies, there was not one informality but many. The debate expanded to include the first conceptualisations of the informal sector (Harris and Todaro 1970) and the definition of informal employment by the ILO (Benanav 2019). In spite of criticisms about the width and thus vagueness of the definition of informal sector that included a vast grey zone overall difficult to measure (Bromley 1978; Lipton 1984). This stream of research eventually led to a revisiting of the concepts of employment and unemployment (Benanav 2019). By the early 2000s, the concept was so widely engaged with that Williams identified a wide variety of adjectives and nouns used to identify the informal economy (Table 3).

This variety had been, and would be, corroborated by studies expanding the meaning of informal employment and labour to the debates on development and governance (Castells and Portes 1989; Gërxhani 2004; Guha-Khasnobis et al. 2006; La Porta and Shleifer 2014) and included the failure of a state to govern a given sector or territory (de Soto 1989; Loyaza 2016), as a consequence of over-regulation and the incapacity of the public sector to create a healthy environment (Johnson et al. 1998; Kucera and Roncolato 2008; Schneider 2011; Schneider and

Table 3 Adjectives and nouns used to denote cash-in-hand work

Adjectives			
Black	*Cash-in-hand*	*Clandestine*	*Concealed*
Dual	Everyday	Ghetto	Grey
Hidden	Invisible	Irregular	Marginal
Moonlight	Non-observed	Non-official	Occult
Off-the-books	Other	Parallel	Peripheral
Precarious	Second	Shadow	Submerged
Subterranean	twilight	Underground	Unexposed
Unobserved	Unofficial	Unorganised	Unrecorded
Unregulated	Untaxed	Underwater	
Nouns			
Activity	Economic activity	Economy	Employment
Sector	Work		

Source Williams (2004)

Medina 2017). Even if employing different definitions, from shadow to informal economy or even informality, these studies had in common the focus on unrecorded or under-recorded economic transactions that eventually affect state capacity and undermine trust in institutions. The overall strategy in this respect was to look for a correlation and bet on what measures to take to ensure that the levels of informal economy in a given context, as classified in Table 4.

Besides the economic ones, two more approaches on informality were developed almost in parallel and the same period in urban studies and then geography (Turner 1977), the other in economic anthropology and subsequently sociology (Hart 1973). The projection of trust network studies onto housing offered a framework to study informal and extra-legal settlements for poor and marginalised people not having access to state support or resources that is at the origins of urban informality studies (Boanada-Fuchs and Fuchs 2018). Moving away from ILO's definition of economic informality, planning scholars started paying attention to the political and social aspects of informality (Innes et al. 2007; Roy and AlSayyad 2004), the management of public areas (de Paula and Scheinkman 2010; Roy 2009; Falla and Valencia 2019) and thus expand the scope of policymaking (Canclini 2019; Davis 2017) to the everyday, bottom and micro actors and the forces underlying urban governance

Table 4 Main causes/indicators determining the shadow economy

Possible correlation	*References*
(1) **Tax and social security contribution burdens**: The bigger the difference between the total labour cost in the official economy and after-tax earnings (from work), the greater the incentive to reduce the tax wedge and work in the shadow economy	Johnson et al. (1998), Tanzi (1998), Dell'Anno (2010), Dell'Anno et al. (2007)
(2) **Quality of institutions or corruption**: The degree of efficiency and discretion with which the tax code is applied, and the effectiveness of government regulation plays a crucial role in the decision to work off the books. The informal sector might shrink in size if institutions can be strengthened, and fiscal policy moves closer to the median voter's preferences	Dreher et al. (2009), Teobaldelli and Schneider (2013), Amendola and Dell'Anno (2010), Hassan and Schneider (2016), Williams and Schneider (2016)
(3) **Regulations**: Regulations and barriers lead to a substantial increase in labour costs in the official economy and thus provide another incentive to work in the shadow economy	Johnson et al.(1998), Kucera and Roncolato (2008), Schneider (2011), Hassan and Schneider (2016)
(4) **Public sector services**: A large shadow economy lowers state revenues and thus state capacity. When the state increases taxes to compensate for lower incomes, it can further incentivise participation in the shadow economy	Johnson et al. (1998), Feld and Schneider (2010)
(5) **Development of the official economy**: The higher the unemployment quota and/or the lower GDP growth, the greater the incentive to work in the underground economy, ceteris paribus	Schneider and Williams (2013), Feld and Schneider (2010)
(6) **Self-employment**: The higher the rate of self-employment, the more activities can be performed in the shadow economy, ceteris paribus	Schneider and Williams (2013), Feld and Schneider (2010)
(7) **Unemployment**: The higher the rate of unemployment, the higher the probability of work in shadow or criminal activities, ceteris paribus	Schneider and Williams (2013), Williams and Schneider (2016)
(8) **Size of the agricultural sector**: The larger the agricultural sector, the more possibilities to work in the shadow or underground economy, ceteris paribus	Hassan and Schneider (2016)
(9) **Use of cash**: The larger the underground economy, the more cash will be used, ceteris paribus (also, the more cash is used the more there are opportunities for shadow transactions)	Hassan and Schneider (2016), Williams and Schneider (2016)
(10) **Share of the labour force**: The higher the shadow economy, the lower the official labour force participation rate, ceteris paribus	Schneider and Williams (2013), Feld and Schneider (2010)

Source Modified and shortened from the original (Schneider and Buehn 2017)

actors and state–citizen relations (Sarmiento and Tilly 2018; Haid and Hilbrandt 2019). Accordingly, it has been remarked that, inasmuch as the state may claim to reduce informality, it may also end up enhancing it in some areas that fall beyond its control, that are neglected or just remain loosely regulated (Davies and Polese 2015; Haid 2017; Polese et al. 2015), not least because the state is not a monolithic entity but produces gaps in communication between institutions (Kasza 2002) and ultimately relies on the agency of its people who populate institutions (Jones 2008).

This governance approach shows some similar points with what has come out of anthropological views. Based on Hart definition of the informal economy, social scientists have sometimes relied on Scott's moral economy and infrapolitics frameworks (Scott 1977, 2019) in that that the most important aspect in state–citizen relation is not how much the state claims to be just, that is its intentions to harmonise and make life better (Scott 1998). It is what citizens perceive as fair that determines their attitude, and possibly devotion and dedication, to the cause of the state (that is, in turn, in charge of mediate single individualities to create a framework that is optimal for the society). When trust in the state is limited, citizens will inevitably resist and this can come as overt confrontation (infrapolitics, war, insurgency) or as silent everyday resistance (Murru and Polese 2020; Scott 1985) leading to the informal renegotiation of political measures in a given aspect of public life (Bayat 2000). This framework has gone as far as to look at the way political negotiations between small and larger states take place and the power dynamics that, although not expressed formally, are key to the completion of a political process at the national and international level. What could be regarded as the missing chain between anthropological and politics and IR approaches can be regarded as inspired by a Wendtian approach and look at the way individual perceptions influence domestic and international politics. Footing on North (1990) understanding of informal organisations, governance and policy scholars have apparently developed parallel concepts of informal governance that are applied to either domestic political dynamics and international politics (Roger 2020: 10). This approach has contributed to consolidating governance as a mode of international cooperation (Stone 2013; Christiansen and Neuhold 2012; Pauwelyn et al. 2013; Kilby 2013; Kleine 2014), ending up also including the relationship of states with non-state actors that could be business (Piattoni 2006; Stafsudd 2009) or insurgent organisations (Polese and

Hanau Santini 2018; Hanau Santini et al. 2020). These views eventually converge in the use of the word informal that is used to refer to either mere economic or also governance and societal phenomena. This is possibly why further governance indicators (for instance by the World Bank) have then also progressively taken into account not only economic indicators but invisible or unnoticed barriers within a variety of geographical settings that are summarised in the Table 5.

Stemming from the above classification, it is possible to propose a distinction between informal and shadow economy as economic phenomena and informality at large, which is better understood through social science lenses. The shadow economy is thus the aggregate of practices that are used to hide revenues, underreport employees or their salaries and overall affect state economic capacity by reducing the amount of taxes paid or payable. Informality, defined as "the art of bypassing the state" (Polese 2021) is larger in scope and may include economic transactions but concentrates more on the social, cultural and contextual settings that make a practice viable, acceptable or even desirable in a given context. If we refer to informality as the activities that are concealed from a state Ruth (Routh 2011), we should consider that they are not only concealed voluntarily but could go unnoticed because enshrined in the everyday behaviour of the interested actors. If some practices are widely embedded in social relations, trust network or a historically rooted mistrust towards the state, not all the behaviours that will be derived are immediately

Table 5 Geography of informality

Discipline	*Main focus*	*Geographical scope*
Economics	Development economics	Developing world
Planning	Development world urban setting	Non-Western world (with an eye for the Western world)
(Human) Geography	Alternative economies	Worldwide
Area studies	Getting things done	Eurasia (expanding)
International relations	Constructing alliances, influence political decisions	International organisations, global politics
Policy and governance	Political negotiations, policymaking	Russia, national politics, relationship between elites
Political science	Corruption, nepotism, neopatrimonialism	The non-Western world with a particular focus on Eurasia

Source Polese (2021)

visible even to the performing actor. Informality then becomes "the space between two formal rules" (Polese 2016) and it is necessary to modify behaviours by reinstalling trust in the state or a dialogue between states and citizens to produce better governance than mere coercion.

The Content of This Book: Diverging Moralities

This volume has been conceived to explore the different meanings informality can take when state and individual morality tend to diverge. We are not interested in single cases where an individual violates state principles while putting at risk the integrity of the people around them. But we are interested in situations where reliance on informal practices becomes sufficiently widespread to talk of a societal tendency that the state can choose to tolerate it or fight it. While it could eradicate the habit, chances are that if they do not address the underlying issue (that is the very reason why people engaged in such practices) in terms of changing societal behaviour or looking after societal needs. The habit will probably just emerge in a different form. This is a way for these groups of societal actors to look after themselves. It is also an indicator that something concrete should be attempted about a given situation to address societal needs.

In no way we expect to exhaust the spectrum of situations where informality is used. However, there should be plenty of material for a critical reflection on the role of informality, the fact that its importance goes well beyond the economic significance of its actions, to highlights some of the tensions that have recently been depicted in informality studies in some of the most important loci of informality. The starting point of this volume is the intention to further explore the relationship between the state and the citizen and, in particular, what happens when individual and state moralities do not overlap or when they even diverge (Polese 2021). There has been a growing amount of interest in this direction, and the increased amount of works devoted to informality and individual or society-centred accounts that contrast with a state-led view on individual morality (see, among others Morris 2012; Morris and Polese 2015; Polese and Rodgers 2011; Van Schendel and Abraham 2005; Wanner 2005). It is generally assumed that the state and citizens are linked by a social contract whose terms are defined by the state in primis. But what happens in the cases where the state fails to deliver what it has promised? And what happens when the citizens are not willing, or able, to accept state morality as their own? This often happens when an externally imposed view on what is "moral", "fair" or "due" is enforced by the state but not understood by the citizens. Post-Weberian definitions look at the state–citizen relationship as "a due ut das" one. The state grants protection, education, medical care and its citizens contribute labour, compliance, taxes. When this does not occur, it is generally accepted that the citizens are deviating from state goals. But citizens and state are united in a mutually interdependent relationship (Migdal 2012) and they influence one another. Lack of compliance with the totality of state-inducted instructions is not necessarily evidence that citizens plot against the state or undermine its symbolic power. It could also be that citizens are not happy with the way the state is proposing, or imposing, its value, that they do not feel protected by formal structures and they rely on informal ones to replace, supplement or even compete with state institutions.

To follow this line, authors have been requested to produce case studies offering broader understanding of the ways informality manifests itself. From established reciprocity practices to under the table agreements to get things done, they have explored a variety of practices that need to be contextualised and located in a broader framework to contribute to a taxonomy of informal practices. This is something that has been implicitly pointed out by previous research (see, for instance, Dixit 2007; Helmke and Levitsky 2004; Stone 2013) and has informed Alena Ledeneva's Encyclopaedia of Informality (Ledeneva 2018a, b) but can be brought to the next level. Accordingly, the book has been divided into three sections. The first one reflects a consolidated field that is the understanding that mobilities and informality are intimately related and explores the way informality is used by marginal actors in a way to integrate into a new context. In the first chapter, Ignacio Fradejas-García, José Molina and Miranda Jessica Lubbers explore the way Romanian migrants in Spain use informal practices evolve when people migrate from one cultural context to another. Their account of the relationship between informality and mobilities is intended to grasp the complexity of cultural practices in everyday life and is informed by the "mobility turn" (Sheller and Urry 2006) that puts the movement of people, things and knowledge at the core of the social research agenda and an understanding of informality as the aggregate of practices that bypass the regulations of states and their institutions (Routh 2011; Polese 2021) that is present globally but adapted to better fit the different cultural context where they evolve. The following chapter by Karolina Bielenin-Lenczowska and Helena Patzer surveys the consolidation of migrant networks in Warsaw's suburb. They notice that recently Poland has emerged as an attractive migration destination and has witnessed a substantial growth of the migrant population, especially coming from Asian countries. Starting from this, their chapter discusses the visible shift to diversity in the character of a suburban neighbourhood of Warsaw and tries to uncover what it means for the different migrant groups in terms of access to the labour market, the formal and informal practices they engage in, and the role it plays in the migrant imaginary of post-socialist Poland. They suggest that using informal migrant networks in order to cope with everyday life in a foreign country is a substitute for the local practice of "getting things done" and illustrate this by showing how migrants are also being pushed into ethnic niches while entering the grey zone economics through undeclared or half-legal work. A similar approach can be found in the third chapter where Ružica Šimić Banović, Vlatka Škokić, Mirela Alpeza investigate networking activities among ethnic entrepreneurs in Croatia, where immigration is still at a lower level than emigration. Their study looks at the ways ethnic entrepreneurs use different network types to access different resources and overcome the additional formal and informal barriers due to the fact that they need to settle in the country. They suggest that informal networks appear to serve as substitutes to missing or insufficiently efficient institutions of the formal networks aimed at foreigners' needs, entrepreneurs in this case. Besides, they look at the role of the "extension" of the expat groups that the ethnic entrepreneurs consider vital for their business establishment and development, but for their private life as well while developing in-group cooperation and support. The final chapter of the section, by Daniel Kashinitsky, analyses the place of undocumented migrants in the society and national economy of modern Russia. Based on life stories of migrants living with HIV in Russia, he reflects on their vulnerabilities and routine practices of the state to control and maintain an untransparent and ostentatious migration policy towards undocumented migrants. He also asks the rationale behind this "performance of illegality" the government seems to be producing with regards to migrants' experiences, access to social entitlements and the biopolitics of the state in state management of migration with particular attention to HIV affected migrants.

The second part of the book explores informality in a more static, and local, meaning. It is about "staying" and explores the tensions between local actors that intend to remain in a space and market and the overarching state. In her study on tax Morale of Small Businesses in Armenia and Georgia, Joanna Paquin explores whether the difference in the size of the informal economy between the two countries can be attributed to discrepancies between formal and informal institutions. Using quantitative and qualitative secondary sources she points at the fact that tensions between formal and informal institutions can be found for Armenian than for Georgian entrepreneurs so to rule out that institutional incongruence can, at least alone, account for the differences in the size of the informal sectors. The following chapter by Aimar Ventsel focuses on semi-informal SME to revisit the concept of a "new entrepreneurship" in Russia. It explores the role of non-monetary or even non-economic factors in the desire to start a business, a thing that ends up influencing the way morality and loyalty are understood internally and in relationship with society. He also illustrates how the chosen strategy of this kind of entrepreneur is to somehow remain unperceived and thus limit their expansion and innovation activities to live a quiet but longer life, given the current Russian business climate. Further in the section, the study on Azerbaijani Meykhana, by Aneta Strzemżalska, explores a musical genre that incarnates both the goals of the state and its society. She examines how the interplay between formal and informal approach can be used in the study on national building process and the way state and individual approaches seem to converge and agree on the development of a marker that becomes important for both sides so to explore the agency bottom actors have and their impact on top-level politics, including cultural policy. The section concludes with an exploration of the "forms of governance" in Uzbekistan and a comparison of practices that are, at different times, considered illegal, Immoral or illegitimate. Rustamjon Urinboyev, Abel Polese, Mans Svensson, Laura Adams, Tanel Kerikmäe look at the way official state narratives are challenged by silent, unorganised, often unawares, gestures of resistance at the bottom of a society thus suggesting incorporating informal practices in a definition of informality that is more inclusive, and better explains the anatomy of a modern state, whose functioning rests on a combination of formal and informal practices. They thus propose that governance practices in some societies may be regarded as a space where formal institutions and citizens (or informal institutions) compete for power and resources and thereby produce informal, alternative "legal orders" and mechanisms that regulate public life in a given area.

The third part of the book is concerned with the competition between formal and informal activities and their influence on public policy at large. It starts with a chapter where Anil Duman explores the link between informal sector diversity and policy preferences and, in particular, the variation of policy preferences among informal sector participants have received less attention. She argues here that people in the upper tier of the informal sector might perceive the formal–informal practices as complementary and sustain their highly rewarding informal activities. She thus tests the impact of the formal–informal divide as well as diversity among the informal sector participants on political and economic opinions at the individual level to suggest that there is no significant relationship between informal employment, political and economic views once country-level fixed effects are included. However, she proposes that diversity within the informal sector is crucial as income mediates the relationship between political opinions and employment. The following chapter by Joseph Helou presents an empirical case study of the Lebanese state's mismanagement of public finances and citizens' adoption of informality in response to Lebanon's sovereign debt crisis. It explores how the consistently negative balance of payments, soaring national debt, dwindling foreign currency reserves and the political system of sectarianism that thrives on government spending to finance budget deficits make Lebanon the economic time bomb it is. The section continues with a chapter by Elena Druică and Rodica Ianole-Călin about the factors that keep, or chase away, healthcare professionals from their workplace in Romania to see opportunities abroad. The study investigates the main reasons behind the health professionals' decision to leave the country based on factors that could act as potential drivers such as wages, working conditions or wide use of informal practices. The results are then correlated with the overall satisfaction the participants feel for their workplace, pointing to different explanatory layers for what may drive a medical doctor or a nurse to leave the country versus what it impacts their professional satisfaction. The following paper introduces the concept of e-nformality as the bodily and spatial relationship between people and the state and various services changes, available online. Focusing on the relationship between ICT technologies and newly emerging forms of informality, Aksana Ismailbekova and Gulzat Baialieva trace informal networks developed around smartphone applications. They look at the different attitudes emerging across various generations of Kyrgyz citizens and how their age, and level of post-Soviet identity, influences the channels that they use to gain access to services and opportunities that are not immediately available through formal channels. The section ends with Petru Negură chapter that examines the employment situation of homeless people in Moldova in the context of the post-Soviet transformation. The paper maintains that most homeless people are excluded from the formal sector of the labour market and get pushed to engage in various informal economic activities which help them survive and integrate them into informal support networks. Based on this, the study argues that informal work in which homeless people are involved reduces the ability of employees to control the labour process and their autonomy to the employer. The informal labour activities, in his view, further reproduce the precarious social status of the homeless person and perpetuate her social and economic vulnerability.

Taken together, these chapters are our attempt to go beyond informality as an economic phenomenon. True that its visible effects are economic, which also depends on the fact that they are the easiest to measure since there is no real "index on non-compliance" (yet). Besides, lack of trust towards the government is difficult to translate into a concrete understanding of the downsizes of a governance strategy or damage or an index that is widely accepted as authoritative. In a world that is used to measure everything in monetary terms, therefore, it becomes difficult to take into account other measurements or indicators to understand the entity of damage. Yet, some of the most important elements that can help governance are not measurable, not in a traditional way. We thus hope that this book will be regarded as an attempt to include in the debates also factors that are more difficult to notice, measure or integrate into evidence-based policies but are not, because of this, less important than measurable elements (that were not measured until some consensus was reached that such measurements are needed). After all, human history is full of elements that are unimportant until a large number of actors agree that they are important and they start permeating debates, societies, cultures and policies and this is a further attempt to suggest new elements, and approaches, that should be taken into account.

Acknowledgement This book benefited from EU funding European Commission [H2020-MSCA-RISE-2018-Shadow, grant no: 778118 and H2020-MSCA-RISE2019-New Markets, grant no: 824027]. The book was finalized during a research stay in Kyoto funded by the Japan Society for the Promotion of Science Visiting Fellowship Scheme (2020/2021).

Bibliography

Amendola, Adalgisio and Roberto Dell'Anno. 2010. "Desigualdad social, desarrollo económico y curva de Kuznets: un análisis en América Latina." *Studiositas* 5 (3): 39–68. ISSN-e 1909-0366.

Ariely, Dan. 2002. "The Honest Truth About Dishonesty." *The Financial Times*.

Bayat, Asef. 2000. "From 'Dangerous Classes' to 'Quiet Rebels'." *International Sociology* 15 (3). https://doi.org/10.1177/0268580000015003005.

Benanav, Aaron. 2019. "The Origins of Informality: The ILO at the Limit of the Concept of Unemployment." *Journal of Global History*. https://doi.org/10.1017/S1740022818000372.

Boanada-Fuchs, Anthony, and Vanessa Boanada Fuchs. 2018. "Towards a Taxonomic Understanding of Informality." *International Development Planning Review*. https://doi.org/10.3828/idpr.2018.23.

Bourdieu, Pierre. 2020. "Outline of a Theory of Practice." In *The New Social Theory Reader*. https://doi.org/10.4324/9781003060963-11.

Bromley, Ray. 1978. "Introduction—The Urban Informal Sector: Why Is It Worth Discussing?" *World Development*. https://doi.org/10.1016/0305-750X(78)90061-X.

Canclini, Néstor García. 2019. "A Culture of Informality." *Urban Studies* 56 (3). https://doi.org/10.1177/0042098018782635.

Castells, Manuel, and Alejandro Portes. 1989. "World Underneath: The Origins, Dynamics, and Effects of the Informal Economy." *The Informal Economy: Studies in Advanced and Less Developed Countries*.

Christiansen, Thomas, and Christine Neuhold. 2012. *International Handbook on Informal Governance*. https://doi.org/10.4337/9781781001219.

Davies, Thom, and Abel Polese. 2015. "Informality and Survival in Ukraine's Nuclear Landscape: Living with the Risks of Chernobyl." *Journal of Eurasian Studies*. https://doi.org/10.1016/j.euras.2014.09.002.

Davis, Diane E. 2017. "Informality and State Theory: Some Concluding Remarks." *Current Sociology* 65 (2). https://doi.org/10.1177/0011392116657301.

Dell'Anno, Roberto. (2010). "Institutions and Human Development in the Latin American Informal Economy." *Constitutional Political Economy* 21 (3): 207–230.

Dell'Anno, Roberto, and Adalgiso Amendola. 2015. "Social Exclusion and Economic Growth: An Empirical Investigation in European Economies." *Review of Income and Wealth* 61 (2): 274–301. https://doi.org/10.1111/roiw.12096.

Dell'Anno, Roberto, Miguel Gómez-Antonio, and Angel Alañon-Pardo. 2007. "The Shadow Economy in Three Mediterranean Countries: France, Spain and Greece. A MIMIC Approach." *Empirical Economics* 33 (1): 191–197. https://doi.org/10.1007/s00181-007-0138-1.

Dixit, Avinash K. 2007. *Lawlessness and Economics: Alternative Modes of Governance*. Princeton: Princeton University Press.

Dreher, Axel, Christos Kotsogiannis, and Steve McCorriston. 2009. "How Do Institutions Affect Corruption and the Shadow Economy?" *International Tax and Public Finance* 16: 6. https://doi.org/10.1007/s10797-008-9089-5.

Falla, Ana Maria Vargas, and Sandra C. Valencia. 2019. "Beyond State Regulation of Informality: Understanding Access to Public Space by Street Vendors in Bogotá." *International Development Planning Review* 41 (1). https://doi.org/10.3828/idpr.2019.3.

Feld, L. and F. Schneider. 2010. "Survey on the Shadow Economy and Undeclared Earnings in OECD Countries." *German Economic Review* 11 (2): 109–149. https://doi.org/10.1111/j.1468-0475.2010.00509.x.

Gaal, Peter, Tamas Evetovits, and Martin McKee. 2006. "Informal Payment for Health Care: Evidence from Hungary." *Health Policy* 77 (1). https://doi.org/10.1016/j.healthpol.2005.07.024.

Gërxhani, Klarita. 2004. "Tax Evasion in Transition: Outcome of an Institutional Clash? Testing Feige's Conjecture in Albania." *European Economic Review* 48 (4). https://doi.org/10.1016/j.euroecorev.2003.08.014.

Guha-Khasnobis, B., R. Kanbur, and E. Ostrom, eds. 2006. *Linking the Formal and Informal Economy: Concepts and Policies*. Oxford: Oxford University Press.

Haid, Christian G. 2017. "The Janus Face of Urban Governance: State, Informality and Ambiguity in Berlin." *Current Sociology* 65 (2). https://doi.org/10.1177/0011392116657299.

Haid, Christian G., and Hanna Hilbrandt. 2019. "Urban Informality and the State: Geographical Translations and Conceptual Alliances." *International Journal of Urban and Regional Research* 43 (3). https://doi.org/10.1111/1468-2427.12700.

Hanau Santini, Ruth, Abel Polese, and Rob Kevlihan. 2020. *Limited Statehood and Informal Governance in the Middle East and Africa*. https://doi.org/10.4324/9780429504570.

Harris, John R., and Michael P. Todaro. 1970. "Migration, Unemployment and Development: A Two-Sector Analysis." *American Economic Review* 60.

Hart, Keith. 1973. "Informal Income Opportunities and Urban Employment in Ghana." *The Journal of Modern African Studies* 11 (1). https://doi.org/10.1017/S0022278X00008089.

Hassan, Mai, and Friedrich Schneider. 2016. "Size and Development of the Shadow Economies of 157 Countries Worldwide: Updated and New Measures." IZA Discussion Paper No. 10281 October 2016. Accessed on 12 October 2021. https://ftp.iza.org/dp10281.pdf.

Helmke, Gretchen, and Steven Levitsky. 2004. "Informal Institutions and Comparative Politics: A Research Agenda." *Perspectives on Politics* 2 (4). https://doi.org/10.1017/S1537592704040472.

Innes, Judith E., Sarah Connick, and David Booher. 2007. "Informality as a Planning Strategy." *Journal of the American Planning Association* 73 (2). https://doi.org/10.1080/01944360708976153.

Johnson, Simon, Daniel Kaufmann, and Pablo Zoido-Lobatón. 1998. "Corruption, Public Finance, and the Unofficial Economy." *ECLAC Conference*.

Jones, Rhys. 2008. *People/States/Territories: The Political Geographies of British State Transformation*. https://doi.org/10.1002/9780470712849.

Kasza, Gregory J. 2002. "The Illusion of Welfare 'Regimes'." *Journal of Social Policy*. https://doi.org/10.1017/S0047279401006584.

Kilby, Christopher. 2013. "An Empirical Assessment of Informal Influence in the World Bank." *Economic Development and Cultural Change* 61 (2). https://doi.org/10.1086/668278.

Kleine, Mareike. 2014. "Informal Governance in the European Union." *Journal of European Public Policy* 21 (2). https://doi.org/10.1080/13501763.2013.870023.

Kucera, David, and Leanne Roncolato. 2008. "Informal Employment: Two Contested Policy Issues." *International Labour Review* 147 (4). https://doi.org/10.1111/j.1564-913X.2008.00039.x.

Ledeneva, Alena V., ed. 2018a. *The Global Encyclopaedia of Informality, Volume 1*. https://doi.org/10.14324/111.9781911307907.

———. 2018b. *The Global Encyclopaedia of Informality, Volume 2*. https://doi.org/10.14324/111.9781787351899.

Lewis, Arthur. 1954. "Economic Development with Unlimited Supplies of Labour." *The Manchester School* 22 (2). https://doi.org/10.1111/j.1467-9957.1954.tb00021.x.

Lipton, Michael. 1984. "Family, Fungibility and Formality: Rural Advantages of Informal Non-farm Enterprise Versus the Urban-Formal State." In *Human Resources, Employment and Development Volume 5: Developing Countries*. https://doi.org/10.1007/978-1-349-17461-4_10.

Loyaza, Norman. 2016. "Informality in the Process of Development and Growth." 7858. Policy Research Working Paper Series. Washington.

Migdal, Joel. 2012. *State in Society Studying How States and Societies Transform and Constitute One Another*. Cambridge: Cambridge University Press.

Morris, Jeremy. 2012. "Unruly Entrepreneurs: Russian Worker Responses to Insecure Formal Employment." *Global Labour Journal* 3 (2). https://doi.org/10.15173/glj.v3i2.1120.

Morris, Jeremy, and Abel Polese. 2015. *Informal Economies in Post-Socialist Spaces: Practices, Institutions and Networks. Informal Economies in Post-Socialist Spaces: Practices, Institutions and Networks*. https://doi.org/10.1057/9781137483072.

Murru, Sarah, and Abel Polese, ed. 2020. *Resistances: Between Theories and the Field*. New York: Rowman & Littlefield Publishers.

Naylor, R. T. 2004. *Wages of Crime: Black Markets, Illegal Finance, and the Underworld Economy*. Ithaca, NY: Cornell University Press.

North, Douglas. 1990. *Institutions, Institutional Change and Economic Performance*. Cambridge: Cambridge University Press.

Patico, Jennifer. 2002. "Chocolate and Cognac: Gifts and the Recognition of Social Worlds in Post-Soviet Russia." *Ethnos* 67 (3). https://doi.org/10.1080/0014184022000031202.

de Paula, Áureo, and José A. Scheinkman. 2010. "Value-Added Taxes, Chain Effects, and Informality." *American Economic Journal: Macroeconomics* 2 (4). https://doi.org/10.1257/mac.2.4.195.

Pauwelyn, Joost, Ramses A. Wessel, and Jan Wouters. 2013. *Informal International Lawmaking*. https://doi.org/10.1093/acprof:oso/9780199658589.001.0001.

Phelps, Nicholas A. 2021. "The Inventiveness of Informality: An Introduction." *International Development Planning Review* 43 (1). https://doi.org/10.3828/idpr.2021.1.

Piattoni, Simona. 2006. "Informal Governance in Structural Policy." *Perspectives on European Politics and Society* 7 (1). https://doi.org/10.1080/15705850600839561.

Polese, Abel. 2006a. "De Jure Oder de Facto? Die Gesundheits- Und Bildungssektoren in Der Ukraine." *Ukraine-Analysen2* 16: 2–4.

———. 2006b. "Paying for a Free Education: Transitions Online." *Transitions2*.

———. 2009a. "'If I Receive It, It Is a Gift; If I Demand It, Then It Is a Bribe': On the Local Meaning of Economic Transactions in Post-Soviet Ukraine." *Anthropology in Action* 15 (3). https://doi.org/10.3167/aia.2008.150305.

———. 2009b. "*A Post-Soviet Syndrome: The State is Public, the Private Sector is Private...the Public Sector is*?" Working Paper, University of Poznan Adam Mickiewicz.

———. 2016. *Limits of a Post-Soviet State: How Informality Replaces, Renegotiates, and Reshapes Governance in Contemporary Ukraine*. Stuttgart: Ibidem Verlag.

———. 2021. The Art of Bypassing the State: Mapping Informality in Eurasian Spaces (and Beyond), *Eurasian Geographies and Economics*.

Polese, Abel, Alessandra Russo, and Francesco Strazzari. 2019a. "Introduction: 'The Good, the Bad and the Ugly': Transnational Perspectives on the

Extralegal Field." In *International Political Economy Series*, 1–26. https://doi.org/10.1007/978-3-030-05039-9_1.
———. 2019b. *The Good, the Bad and the Ugly: Transnational Perspectives on the Extralegal Field. International Political Economy Series*. https://doi.org/10.1007/978-3-030-05039-9_1.
Polese, A., J. Morris, and B. Kovács. 2015. "Introduction: The Failure and Future of the Welfare State in Post-Socialism." *Journal of Eurasian Studies* 6 (1). https://doi.org/10.1016/j.euras.2014.11.001.
Polese, A., J. Morris, B. Kovács, and I. Harboe. 2014. "'Welfare States' and Social Policies in Eastern Europe and the Former USSR: Where Informality Fits In?" *Journal of Contemporary European Studies* 22 (2): 184–198. https://doi.org/10.1080/14782804.2014.902368.
Polese, A., and P. Rodgers. 2011. "Surviving Post-socialism: The Role of Informal Economic Practices." *International Journal of Sociology and Social Policy* 31 (11–12). https://doi.org/10.1108/01443331111177896.
Polese, Abel, and Rob Kevlihan. 2013. "Locating Insurgency, Informality and Social Movements on a Spectrum: Is There a Theory Linking Them All?" In *9Th Pan-European Conference on International Relations of EISA*, Naxos 23–26.
Polese, Abel, and Ruth Hanau Santini. 2018. "Limited Statehood and Its Security Implications on the Fragmentation Political Order in the Middle East and North Africa." *Small Wars and Insurgencies* 29 (3). https://doi.org/10.1080/09592318.2018.1456815.
La Porta, Rafael, and Andrei Shleifer. 2014. "Informality and Development." *Journal of Economic Perspectives* 28 (3): 109–26.
Rivkin-Fish, Michele. 2005. "Bribes, Gifts and Unofficial Payments: Rethinking Corruption in Post-Soviet Russian Health Care." In *Corruption: Anthropological Perspectives*. London: Pluto Press.
Roger, Charles. 2020. *The Origins of Informality: Why the Legal Foundations of Global Governance Are Shifting, and Why It Matters*. New York: Oxford University Press. https://doi.org/10.1007/s11558-020-09394-x.
Routh, Supriya. 2011. "Building Informal Workers Agenda: Imagining 'Informal Employment' in Conceptual Resolution of 'Informality'." *Global Labour Journal* 2 (3). https://doi.org/10.15173/glj.v2i3.1106.
Roy, Ananya. 2009. "Why India Cannot Plan Its Cities: Informality, Insurgence and the Idiom of Urbanization." *Planning Theory* 8 (1). https://doi.org/10.1177/1473095208099299.
Roy, Ananya, and Nezar AlSayyad, eds. 2004. *Urban Informality: Transnational Perspectives from the Middle East, Latin America, and South Asia*. Lanham: Lexington Books.

Sarmiento, Hugo, and Chris Tilly. 2018. "Governance Lessons from Urban Informality." *Politics and Governance*. https://doi.org/10.17645/pag.v6i1.1169.

Schneider, Friedrich. 2011. *Handbook on the Shadow Economy. Handbook on the Shadow Economy*. https://doi.org/10.4337/9780857930880.

Schneider, Friedrich, and Andreas Buehn. 2017. "Shadow Economy: Estimation Methods, Problems, Results and Open Questions." *Open Economics* 1 (1). https://doi.org/10.1515/openec-2017-0001.

Schneider, Friedrich, and Dominik H. Enste. 2003. *The Shadow Economy: An International Survey*. Cambridge: Cambridge University Press. https://doi.org/10.1017/CBO9780511493591.

Schneider, Friedrich, and Désirée Teobaldelli. 2012. "Beyond the Veil of Ignorance: The Influence of Direct Democracy on the Shadow Economy." 3749. CESifo Working Paper Series. CESifo.

Schneider, Friedrich G. and Colin Williams. 2013. *The Shadow Economy*. Institute of Economic Affairs, June 1, 2013. Available at SSRN: https://ssrn.com/abstract=2286334.

Schneider, Friederich, and Leandro Medina. 2017. "*Shadow Economies Around the World: New Results for 158 Countries over 1991–2015*." 6430. CESifo Working Paper Series. https://papers.ssrn.com/sol3/papers.cfm?abstract_id=2965972.

Schneider, Friedrich, Konrad Raczkowski, and Bogdan Mróz. 2015. "Shadow Economy and Tax Evasion in the EU." *Journal of Money Laundering Control* 18 (1). https://doi.org/10.1108/JMLC-09-2014-0027.

Scott, James C. 1977. *The Moral Economy of the Peasant: Rebellion and Subsistence in Southeast Asia*. Yale: Yale University Press.

———. 1985. *Weapons of the Weak: Everyday Forms of Peasant Resistance*. New Haven: Yale University Press.

———. 1998. *Seeing Like a State: How Certain Schemes to Improve the Human Condition Have Failed*. Yale: Yale University Press.

———. 2019. *Two Cheers for Anarchism. Two Cheers for Anarchism*. https://doi.org/10.1515/9781400844623.

Sheller, Mimi, and John Urry. 2006. "The New Mobilities Paradigm." *Environment and Planning A: Economy and Space* 38 (2): 207–26. https://doi.org/10.1068/a37268.

de Soto, Hernando. 1989. *The Other Path: The Invisible Revolution in the Third World*. New York: Basic Books.

Stafsudd, Anna. 2009. "Corporate Networks as Informal Governance Mechanisms: A Small Worlds Approach to Sweden." *Corporate Governance: An International Review* 17 (1). https://doi.org/10.1111/j.1467-8683.2008.00721.x.

Stone, Randall W. 2013. “Informal Governance in International Organizations: Introduction to the Special Issue.” *Review of International Organizations*. https://doi.org/10.1007/s11558-013-9168-y.

Tanzi, Vito. 1998. “Corruption Around the World: Causes, Consequences, Scope, and Cures.” *IMF Staff Papers* 45 (4): 559–594. https://doi.org/10.2307/3867585.

Teobaldelli, Désirée, and Friedrich Schneider. 2013. “The Influence of Direct Democracy on the Shadow Economy.” *Public Choice* 157 (3–4): 543–567. https://doi.org/10.1007/s11127-013-0098-2.

Turner, John. 1977. *Housing by People: Towards Autonomy in Building Environments*. New York: Pantheon Books.

Van Schendel, Willem, and Itty Abraham. 2005. *Illicit Flows and Criminal Things: States, Borders, and the Other Side of Globalization*.

Wanner, Catherine. 2005. “Money, Morality and New Forms of Exchange in Postsocialist Ukraine.” *Ethnos* 70 (4). https://doi.org/10.1080/00141840500419782.

Williams, Colin C. 2004. *Cash-in-Hand Work: The Underground Sector and the Hidden Economy of Favours: Cash-in-Hand Work: The Underground Sector and the Hidden Economy of Favours*. https://doi.org/10.1057/9780230506190.

Williams, Colin and Friedrich Schneider. 2016. *Measuring the Global Shadow Economy*. London: Edward Elgar Publishing.

Coming

(Im)Mobilities and Informality as Livelihood Strategies in Transnational Social Fields

Ignacio Fradejas-García, José Luis Molina, and Miranda Jessica Lubbers

Introduction

Mobility and informality have become popular concepts in the social sciences as a way of attempting to grasp the complexity of cultural practices in everyday life. First, the "mobility turn" (Sheller and Urry 2006) put the movement of people, things, and knowledge at the core of the social research agenda, encompassing the mobilities associated with migration and transnationalism (Salazar 2019). Second, "informality," understood broadly as the aggregate of practices that bypass the regulations of states and their institutions (Routh 2011; Polese 2021), is a global phenomenon that inherently co-exists with any formal system to "get things done" (Ledeneva 2018; Polese et al. 2017). However, as the precise practices are culture-specific (Ledeneva 2018; Lomnitz 1988), we

I. Fradejas-García (✉)
University of Iceland, Reykjavik, Iceland
e-mail: Ignacio@hi.is

J. L. Molina · M. J. Lubbers
Department of Social and Cultural Anthropology,
Universitat Autònoma de Barcelona (UAB), Bellaterra, Spain

A. Polese (ed.), *Informality, Labour Mobility and Precariousness*, International Political Economy Series,
https://doi.org/10.1007/978-3-030-82499-0_2

can expect the informal practices that individuals employ to evolve when they migrate to a different cultural context. Despite this, the connection between transnational mobility and informality has hardly been explored in the literature on informality. Therefore, in this chapter, we explore the relationship between informality and the mobilities of Romanian migrants in Spain, analyzing how their informal practices evolve when people migrate from one cultural context to another.

The connection between informality and migration was first established by Keith Hart, who investigated the informal activities of illiterate, unskilled migrants from rural areas of northern Ghana in Accra (Hart 1973). Hart is frequently credited for having pioneered the first typologies of informal economic activities (Ledeneva 2018) and for coining the term "informal economy" as the ensemble of productive activities outside the "organized labor force" (Morris and Polese 2014). Hart's work also shows that both formal and informal activities were part of migrants' mobility and immobility patterns—hereafter (im)mobilities—which were facilitated by their social networks, based on kin ties and ethnic membership.

Since then, migration scholars have focused on the relations between undocumented migrants and their employment in the informal sector (Baldwin-Edwards and Arango 1999; Berggren et al. 2007; Likic-Brboric et al. 2013), on transnational entrepreneurs able to obtain a competitive advantage by relying on informal economies and networks (Portes et al. 2002; Turaeva 2014), and on transnational practices of resistance to state control (Garapich 2016). Furthermore, scholarship on informality has paid attention to informal border-crossing practices such as smuggling and trafficking (van Schendel and Abraham 2005; Bruns et al. 2011; Kalir and Sur 2012) and the shadow economies of migrant workers in post-socialist contexts (Urinboyev and Polese 2016; Cieslewska 2014; Yalcin-Heckmann 2014).

However, academic contributions exploring whether and how individuals' informal practices evolve when people migrate from one cultural context to another are scarce. The "sedentarist metaphysics" (Malkki 1992) that identifies the relationship between peoples and places usually through migrants' countries of origin and destiny or that focus on informal exchanges *within* physical borders has been dominant so far (Spyer 1988; Bruns and Miggelbrink 2012). When people move between cultural contexts, the informal practices that they learned may not work or become superfluous in the new context, while new needs, opportunities, or limitations may arise. Besides, transnational migrants are not confined to a single nation-state (Wimmer and Glick Schiller 2002), but live at the crossroads of two or more nation-states, influenced by multiple sets of laws and institutions (Levitt and Glick Schiller 2004), which implies they navigate different "mobility regimes" (Glick Schiller and Salazar 2013), as well as different administrative, legislative and cultural systems. Consequently, their informal practices to "get things done" may also respond to or exploit more than one formal system.

In this chapter, we adopt a livelihood perspective, which sees both transnational migration and informal practices as strategies that households employ to make a living. This perspective suggests that, for households, the two complementary activities have the common goal of reproduction. Moreover, we assume that migrants contribute to the creation of a transnational social field—hereafter TSF—i.e., the networks of personal relationships that extend across national borders "through which ideas, practices, and resources are unequally exchanged, organized, and transformed" (Levitt and Glick Schiller 2004: 1009). TSFs may start as nothing more than networks of personal relationships, but the migrant enclaves at the destination may gradually attain an "institutional completeness" (Breton 1968) reproducing the institutions of society at large. In this case, the emergence of Romanian welfare organizations, political organizations, cultural associations, churches, schools, language classes, newspapers, and transport companies in Spain catering specifically to the needs of the migrants (Molina et al. 2018). Thus, the TSF perspective helps us identify three types of interdependent actors (migrants, return migrants, and Romanians living in Romania) and allows us to explore the agency of household members in the context of collective and institutional processes.

In sum, this chapter poses the research question of how the informal practices of Romanians evolve in the process of their migration to Spain, whether individually or collectively? This approach focuses on westward migration and mobilities from Romania to Spain, extending previous accounts of "transnational informality" in post-socialist spaces (e.g., Urinboyev 2016), and enhances our understanding of the relationship between informality and mobility.

The data presented in this chapter are based on a research project[1] that analyzes the TSFs created by Romanian immigrants in Spain. To investigate their livelihood strategies, we used a mixed-methods approach combining a binational survey[2] ($N = 303$ for the field connecting Dâmbovița to Castelló de la Plana) with ethnographic fieldwork. The survey inquired about migration and mobility trajectories, family situations, formal and informal economic activities—e.g., remittances, flows of goods, work contracts—homeownership, and social networks. It was conducted between November 2017 and July 2018 in Spain (Castelló de la Plana) and Romania (Dâmbovița), using a novel sampling methodology called "binational link tracing" (Mouw et al. 2014), a technique closely related to respondent-driven sampling (Heckathorn 1997, 2002). Also, after administering the survey to respondents, the interviewers wrote brief field notes highlighting ethnographic information on people's livelihood strategies that have complemented the survey data. In addition, between 2017 and 2020 we conducted multi-sited ethnographic fieldwork (Marcus 1995; Falzon 2009) in Dâmbovița and Bistrița-Năsăud (Romania) and Castelló de la Plana and Roquetas de Mar (Spain). The ethnographic fieldwork consisted of a combination of four methods: (1) participant observation in the daily activities of churches, associations, and companies, as well as carpooling trips with Romanian migrants to and from Romania; (2) dozens of informal interviews with migrants, politicians, and religious and community leaders; (3) four focus groups on informal practices (three with Spaniards on informal practices in Spain, and one with a Romanian family living in Spain to identify and list

[1] In this chapter we present data from the first phase of the ORBITS project, "The Role of Social Transnational Fields in the Emergence, Maintenance and Decay of Ethnic and Demographic Enclaves", funded by the Spanish government (MINECO-FEDER-CSO2015-68,687-P). In the second phase of the project, we are studying a second TSF field between Roquetas de Mar (Spain) and Bistrița-Năsăud (Romania). Quantitative data from this second phase it is not included in this chapter. More information: https://pages.uab.cat/orbits/en.

[2] The sample includes three types of respondent: Romanian citizens currently residing in Spain ($N = 147$), Romanian return migrants living in Romania ($N = 19$), and non-migrants living in Romania ($N = 138$). In Castelló, 73% of the respondents were female, while gender was more equally distributed in Dâmbovița (with 47% females). Respondents' ages varied from 19 to 72 in Castelló, with an average of 43 years, and between 18 and 75 years in Dâmbovița, with an average of 36 years. In both places, about a quarter of the respondents had post-high school and higher education.

informal practices in both countries); and (4) ten semi-structured interviews with Romanian migrants about informal practices. Anonymity and confidentiality were guaranteed and all the participants in this research have signed informed consent forms. The results presented in this chapter are based on these data.

The chapter is structured as follows. The following two sections discuss the theoretical intersections of informality, transnational mobilities, and livelihood strategies, as well as providing a brief overview of Romanian migration to Spain in general and to Castelló de la Plana and Roquetas de Mar in particular. Thereafter, in Sect. "Informality Adaptation: A Schema of Informal Practices and Transnational Migration", we propose a schema of the adaptation of informal practices by TSFs based on our findings and state our expectations. In Sects. "Informal (Im)Mobilities of Romanian Migrants in Spain", "Navigating Processes of Formalization and Informalization: Regularization, Immobility, and Institutionalization" and "(Im)Mobilities and Informality of Non-migrants and Returnees", we present our results on how informal practices change during different phases of migration. Finally, we conclude with some findings on the adaptation of informal practices as an overture to our future research on the topic.

Informality and (Im)Mobilities as Livelihood Strategies in Transnational Social Fields

Livelihood strategies are the repertoire of economic and non-economic strategies through which people strive to make a living (De Haan 2012), a notion that extends to both informal practices and mobilities. Informal practices, defined as "regular strategies to manipulate or exploit formal rules by enforcing informal norms and personal obligations in formal contexts" (Ledeneva 2008: 119), penetrate all aspects of public life globally, including economic, social, and political practices (Polese, Morris et al. 2016). They are embedded in market exchanges, but also in non-economic dimensions such as non-profit activities and in exchanges within personal relationships (Ledeneva 1998). Their pervasiveness suggests that they are adopted irrespective of the economic circumstances of citizens or countries (Morris and Polese 2014: 14). Ledeneva stresses the importance of unwritten rules, or "the know-how needed to 'navigate' between formal and informal sets of constraints" (2011: 722). Informal practices vary across time and space, responding to cultural, political, and economic transformations (Ledeneva 2018; Yalcin-Heckmann 2014). They are also embedded in grey zones "associated with in-betweenness, liminality, marginality and ambiguity" (Ledeneva 2018: 2), as they are "neither hidden nor fully articulated" (p. 11).

In this chapter, we study informal practices within the TSFs through the personal relationships that migrants maintained with one another and with non-migrants in Romania. The TSF concept allows empirical research not only on individual responses and migrant processes, but also on the collective and institutional responses to mobility regimes (Glick Schiller and Salazar 2013) that are constituted by existing regulations, institutions, and infrastructure (Baker 2016) and that either limit or facilitate (im)mobilities, depending on the power relationships within a social field (Glick Schiller and Salazar 2013). From our perspective, mobility within a TSF is not just an individual but a household decision that is also driven by ties beyond the household that follow the chain of migration (MacDonald and MacDonald 1964), as the costs of migration fall with each new wave. Network externalities start to emerge once a certain lower threshold of migrants is reached, which includes a growing institutionalization in the TSF. Moreover, specialized roles can emerge of immobile people who coordinate and optimize mobility across the field (Dahinden 2010; Bashi 2007; Molina et al. 2015).

Although the vast body of literature on informality acknowledges its ubiquity, it is mostly based on evidence from post-socialist countries, the Global South, and developing countries. In Western Europe informality has been studied less often than in post-socialist countries because it is embedded in formality in more complex ways there (Morris and Polese 2014). Thus, the focus on livelihood strategies and TSFs allows us to analyze the complex processes of both formalization and informalization (Boudreau and Davis 2017) along with the phases of Romanian migration in Spain. The purpose is twofold. On the one hand, it allows formal and informal labor practices to be reexamined as an overlapping continuum that goes from formal employment to self-provisioning (Williams and Onoshenko 2014), performed as forms of either resistance or exploitation (Round et al. 2008). This phenomenon can be explained by the decline in formal employment and the process of informalization of previous formal relations (Likic-Brboric et al. 2013; Williams and Onoshenko 2014). Trying to find comparative dimensions of informal economies in the UK, Pahl suggested the expression "forms of informal work" (1990) to describe productive activities that are embedded in social relations, using the household as a unit of analysis to include the domestic economy as the locus on both formal and informal sources of income (Pahl 1984; Martinez Veiga 2005; Molina and Valenzuela 2007).

On the other hand, as developed in Sect. "Informality Adaptation: A Schema of Informal Practices and Transnational Migration", informal practices are culturally dependent and migrants might learn how to get things done at a migrant destination and along TSFs. Thus, migrant adaptation to informality is a process of learning the strategies, mastering the practices, and developing the relationships required to manipulate or exploit the formal rules and context of a given destination.

Overview of Romanian Migration to Spain: The Romanian Enclaves of Castelló de la Plana and Roquetas de Mar

One consequence of European integration is the growing number of European Union (EU) citizens who live permanently in a different country than where they were born. In 2017, 19.3 million (Eurostat 2018b) of the approximately 500 million inhabitants of the EU were living in another EU country, of whom roughly nine million are EU citizens of working age who are active in the labor market (Fries-Tersch et al. 2018). The post-socialist states are particularly active as sending countries: Romania and Poland are the largest sending countries in the EU, with more than three and two million intra-EU emigrants, respectively (Eurostat 2018b).

In Romania, the transition to the market economy has deepened the subalternization and fragmentation of labor (Kideckel 2008; Verdery 2009), which has increased poverty and unemployment, reinforced informal activities as survival strategies (Ciupagea 2002; Neef 2002), and pushed people to move abroad (Sandu 2005; Marcu 2009). In contrast to migrants from other east European countries, the main internal European destinations for Romanians are Italy[3] and Spain, which also speak Romance languages and have a relatively low cost of living.

The first phase of Romanian migration to Spain comprises the period between the first arrivals at the beginning of the 1990s and the lifting

[3] For a description of the formation of the transnational social spaces of Romanians in Italy, see Remus (2008).

of visa requirements in 2002, which smoothed movement within the Schengen area. The second phase was a transitory period between 2002 and the entry of Romania in the EU in 2007, when Romanians could live legally in other EU countries, but still without the legal right to work (Marcu 2009). The third phase stretches from 2007 to today and consists of circular migrations, with a young generation of highly skilled Romanians who are willing to work abroad within the EU and are considered "mobile European citizens" (Marcu 2015).

Over these years, the Romanian population in Spain increased sharply, from a few thousand in 1998 to almost 900,000 in 2012 (see Fig. 1), when it became the largest foreign population in Spain. Although the economic crisis of 2008–2014 drove many Romanians back out of Spain, they continue to be the second-largest foreign population, with 671,985 Romanian nationals living in Spain in 2019 (INE 2020).

Romanians were attracted by the expanding labor markets, both formal and informal, and were supported by social and religious migration networks (Paniagua 2007; Elrick and Ciobanu 2009; Bernat and Viruela

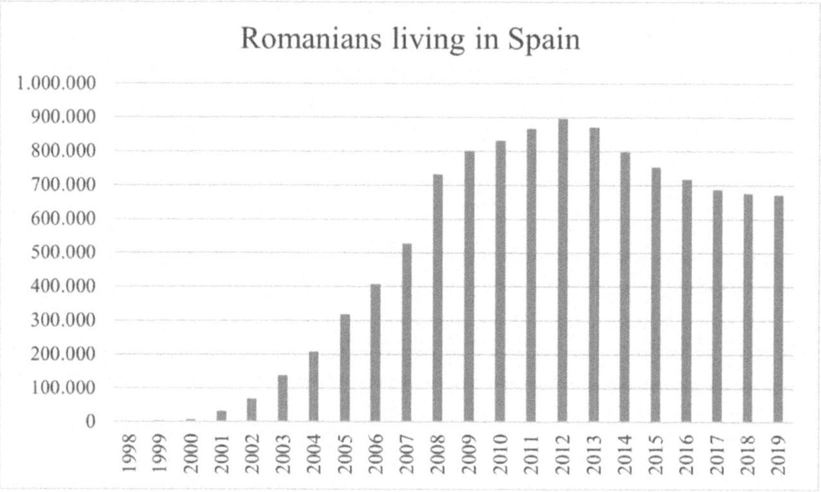

Fig. 1 Population with Romanian nationality in Spain. Own elaboration based on the Padrón continuo. October 21, 2020. www.ine.es

2011; Molina et al. 2018; Marcu 2009), as well as by the growing institutionalization resulting from these networks (De Haas 2010). Geographically, the Romanian population is not homogeneously distributed, but rather forms demographic enclaves within Spain where the percentage of Romanians is particularly high. The Mediterranean cities of Castelló de la Plana and Roquetas de Mar are paradigmatic Romanian enclaves.

At the start of fieldwork in Castelló de la Plana in 2017, the number of inhabitants of Romanian nationality accounted for 15,748 out of the total population of 169,498, or roughly 10% of the total population of the city (INE 2020). Many of them came from a bounded geographical area in Romania: Dâmbovița, a county northwest of Bucharest. The development of this migrant enclave is described elsewhere (Molina et al. 2018). In the case of the city of Roquetas de Mar, in 2017 the total population accounted for 93,363 inhabitants, of whom 24,948 (27.3%) were of foreign nationality, including 8939 Romanians. This means that Romanians are by far the largest population of foreign nationality (35.8% of all foreigners) and 9.5% of the total population (INE 2020), many of whom come from Bistrița-Năsăud, a county in Transylvania, Romania (Fig. 2).

These migration corridors, from Dâmbovița to Castelló de la Plana and from Bistrița-Năsăud to Roquetas de Mar, constitute TSFs where people's permanent and temporal mobilities are facilitated by transnational networks of kinship, friendship, and acquaintanceship, as well as regular channels of communication, through which people move, and goods, services, and information are exchanged.

Informality Adaptation: A Schema of Informal Practices and Transnational Migration

During the socialist period in Romania, instrumental social relations were necessary to overcome scarcities, obtain access to good quality services, or resolve legal issues. Despite the fall of socialism, informal networks and practices are still fundamental to obtaining access to education, health, business, and the labor market (Stoica 2012). In this context, neoliberal reforms amplified the competition for scarce resources, increasing the inequalities of power in patron–client relations in basic sectors such as the health-care system (Stan 2012). In Romania, the "widespread networks of personal exchange and favors [similar to Russian *blat*] have been '*relatii*' (relations), '*cunostinte*' (acquaintances), and '*pile*'" (Stoica 2012: 173), where *pile*—or "*A avea o pilă*"—refers to connections that can smooth

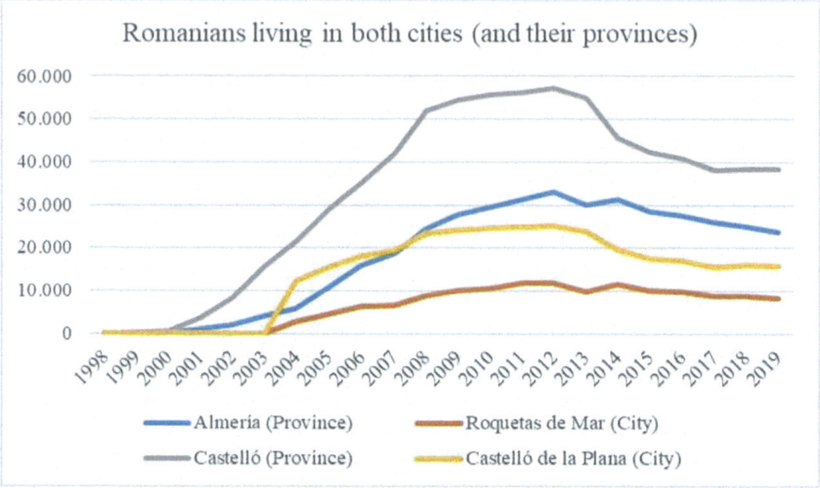

Fig. 2 Population of Romanian nationality living in the Spanish cities of Castelló de la Plana (capital of Castelló province), and Roquetas de Mar (in the province of Almería). Own elaboration based on the Padrón continuo. October 26, 2020. www.ine.es

things out.[4] As Ledeneva shows (2018), the instrumentality of sociability exists with similar patterns under different names all over the world.

In Spain, the informal practice of using social networks to get things done is called *enchufismo*,[5] translated directly as "to plug in" (*enchufar*), a figurative way of denoting the practice of "pulling strings." The verb *enchufar* means "to give a position or appointment to someone who does not merit it, through friendship or political influence" (RAE 2019), while *enchufismo* has been defined as "political and social corruption" (ibid.). It is nonetheless common practice within the endogenous Spanish labor market and in Spanish politics, providing opportunities for corrupt practices. No fewer than 40% of the Spanish population finds work through

[4] "Romanians joked that the acronym for the Romanian Communist Party (Partidul Comunist Român, PCR) stood for 'Props [or Files], Acquaintances, and Relations' (or '*Pile, Cunostinte si Relatii*' in Romanian)" (Stoica 2012: 172).

[5] Also, *amiguismo*, which "indicate(s) a specifically instrumental use of friendship ties" (Giordano, 2018: 102).

informal channels of relatives, friends, and acquaintances, a much higher percentage than in northern European countries such as the Netherlands, Denmark, or Finland (e.g., Pellizari 2010, cited in Vacchiano et al. 2018). Thus, as our participants noted, the Romanian term "*avea o pilă*" translates directly, both in theory and everyday practice, as to have "*enchufe.*"

Other informal practices taken from Romania have been adapted to the destination context in Spain, a country with a large tradition of informality already. For instance, Pitt-Rivers' ethnographic investigation in the 1950s showed how people from a Spanish village made a living relying on undeclared crops and illegal trade (Pitt Rivers 1971). Benton demonstrated the relevance of off-the-books workers in industrial development in Spain after the dictatorship (1990), a practice that has lasted until today (European Commission 2014). Informal economies in Spain have been reported in mining (García 1996), industry (Narotzky 1988), agriculture (Du Bry 2015; Martinez Veiga 2005), and of course domestic service (Viruela 2013), among other sectors. Indeed, the pervasiveness of the hidden economy in Spain is estimated at 23% of national income or 6% of GDP lost to the exchequer (Serrano and Gadea 2005; Lago 2018).

A good practical example of informal work practices in Spain is making *chapuzas*, meaning minor repairs, especially if done shoddily, for which undeclared workers are often hired. Romanians are well known in Spain as *manitas* (handymen) who perform good work cheaply. The worker does not declare the work and the client does not pay 21% VAT, a practice called "to pay in *B* or *en negro*" ("in black" or *la negru* in Romanian). In many cases, the line between informal and corrupt practices is blurred, and the importance of corruption as a subcase of informality (Baez-Camargo and Ledeneva 2017; Polese 2021) should not be underestimated. Indeed, any casual conversation with Romanian migrants in Spain easily includes strong criticism of "Romanian institutions and politicians because of corruption" (Paniagua 2007: 167), which is sometimes pointed out as one of the reasons for leaving—and/or not returning—to Romania. When people are asked about specific cases of corruption, the replies are unclear but sometimes distinguish daily informal practices in making a living—e.g., informal bribery of civil servants, known as *mită* or *şpaga*, that is, to have to pay a bribe to obtain access to health services (Stan 2012) and the diversion of public resources by rent-seeking political and economic elites. For example, one research participant stated that "we have the most expensive cost per kilometer of the highway in Europe because of corruption,"[6] a view confirmed in a report.[7]

According to Zerilli (2005), there is a naturalized and stereotypical view of corruption associated with Romanian-ness that is grounded in the rhetorical devices of history—the Ottoman legacy, the influence of communism, and/or folklore—which is used as "passive resistance" to the unwritten rules of the system, spoken about with irony and jokes, as a form of resistance by the powerless (Scott 1985). Similarly, the stereotyped vision of corruption in Spain is commonly associated with the term *picaresca*,[8] which refers to taking advantage of others—or exploiting the formal system—for one's own benefit. This is also an (uncritically) naturalized cultural characteristic of Spaniards that includes several informal practices such as *gorroneo* (Fradejas-García 2021a), *chanchullo*, or *triquiñuela*, among others.

Both countries' populations also have in common higher perceptions of corruption than other EU countries. The 2017 Eurobarometer report indicates that Romania (68%) and Spain (58%) are at or near the top—first and third, respectively—in the EU whose respondents say that they have been personally affected by corruption (European Commission 2017). For our research participants, the main difference is that practices of informal governance are being prosecuted in Spain,[9] even forcing some politicians to resign, while in Romania this is still unusual. As one interviewee stated, the Romanian population "do this precisely because politicians do it, they already grow up in a culture of robbery. Well, if the president of the government does it, why shouldn't I make a living[10] as

[6] A middle-aged man who has lived for more than twenty years in Spain. Unrecorded informal interview, 19.08.2017.

[7] http://www.consiliulconcurentei.ro/uploads/docs/items/bucket8/id8693/raport.pdf.

[8] In the sixteenth century, so-called picaresque novels depicted a rogue and or anti-hero overcoming the daily life struggles of poor people, wheeling and dealing by creative tricks. Nowadays, this old literature genre coexists in Spain with the Spanish picaresque as a cultural set of deceiving practices.

[9] In 2018, a case of political corruption forced a change of government in Spain after a motion of censure (País 2018).

[10] In Spanish, "*buscarse la vida*" means to do whatever is necessary to survive in a limited situation.

well?"[11] This top-to-bottom continuum of corruption was also described as a survival strategy by another participant:

> People have to do whatever it takes to eat. And if you have a measly salary you have no alternative: people are not going to rob a neighbor, put their hand in his pocket. But someone who can use his/her job to complete his salary is going to do it, and I see that as legit. (...) They have to survive: if you limit them to a miserable salary, obviously corruption will continue to exist at all levels.[12]

Indeed, some exchanges that political sciences and economics would see as corruption are interpreted by participants as legitimate (Polese et al. 2016). Romanian migrants learn very quickly which practices are not welcome, less explicit, or less accepted in Spain.[13] For instance, overall informal practices like attempts to bribe police officers or making informal economic exchanges to obtain access to public resources are not just illegal but are deemed unacceptable by the local population, which may limit its use to dealing with Romanian compatriots or institutions transnationally. Moreover, the experience of learning how informal governance and informal practices of corruption among public servants, politics, and economic elites[14] function in another country produces a reconsideration of the harmful consequences of diverting public resources. Young migrants who came to Spain as children are very clear about this, as in this example:

> In the end, you enter a dynamic as a whole loop. Because you are paying that money to the police and not the state (...) you cannot improve the service. So, you think that the service is crap and to be better served you

[11] A 24-year-old man who migrated to Spain with his family when he was seven. Recorded focus group, 19.06.2019.

[12] A middle-aged woman who has lived in Spain for more than twenty years and is very well-connected transnationally. Recorded interview, 03.03.2020.

[13] This comparison entails ambiguity, since informal practices are more primitive in post-socialism (I pay to get access to a service) and more subtle in neoliberalism (I buy private insurance to get access to a service). We thank Abel Polese for suggesting this insightful distinction.

[14] Poenaru argues that mass mobilizations against corruption in Romania at the beginning of 2017 (e.g., https://www.theguardian.com/world/2017/feb/01/romanians-protests-emergency-law-prisoner-pardons-corruption) have made politicians synonymous with corruption while business practices have been exonerated (2017).

> pay. But then we enter the same thing again. And if you don't get out of the loop, you never end. (…) [T]hat's it, [when visiting Romania] I'm not going to pay a policeman, or a doctor or anything.[15]

In this framework of how migrant adapt to a new context at both the individual and community levels, we contend that selected informal practices that exist in the (post-socialist) sending country are preserved and adapted during the process of migration, while other, mostly illegal, illicit, and harmful practices are abandoned (see this book's introduction). Indeed, transnational migrants also learn to deal with almost two-state legal systems in a coexistence of laws defined as "legal pluralism" that may include others such as customary laws or religious laws (von Benda-Beckmann and von Benda-Beckmann 2016). Thus, some of the practices that are maintained may be used without changes between individuals of the same nationality across the TSF, whereas others may be adapted to the local context of the destination (see Fig. 3).

A constitutive aspect of migrants' sociocultural adaptations and livelihood strategies is learning the local formal rules that are part of the formalization process, as well as the unwritten rules and informal practices that provide contacts and facilitate access to employment, work and training opportunities, schooling, health care, economic investments, or housing, among others. Indeed, the actor's strategic actions and behavior are guided by a toolkit composed of a repertoire of habits, skills, and styles (Swidler 1986), but that also needs adaptation to the new cultural milieu in which the old and the new contexts coexist in a TSF. As Ledeneva puts it, following Wittgenstein, "certain mastery and expertise can only be achieved by dealing with constraints in practice" (Ledeneva 2011: 722).

Thus, from a livelihood perspective, we would expect that migrant households adopt informality when their access to legal residence, employment, and housing through formal channels is restricted. We would further expect them to rely on compatriots in the first phase after migration, when the migrants lack contacts with the indigenous population at the destination, but they can gradually incorporate Spanish contacts that can prove to be instrumental as well if they simultaneously learn the unwritten Spanish rules of informality. From the TSF perspective, we would expect that the informal practices adopted by migrants

15 A 24-year-old man who migrated to Spain with his family when he was seven. Recorded focus group, 19.06.2019.

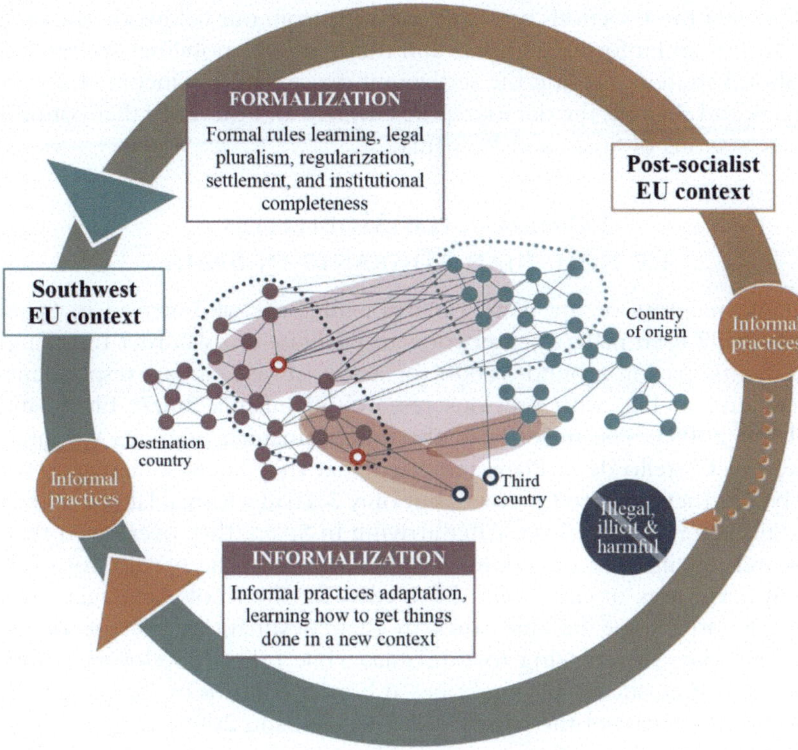

Fig. 3 Schema of adaptation: informalizaton and formalization in a TSF between post-socialism and southwest EU

are diffused socially through networks and that migrants can mobilize local contacts, as well as people in the country of origin (e.g., for the construction of a house in Romania, for entrepreneurship), to get things done. Higher institutional completeness in the TSF can compensate for restricted access to formal channels in Spain and thus reduce informality, but it can also introduce more informality into how migrants deal with these institutions. In this latter case, as the institutions come from the same cultural context, we would expect certain practices to be imported from the country of origin without major adaptations, though some settled migrants are critical of such naturalized informal practices.

The next three sections present the results from our fieldwork, showing how formal and informal activities and transnational mobilities intersect as livelihood strategies easing the settlement process of low-income migrant workers and shaping the demographic enclaves and the TSFs that connect specific regions of Spain and Romania.

Informal (Im)Mobilities of Romanian Migrants in Spain

The vast majority of the first Romanian migrants who arrived in Spain before 2002 used mafia-like networks to facilitate cross-border travel and documents, paying around $1000 for a tourist visa, as our respondents stated, and in line with previous research (Paniagua 2007; Elrick and Ciobanu 2009). The majority of the 147 Romanian migrants we interviewed in Castelló de la Plana indicated that they knew someone in the city before they came to the town, but only 3% had a formal labor contract in Spain before they arrived. After arriving in Spain, they overstayed their visas and remained in the country undocumented. The penalty for overstaying a visa was an entry ban of five years for the whole Schengen area. Consequently, some became stuck in their destinations because of the costs and risks of returning to Romania, while others developed various travel strategies, such as paying bribes at border controls or changing their travel routes, to avoid the ban (Elrick and Ciobanu 2009).

During this phase of migration, as happens in many cases of migration (Mahler 1995; Menjívar 2000), some were supported by informal networks of other Romanians who had settled before them, but many others were left on their own by their contacts. Some were forced to scavenge and to live in abandoned houses, train stations, or squares. After the difficulties with travel and arrival, many respondents indicated that they started to work irregularly[16] without a residence permit until 2002, when the visa requirements changed and the costs of migration fell, opening the door to migration by people without the capital and/or social networks previously required (Elrick and Ciobanu 2009).

[16] As some of our research participants stated, the informal economic practices of using the permits and working papers of another person, lent to a friend or family member, or rented for approximately €150 per month, were common at that time.

The entry of Romania into the EU in 2007 eased access to formal labor markets within the Schengen area and facilitated transnational mobilities. It also increased Romanians' mobility capital or "motility," a term defined as the capacity and potential to be mobile within a social field (Kaufmann et al. 2004). Cheap flights started to replace the two-day bus trips between Romania and Spain, although the fares for the latter are still very low—most affordable one-way ticket cost €69 in 2020. The appearance of numerous formal and informal Romanian road transport companies since the end of the 1990s in Spain favored the arrival of more Romanians, as well as an informal influx of products from Romania to Spain and vice versa—in the beginning costing merely €2 a kilo—some of which are handmade, as well as unlabeled food and alcohol (Petrescu and Rodriquez 2006). This flow of products for trading, gifts, or self-consumption continues and is now even cheaper at €1 a kilo in 2020, facilitating social remittances that reinforce transnational relations (Levitt and Lamba-Nieves 2011) and transnational networks of trust (Tilly 2007). In many cases, things, documents, and money, normally small amounts of cash, are sent via the international passenger buses that ply in both directions between Spain and Romania. This service is widely used, being faster, safer, and cheaper than the regular post, and offering hand delivery. In fact, it can be suggested that the informal Romanian practice of sending small packages via local and regional passenger bus drivers for hand delivery has become transnational. These practices are combined with travel from Spain to Romania by air, bus, or private cars back and forth for holidays, social events such as weddings, and arranging birth and marriage certificates and other bureaucratic necessities (Fradejas-García 2021b). Consequently, 77% of our respondents had traveled to Romania at least once in the last two years, and they spent on average more than five weeks in their home country over those two years.

Thus, to navigate the TSF, mobility and informal activities intersect as strategies to "manipulate or exploit the formal rules" (Ledeneva 2008: 119). For example, some of our respondents complained that fellow citizens received unemployment payments from Spain while they were living in Romania, it being possible to draw these benefits via the internet. A second example is that a few respondents who had acquired Spanish nationality maintained both nationalities and passports—Romanian and Spanish—even though dual citizenship is not allowed in either country. A third example is civil registration strategies in the places of both destination and origin. As a livelihood strategy, international mobility can be combined with certain strategies for obtaining documents and meeting the requirements of the various administrative systems. Indeed, as one interviewee told us, "there is a lot of *trapicheo* (scheming) because people ask for €300 or €400 to register you at their houses [in the Padrón]."[17] Though they do not live there, being registered in the Padrón, a census of inhabitants conducted by local governments, is needed to start the process of obtaining a residence permit in Spain, among other local and regional social and economic benefits.

Summing up, migrants' (im)mobilities and informal practices are intertwined as livelihood strategies for those who live or participate in the TSF. The ability to move due to the regularization of intra-EU mobility by workers, good infrastructure, cheap travel, smooth transnational connections, and social support enables various mobilities—settled, circular, temporal, open-ended, etc.,[18]—that permit people to exploit the formal rules of various states and their institutions.

Navigating Processes of Formalization and Informalization: Regularization, Immobility, and Institutionalization

The parallel processes of formalization and informalization examined here involve administrative regularization, transnational institutionalization, adaptation to formal and informal practices and economies, and moving from undeclared work to the informalization of formal labor, defined by precarity, exploitation, and flexibilization (Likic-Brboric et al. 2013). Our interviewees reported an average of 13.4 years of residence in Spain at the time of the interview, ranging from nine months to 25 years, and around 14% had lived in countries other than Romania and Spain. Their main motivations for migration were seeking better economic conditions or family reunification or both. What is striking is that only 10% had been unemployed in Romania, where 55% were in formal—employment, 45% with a full-time fixed contract, before they migrated to Spain. Consequently, they moved from a formal context of

[17] Middle-aged female who has lived in Spain for more than twenty years and is very well-connected transnationally. Recorded interview, 03.03.2020.

[18] Within the TSFs, we have analyzed various types of international migration mobility: permanent (A → B); circular (A ↔ B); returnees (A → B → A); re-emigration (A → B → C); returnees to the previous enclave (A → B → C → B); and immobile (A ↮ B).

labor in Romania to a situation in Spain of higher salaries but often undeclared jobs. A survey conducted at the end of 2007, a year after the entry of Romania into the EU, estimated that 45% of Romanians of working age living in Spain were working irregularly, 15% were combining regular and irregular work, and 28% working regularly; the other 12% did not respond to the survey (Marcu 2009: 176–177). These data suggest that migration was a subsistence strategy because work formalization was not a motivation for migration.

For some respondents, their arrival was the starting point of a long parallel process of formalization, with access to formal jobs and administrative regularization, and informalization, that is, adapting to new informal practices and learning the new rules of "informality." Formalization went hand in hand with the bilateral and EU policy agreements of 2002–2007, which smoothed the path toward residence regularization, work permits, and family reunifications, as well as in learning local informal practices and developing local informal relations.[19]

The steps from undeclared work to formal job contracts are paradigmatic of these two processes. These usually started with an informal agreement to do an undeclared job. It was seen as a test period and could last several years. Then the employer had the option of formalizing the contract and thus facilitating the regularization. We take an example of this process from our field notes:

> Ironim (a pseudonym) says he was very lucky when he arrived in Castelló in 2000. After six days waiting in Plaza Maria Agustina, the immigrant location for informal work-seekers, someone asked them who wanted to work in a bakery. Nobody was interested, but Ironim boldly accepted, and he is still working in this bakery. In the first two years, he worked without a contract, but the firm supported him in regularizing the situation with a temporary contract in 2002. Since 2006, he has had a full-time contract and has become a pastry chef expert in local sweets.[20]

[19] As some research participants stressed, bars and restaurants in Spain played an important role as informal social spaces where Romanians could develop their (local) personal networks. This relational work has been instrumentalized to access jobs, accommodation, and other basic needs by word of mouth. Also, in comparison with Spaniards, Romanians tip better tips (*baksheesh* in Romanian, *propina* in Spanish), an informal practice which has been proudly maintained in Spain by Romanian migrants.

[20] A male participant from Romania. Brief fieldnotes, CAS032, February 2018.

It is interesting to note that many Romanian employees were sponsored by their Spanish employers in obtaining documents and regularizing their employment situation, although some informal features might remain. For example, employees worked for the formal minimum wage but informally received the money in B ("in black") to complete the salary. In some cases, the relation between employee and employer became blurred, as they became friends, mostly among those who had daily close contact, as in the case of waitresses and domestic workers, in line with Kovács' research among informal childcare workers in Romania (2014).

Permanent contracts provide stability and encourage the formation of a long-term life project. As a result, 32% of our respondents now have a full-time permanent contract, 12% a part-time one. In Castelló de la Plana, the ceramics industry also gives formal contracts that provide an anchor for permanent settlement (Molina et al. 2018), the same role that agribusiness has played in Roquetas de Mar (Fradejas-García et al. 2022). This process has an immobility effect because formal labor is a precious asset, and even with low salaries, hard work, and unpaid extra hours, people tend to maintain their formal jobs in Spain. Indeed, 54% of employed respondents felt that what they earned in Spain was much better than they could in Romania, with a further 29% indicating that it was a little better.

In addition, job stability means meeting new informal contacts at the workplace who can mitigate future uncertainties over employment. One research participant told us that he was not worried about jobs in the future because he already has the contacts, both Romanians and Spaniards, to *enchufarme* in something, that is, to pull strings to find employment if needed, as explained in Sect. "Informality Adaptation: A Schema of Informal Practices and Transnational Migration". In this regard, 71% of our working respondents reported that they had relied on family, friends, and acquaintances to find their current jobs. This percentage is higher than among Spain's general population (see Sect. "Informality Adaptation: A Schema of Informal Practices and Transnational Migration"), and it suggests that Romanian migrants in Spain are using their informal networks more than Spaniards for seeking employment. Finally, when asked whether they had to pay brokers or middlemen to find a job, some participants knew of cases in Romania as well as in Spain. Asked whether she knew anyone who had paid an intermediary to get a job, one interviewee pointed out that "Everyone who goes to a temporary employment agency has to make it [pay to get a job],"[21] showing how the informalization of labor works in the current formal labor market as well.

Along with the process of job regularization and stabilization described earlier, our ethnographic data reveal a parallel process in which household income and reproduction are made up with undeclared jobs and informal economic activities such as house cleaning or temporary or one-off jobs in agriculture, construction, and services, as well as childcare, baking cakes and sweets for parties, renting out rooms in their homes, working as a DJ at social events, and even collaborating in transnational enterprises that import and export cars (Fradejas-García 2021b). The strategy of combining declared and undeclared jobs avoids dependence on a single source of income and can be a buffer against unemployment (Hart 1973). This finding recalls that of Pahl, that families and households with some protected wage labor are better placed to have a surplus in informal forms of work (1984).

It is also important to note that many Romanians have settled in Spain to provide their children with stability. More than 100,000 Romanians with formal residence in Spain in 2016 were under sixteen years old (Ministerio de Trabajo y Economía Social de España 2016), meaning that a young generation of Romanians is growing up in Spain. Indeed, some respondents lived in Romania until their parents regularized their residence in Spain and brought them to Spain after the large-scale family reunifications of 2007 (Marcu 2015). Family reunification is part of a process of settlement that is also accompanied by the institutionalization of Romanian diasporic formations and demographic enclaves and that ends with institutional completeness (Molina et al. 2018) in the form of more favorable Romanian legislation for citizens abroad, bilateral agreements, church construction, the foundation of ethnic associations, and the opening of consulates and cultural centers, such as the Ministry of Romanian Citizens Living Abroad (*Ministerului Pentru Românii de Pretut Indeni*), set up at the end of the 1990s.[22] Locally, institutions like the city council also played a role in supporting migrants with intercultural, social, and health services, and even subsidies to rent houses. In this

[21] A middle-age female living in Spain for more than 20 years and very connected transnationally. Recorded interview, 03.03.2020.

[22] Strategia Națională pentru Românii de Pretut indeni pentru perioada 2017–2020 http://www.mprp.gov.ro/web/strategia-privind-relatia-cu-romanii-de-pretutindeni-2/.

regard, the twinning agreement[23] signed between Castellón de la Plana and Târgovişte, the capital of Dâmboviţa County, in 2017 facilitates the relations between local institutions that are rarely connected politically at the translocal level.

Along with these institutions, other non-governmental, charity organizations also played an important role by providing informal support, such as paying bills, providing food, clothes, books, and language courses, backing up registration processes, and even helping Romanian migrants find jobs. Some respondents highlighted the support of local charitable organizations like Caritas, the Red Cross, and the Orthodox and Adventist churches, whose respective clergy were key community actors, as well as some Romanian associations. One respondent, however, believed that formal and informal Romanian institutions in Spain "have set up their *chiringuitos* in Spain to receive public funding to line their own pockets".[24] In Spain, *chiringuito* means kiosk or beach bar in the street or on the beach, but colloquially it refers to a shady company organized to obtain informal economic benefits. Romanian communities abroad express their horizontal solidarity in other ways. For example, when a migrant passes away and has no repatriation insurance, nor the money to send the body back to be buried in Romania, money boxes are placed in Romanian bars, restaurants, associations, and churches to raise the money and help the family with the costs.

However, as discussed in Sect. "Informality Adaptation: A Schema of Informal Practices and Transnational Migration", the discourses about corruption are somehow naturalized and accepted uncritically. We do not have evidence about informal forms of governance, but our ethnographic work does confirm that these "conspiracy" theories are widespread and limit the participation of Romanians in some of their institutions, such as associations, churches, consulates, cultural centers, etc., while some practices of disruption, such as the failed organization of a referendum

[23] In the same vein, a twinning agreement between Almería (the capital of Almería province, in which Roquetas de Mar is located), and Bistriţa (the capital of Bistriţa-Năsăud) was suggested by politicians from both cities during the commemoration of the Great Union Day of Romania in Spain.

[24] Man of 39 years old living more than 20 years in Spain. Recorded interview, 18.01.2019.

for Romanians living out of the country in 2019,[25] create little trust in Romanian institutions.

The process of regularization, adaptation, and institutionalization of migrants in TSFs analyzed in this section would be incomplete without analyzing its consequences for non-migrants and returnees. In the following section, we analyze the "stuckness" (Cresswell 2012) or immobility effects that are associated with transnational mobilities and informality.

(Im)Mobilities and Informality of Non-migrants and Returnees

During the hardest times of the economic crisis and its aftermath—approximately 2008–2016,—some migrants have returned to Romania or have moved to other EU countries (Viruela and Marcu 2015). Although some have returned to Spain afterward, the decline of the Romanian population in Spain continuous (see Figs. 1 and 2). To be attuned to various types of mobilities we asked the Romanian respondents who resided in Spain to refer us to people who had returned to Romania after living in Spain. After interviewing 19 of those returnees in Romania, we identified three types: (1) highly mobile people who had experienced circular migration or moved to third countries, (2) people who tried migration unsuccessfully and went back, (3) people who have returned for work, care for the family, or retirement. Some had been living for nearly 20 years in Spain and they arrived at the age of retirement. Others had saved money and have returned as entrepreneurs, opening small businesses like bakeries, restaurants, or pensions, sometimes supported by formal programs from the EU and the Romanian government to promote the return of migrants, granting them 40,000€ to fund a start-up.[26]

However, returning to Romania is not necessarily easy. Many migrants have children, mortgages,[27] and properties in Spain—15% of our respondents own a house in Spain. Their networks of support in Romania are

[25] As example, in 2019, many Romanian citizens in Spain were left without depositing their ballot in a referendum to reform the judicial system due to the lack of facilities https://www.lasprovincias.es/politica/ocho-horas-votar-20190531003845-ntvo.html.

[26] E.g., "Romania din Spania" http://romania.startupeuropeaccelerator.eu/main_21/ or "Acasa Entrepreneur" www.romania.startupeuropeaccelerator.eu/main_21/.

[27] Spanish legislation does not facilitate returning a property to the bank as compensation for the mortgage. However, some Romanians have negotiated with the banks, giving back their properties in payment.

generally small after they have spent years abroad, as returned migrants indicated in the interviews and as was confirmed by our survey, which revealed an average of 3.2 family members and friends in the country of origin. Some migrants were also aware that in Romania *informality* may jeopardize their entrepreneurial projects, hampering their return through bribery generally and the lack of contacts, but also blackmail. Still, 58% of our respondents in Spain declared they own a house in Romania, keeping alive the prospect of return after their retirement (Werbner 2013), fed by constructing and maintaining houses in Romania, saving money, and working hard to get a pension from Spain in euros instead of the Romanian official currency, the leu, which is less stable. Along with those who are planning to go back when they retire, many others feel stranded in Spain, living what Sayad (2010) called a *double absence*, not being fully satisfied in either their country of destination or origin.

Returnees and non-migrants both reflected on the difficulties of living in Romania because the cost of living has risen, while salaries and wages are low. When asked about life satisfaction, one non-migrant who was a medical nurse told us that:

> If you want to buy something you like, you should weigh very well the situation and hierarchize the priorities you have. Here in Romania, you can always think about tomorrow and about the fact that you have nothing to eat.[28]

This is consistent with statistics: in 2014, 25.9% of Romanians were living with severe material deprivation, and 25% were at risk of poverty (Eurostat 2018a). Furthermore, the price level for consumer goods and services in Romania is 45% below the average of EU member states in 2019—in Spain the figure is 3.4% below—(Eurostat 2020), but that is not enough to guarantee one's daily maintenance because Romania has also the second-lowest median gross hourly earnings in the EU (€2), when in Spain the figure is €9.80, and the median gross hourly pay in the EU is €13.20 (Eurostat 2014).

The difficulties in making ends meet is a push factor for the poorest segments of the population, who are still thinking of leaving the country. In theory, nowadays it would be easier to migrate within the EU because

[28] A middle-aged female living in Romania. Brief fieldnote, February 2018.

many have family, friends, and acquaintances living abroad, and therefore transnational connections. Yet international mobility is the exception. Most people connected with TSFs who are struggling to live in Romania prefer to stay and wait for better times, managing to compensate for the low wages through the informal economy and remittances—15% of our respondents in Spain send money regularly to Romania.[29] Here, age is an important factor in mobility. A new generation of young non-migrants born after socialism has other forms of cultural capital and takes a different approach to migration and mobility (Marcu 2018). Some have been living in the EU Schengen area for most of their lives, have gone on holidays to other countries, speak foreign languages, and want to move to look for a better quality of life, open values, and more high-skilled career opportunities through mobility within the EU. However, their motivations and imaginaries are focused on northern European countries instead of the informalized and precarious labor markets in Italy or Spain.

In short, most non-migrants and returnees in Romania who are connected to TSFs are experiencing economic difficulties. However, the effects can be mitigated by informal remittances, receiving savings and pensions from their period of residence in Spain, and using their potential to be internationally mobile (e.g., seasonal work, studying abroad, etc.,), thus instrumentalizing their transnational social relations abroad. How informal practices from Spain are used by return migrants in Romania or in other, third countries connected with the TSF has yet to be investigated.

Conclusion

In this chapter, we have explored the relationship between mobility and informality by analyzing how informal practices evolve when people migrate and move within TSFs. The livelihood perspective allows us to analyze informality and (im)mobility as strategies that individuals and households perform to make a living, including the role played by institutions transnationally. The chapter shows that transnational migrants learn

[29] Although the economic crisis had a deep negative impact, workers' remittances from Spain to Romania remain one of the major financial corridors in the EU, amounting to €430 million in 2019 according to Eurostat: https://ec.europa.eu/eurostat/statistics-explained/pdfscache/39326.pdf.

how to navigate and exploit formal rules to get things done by adapting their informal practices to their new context of living.

In the migratory process described in this chapter, we developed two parallel processes. On the one hand, the process of adapting informality entails learning the unwritten rules and selecting, preserving, and adjusting the informal practices that exist in the (post-socialist) sending country to the new context, while other practices are abandoned, mostly illegal, illicit, and harmful ones. On the other hand, the formalization process involves learning the formal rules, the regularization of residence and working permits, and institutional completeness. Thus, transnational networks and geographical mobilities allowed migrants to exploit the grey areas of various formal systems and their institutions in making a living.

As we expected, informality was adopted to cope with formal restrictions on legal residence, employment, and housing in the first phase of migration. At the time, the instrumental use of personal networks was very necessary, relying on those who had also come from their home cities and towns in Romania. This instrumental sociability contributed to the creation of transnational social fields in certain demographic enclaves and facilitated the diffusion of informal practices through social networks in order to obtain access to resources and get things done. The formalization process was eased by the entry of Romania into the EU, which permitted regularization (residence permits) and geographical mobility within the Schengen area, as well as the more institutional support of local and Romanian organizations. Moreover, the creation of demographic enclaves was smoothed by the formal labor markets associated with robust industrial districts, such as the ceramic industry in Castelló de la Plana and agribusiness in Roquetas de Mar, which also provided several formal and informal forms of work and employment opportunities in agriculture, construction, and services. Immobility at the destination (settlement) is thus a livelihood strategy for Romanian migrants, who thereby strengthened their informal networks and learned how to master both local and transnational informal practices. This practical knowledge of informal practices provided livelihood resources to their transnational families and friends, allowing the latter to move internationally if they wished and helping them to mitigate economic uncertainties by providing other forms of work, new jobs, and advantages in order to get things done when needed.

Finally, the existence of a TSF does not just open up new avenues for the migration and adaptation of informal practices, it might also contribute to the development of new ones, especially those that take advantage of the bridges and grey areas between different formal systems, thus creating new values and allowing people to get ahead.

Acknowledgements This work was funded by a doctoral grant to the first author (FPI grant number BES-2016-076859) and by the ORBITS project, both funded by the Ministry of Economy and Competitiveness, Government of Spain (MINECO-FEDER: CSO2015-68687-P, 2016–2020). We are grateful for the support of Renáta Hosnedlová and the work of Marian-Gabriel Hâncean and his team in Romania (Bianca Mihăilă, Adelina Stoica and Iulian Oană).

References

Baez-Camargo, Claudia, and Alena Ledeneva. 2017. "Where Does Informality Stop and Corruption Begin? Informal Governance and the Public/Private Crossover in Mexico, Russia and Tanzania." *The Slavonic and East European Review* 95 (1): 49–75. http://www.jstor.org/stable/10.5699/slaveasteurorev2.95.1.0049.

Baker, Beth. 2016. "Regime." In *Keywords of Mobility: Critical Engagements*, edited by Noel B. Salazar and Kiran Jayaram. New York: Berghahn Books.

Baldwin-Edwards, Martin, and Joaquín Arango. 1999. *Immigrants and the Informal Economy in Southern Europe*. London: Frank Cass.

Bashi, V. 2007. *Survival of the Knitted: Immigrant Social Networks in a Stratified World*. Stanford: Stanford University Press.

von Benda-Beckmann, Franz, and Keebet von Benda-Beckmann, eds. 2016. *Mobile People, Mobile Law Expanding Legal Relations in a Contracting World*. London: Routledge.

Benton, Lauren. 1990. *Invisible Factories: The Informal Economy and Industrial Development in Spain*. New York: State University on New York Press.

Berggren, Erik, Branka Likić-Brborić, Gülay Toksöz, and Nicos Trimiklinotis. 2007. *Irregular Migration, Informal Labour and Community: A Challenge for Europe*. Maastricht: Shaker Publishing.

Bernat, J. S., and R. Viruela. 2011. "The Economic Crisis and Immigration: Romanian Citizens in the Ceramic Tile District of Castelló (Spain)." *Journal of Urban and Regional Analysis* 3 (1): 45–65.

Boudreau, Julie-Anne, and Diane E. Davis. 2017. "Introduction: A Processual Approach to Informalization." *Current Sociology* 65 (2): 151–66. https://doi.org/10.1177/0011392116657286.

Breton, R. 1968. "Institutional Completeness of Ethnic Communities and the Personal Relations of Immigrants." In *Canadian Society: Sociological Perspectives*, edited by B. R. Blishen, F. E. Jones, K. D. Naegele, and J. Porter,

77–94. London: Palgrave Macmillan UK. https://doi.org/10.1007/978-1-349-81601-9_5.

Bruns, Bettina, and Judith Miggelbrink. 2012. "Introduction." In *Subverting Borders: Doing Research on Smuggling and Small-Scale Trade*, edited by Bettina Bruns and Judith Miggelbrink. Wiesbaden: VS Verlag.

Bruns, Bettina, Judith Miggelbrink, and Kristine Müller. 2011. "Smuggling and Small-Scale Trade as Part of Informal Economic Practices." Edited by Abel Polese. *International Journal of Sociology and Social Policy* 31 (11/12): 664–80. https://doi.org/10.1108/01443331111177869.

Du Bry, Travis. 2015. "Agribusiness and Informality in Border Regions in Europe and North America: Avenues of Integration or Roads to Exploitation?" *Journal of Borderlands Studies* 30 (4): 489–504. https://doi.org/10.1080/08865655.2016.1165132.

Cieslewska, Anna. 2014. "From Shuttle Trader to Businesswomen: The Informal Bazaar Economy in Kyrgyzstan." In *The Informal Post- Socialist Economy: Embedded Practices and Livelihoods*, edited by Jeremy Morris and Abel Polese, 121–34. London and New York: Routledge.

Ciupagea, Constantin. 2002. "Economic Functions of Informal Activities in Romania." In *The Social Impact of Informal Economies in Eastern Europe*, edited by Rainer Neef and Manuela Stanculescu, 188–210. Aldershot, Hants: Ashgate.

Cresswell, Tim. 2012. "Mobilities II: Still." *Progress in Human Geography* 36 (5): 645–53.

Dahinden, Janine. 2010. "The Dynamics of Migrants' Transnational Formations: Between Mobility and Locality." *Diaspora and Transnationalism: Concepts, Theories and Methods*, no. December: 51–71. https://doi.org/10.5744/florida/9780813044583.003.0003.

Elrick, Tim, and Oana Ciobanu. 2009. "Migration Networks and Policy Impacts: Insights from Romanian-Spanish Migrations." *Global Networks* 9 (1): 100–116. https://doi.org/10.1111/j.1471-0374.2009.00244.x.

European Commission. 2014. "Undeclared Work in the European Union. Special Eurobarometer 402." https://doi.org/10.2767/37041.

———. 2017. "'Corruption' Report—Special Eurobarometer 470." https://doi.org/10.2837/513267.

Eurostat. 2014. "Earnings Statistics." https://ec.europa.eu/eurostat/statistics-explained/index.php/Earnings_statistics#General_overview.

———. 2018a. "Living Conditions in Europe." https://ec.europa.eu/eurostat/documents/3217494/9079352/KS-DZ-18-001-EN-N.pdf/884f6fec-2450-430a-b68d-f12c3012f4d0.

———. 2018b. "Migration and Migrant Population Statistics." Eurostat. http://epp.eurostat.ec.europa.eu/statistics_explained/index.php/Migration_and_migrant_population_statistics#Foreign_and_foreign-born_population.

———. 2020. "Comparative Price Levels of Consumer Goods and Services." https://ec.europa.eu/eurostat/statistics-explained/index.php/Comparative_price_levels_of_consumer_goods_and_services.

Falzon, Mark Anthony. 2009. *Multi-sited Ethnography: Theory, Praxis and Locality in Contemporary Research*. Farnham: Ashgate.

Fradejas-García, Ignacio. 2021a. "Gorroneo." In *The Global Encyclopaedia of Informality: Understanding Social and Cultural Complexity*. Volume III, edited by Alena Ledeneva. UCL Press.

———. 2021b. "Informality on Wheels: Informal Automobilities Beyond National Boundaries." *Migration Letters*, 18 (2): 149–163. https://doi.org/10.33182/ml.v18i2.1175

Fradejas-García, Ignacio, José Luis Molina, and Miranda J. Lubbers. 2022. "Migrant Entrepreneurs in the 'Farm of Europe': The Role of Transnational Structures." *Globalizations*, in press.

Fries-Tersch, Elena, Tugce Tugran, Rossi Ludovica, and Bradley Harriet. 2018. "2017 Annual Report on Intra-EU Labour Mobility." https://doi.org/10.2767/974797.

Garapich, Michał P. 2016. "'I Don't Want This Town to Change': Resistance, Bifocality and the Infra-Politics of Social Remittances." *Central and Eastern European Migration Review* 5 (2): 155–66. https://doi.org/10.17467/ceemr.2016.17.

García, José Luis. 1996. *Prácticas Paternalistas. Un Estudio Antropológico Sobre Los Mineros Asturianos*. Barcelona: Ariel.

Giordano, Christian. 2018. "Amici, Amigos (Mediterranean and Latin America)." In *The Global Encyclopaedia of Informality: Understanding Social and Cultural Complexity*. Volume 1, edited by Alena Ledeneva, 102–6. London: UCL Press.

Glick Schiller, Nina, and Noel B. Salazar. 2013. "Regimes of Mobility Across the Globe." *Journal of Ethnic and Migration Studies* 39 (2): 183–200. https://doi.org/10.1080/1369183X.2013.723253.

De Haan, Leo J. 2012. "The Livelihood Approach: A Critical Exploration." *Erdkunde* 66 (4): 345–57. https://doi.org/10.3112/erdkunde.2012.04.05.

De Haas, Hein. 2010. "The Internal Dynamics of Migration Processes: A Theoretical Inquiry." *Journal of Ethnic and Migration Studies* 36 (10): 1587–617. https://doi.org/10.1080/1369183X.2010.489361.

Hart, Keith. 1973. "Informal Income Opportunities and Urban Employment in Ghana." *The Journal of Modern African Studies* 11 (1): 61–89.

Heckathorn, Douglas D. 1997. "Respondent-Driven Sampling: A New Approach to the Study of Hidden Populations." *Social Problems* 44 (2): 174–99. https://doi.org/10.2307/3096941.

———. 2002. "Respondent-Driven Sampling II: Deriving Valid Population Estimates from Chain-Referral Samples of Hidden Populations." *Social Problems* 49 (1): 11–34.

INE. 2020. "Spain Statistics 2020." Madrid. www.ine.es/en/.

Kalir, Barak, and Malini Sur, eds. 2012. *Transnational Flows and Permissive Polities*. Amsterdam: Amsterdam University Press. https://doi.org/10.2307/j.ctt45kfk8.

Kaufmann, Vincent, Manfred Max Bergman, and Dominique Joye. 2004. "Motility: Mobility as Capital." *International Journal of Urban and Regional Research* 28 (4): 745–56. https://doi.org/10.1111/j.0309-1317.2004.00549.x.

Kideckel, David A. 2008. *Getting by in Postsocialist Romania*. Bloomington and Indianapolis: Indiana University Press.

Kovács, Borbala. 2014. "Nannies and Informality in Romanian Local Childcare Markets." In *The Informal Post-Socialist Economy: Embedded Practices and Livelihoods*, edited by Jeremy Morris and Abel Polese, 67–. New York: Routledge.

Lago, Santiago. 2018. *Economía sumergida y fraude fiscal en España: ¿qué sabemos? ¿Qué debemos hacer?*. Madrid: Estudios de la fundación FUNCAS. Serie economía y sociedad.

Ledeneva, Alena. 1998. *Russia's Economy of Favours: Blat, Networking and Informal Exchange*. Cambridge: Cambridge University Press.

———. 2008. "Blat and Guanxi: Informal Practices in Russia and China." *Comparative Studies in Society and History* 50 (1): 118–44. https://doi.org/10.1017/S0010417508000078.

———. 2011. "Open Secrets and Knowing Smiles." *East European Politics and Societies* 25 (4): 720–36. https://doi.org/10.1177/0888325410388558.

———. 2018. *The Global Encyclopaedia of Informality: Understanding Social and Cultural Complexity*. Volume 1 & 2. London: UCL Press. https://doi.org/10.14324/111.9781911307907.

Levitt, Peggy, and Deepak Lamba-Nieves. 2011. "Social Remittances Revisited." *Journal of Ethnic and Migration Studies* 37 (1): 1–22. https://doi.org/10.1080/1369183X.2011.521361.

Levitt, Peggy, and Nina Glick Schiller. 2004. "Conceptualizing Simultaneity: A Transnational Social Field Perspective on Society." *International Migration Review* 38 (3): 1002–39. https://doi.org/10.1111/j.1747-7379.2004.tb00227.x.

Likic-Brboric, Branka, Zoran Slavnic, and Charles Woolfson. 2013. "Labour Migration and Informalisation: East Meets West." *International Journal of Sociology and Social Policy* 33 (11/12): 677–92. https://doi.org/10.1108/IJSSP-10-2012-0087.

Lomnitz, Larissa Adler. 1988. "Informal Exchange Networks in Formal Systems: A Theoretical Model." *American Anthropologist* 90 (1): 42–55. https://doi.org/10.1525/aa.1988.90.1.02a00030.

MacDonald, John S., and Leatrice D. MacDonald. 1964. "Chain Migration Ethnic Neighborhood Formation and Social Networks." *The Milbank Memorial Fund Quarterly* 42 (1): 82–97. https://doi.org/10.2307/3348581.

Mahler, Sarah J. 1995. *American Dreaming: Immigrant Life on the Margins*. Princeton, NJ: Princeton University Press.

Malkki, Liisa H. 1992. "National Geographic: The Rooting of Peoples and the Territorialization of National Identity among Scholars and Refugees." *Cultural Anthropology* 7 (1): 22–44.

Marcu, Silvia. 2009. "Del Este Al Oeste. La Migracion de Rumanos En La Unión Europea: Evolucion y Caracteristicas." *Migraciones Internacionales* 5 (1): 155–91.

———. 2015. "From the Marginal Immigrant to the Mobile Citizen: Reconstruction of Identity of Romanian Migrants in Spain." *Population, Space and Place* 21 (6): 506–17. https://doi.org/10.1002/psp.1845.

———. 2018. "Mobility as a Support Strategy: Linked Lives Through the Life Course Among Eastern Europeans in Spain." *Geoforum* 97: 335–42. https://doi.org/10.1016/j.geoforum.2018.09.034.

Marcus, George E. 1995. "Ethnography in/of the World System: The Emergence of Multi-Sited Ethnography." *Annual Review of Anthropology* 24: 95–117.

Martinez Veiga, Ubaldo. 2005. "El Otro Desempleo: El Sector Informal." In *Entre Las Gracias y El Molino Satánico: Lecturas de Antropología Económica*, edited by Paz Moreno Feliú, 411–62. Madrid: UNED.

Menjívar, Cecilia. 2000. *Fragmented Ties: Salvadoran Immigrant Networks in America*. Berkeley: University of California Press.

Ministerio de Trabajo y Economía Social de España. 2016. "Anuario de Estadísticas." http://www.mites.gob.es/es/estadisticas/anuarios/index.htm.

Molina, José Luis, and Hugo Valenzuela. 2007. *Invitación a La Antropología Económica*. Barcelona: Bellaterra.

Molina, José Luis, Luis Martínez-Cháfer, Francesc Xavier Molina-Morales, and Miranda J. Lubbers. 2018. "Industrial Districts and Migrant Enclaves: A Model of Interaction." *European Planning Studies* 26 (6): 1160–80. https://doi.org/10.1080/09654313.2018.1455808.

Molina, José Luis, Sören Petermann, and Andreas Herz. 2015. "Defining and Measuring Transnational Social Structures." *Field Methods* 27 (3): 223–43. https://doi.org/10.1177/1525822X14556254.

Morris, Jeremy, and Abel Polese. 2014. "Introduction: Enduring Practices, Entwined Livelihoods." In *The Informal Post-Socialist Economy: Embedded*

Practices and Livelihoods, edited by Jeremy Morris and Abel Polese, 1–18. London and New York: Routledge.

Mouw, Ted, Sergio Chavez, Heather Edelblute, Ashton Verdery, North Carolina, and Chapel Hill. 2014. "Binational Social Networks and Assimilation: A Test of the Importance of Transnationalism." *Social Problems* 61 (3): 329–59. https://doi.org/10.1525/sp.2014.12192.

Narotzky, Susana. 1988. *Trabajar En Familia. Mujeres, Hogares y Talleres*. Valencia: Edicions Alfons El Magnànim.

Neef, Rainer. 2002. "Aspects of the Informal Economy in a Transforming Country: The Case of Romania." *International Journal of Urban and Regional Research* 26 (2): 299–322. https://doi.org/10.1111/1468-2427.00381.

Pahl, Ray E. 1984. *Divisions of Labour*. Oxford [Oxfordshire]; New York, NY: B. Blackwell.

———. 1990. "De l'«economia Informal» a «formes de Treball»models i Tendències Transnacionals." *Papers: Revista de Sociología* 34: 63–95. http://dx.doi.org/10.5565/rev/papers/v34n0.1561.

País, El. 2018. "Moción de Censura." *El País*. https://elpais.com/tag/mociones_censura/a.

Paniagua, Julián. 2007. *Red Migratoria de Los Rumanos En España. Adventistas, Ortodoxos y Gitanos*. Torres de Alameda: Fugaz Ediciones.

Petrescu, Alexandru, and Victor Rodriquez. 2006. "Los Rumanos Ya Tienen Su Propio «SEUR»." *EL MUNDO*. https://www.elmundo.es/suplementos/cronica/2006/557/1151791204.html.

Pitt Rivers, Julian A. 1971. *Los Hombres de La Sierra. Ensayo Sociológico Sobre Un Pueblo de Andalucía*. Barcelona: Grijalbo.

Poenaru, Florin. 2017. "Romanian Protests: A Cake with Three Layers (and a Cherry on Top)." *FocaalBlog*, 27. www.focaalblog.com/2017/02/27/florin-poenaruromanian-%0Aprotests-a-cake-with-three-layers-and-a-cherry-on-top.

Polese, Abel. 2021. "What Is Informality? About 'the Art of Bypassing the State' and Its Theorizations in Post-Socialist Spaces (and Beyond)." *Eurasian Geography and Economics*, in press.

Polese, Abel, Borbála Kovács, and David Jancsics. 2016. "Practici Informale Nocive Sau Inofensive?: 'În Pofida' Și 'Dincolo de' Stat; Cazul Ungariei Și României." *Studia Politica: Romanian Political Science Review* 16 (2): 219–41. https://www.ssoar.info/ssoar/handle/document/51763.

Polese, Abel, Colin C. Williams, Ioana A. Horodnic, and Predrag Bejakovic. 2017. *The Informal Economy in Global Perspective: Varieties of Governance*. Cham: Palgrave Macmillan. https://doi.org/10.1007/978-3-319-40931-3.

Polese, Abel, Jeremy Morris, and Borbala Kovács. 2016. "'States' of Informality in Post-Socialist Europe (and Beyond)." *Debatte* 24 (3): 181–90. https://doi.org/10.1080/0965156X.2016.1261216.

Portes, Alejandro, Luis Eduardo Guarnizo, and William J. Haller. 2002. "Transnational Entrepreneurs: An Alternative Form of Immigrant Economic Adaptation." *American Sociological Review* 67 (2): 278. https://doi.org/10.2307/3088896.

RAE. 2019. "Gorronear; Gorrón; Pegote; Godería; Rozavillón." In *Diccionario de La Lengua Española*. https://dle.rae.es/?id=JMrUqze.

Remus, Gabriel Anghel. 2008. "Changing Statuses: Freedom of Movement, Locality and Transnationality of Irregular Romanian Migrants in Milan." *Journal of Ethnic and Migration Studies* 34 (5): 787–802. https://doi.org/10.1080/13691830802106069.

Round, John, Colin C. Williams, and Peter Rodgers. 2008. "Everyday Tactics and Spaces of Power: The Role of Informal Economies in Post-Soviet Ukraine." *Social & Cultural Geography* 9 (2): 171–85. https://doi.org/10.1080/14649360701856110.

Routh, Supriya. 2011. "Building Informal Workers Agenda: Imagining 'Informal Employment' in Conceptual Resolution of 'Informality'." *Global Labour Journal* 2 (3): 208–27.

Salazar, Noel B. 2019. "Mobility." *REMHU: Revista Interdisciplinar Da Mobilidade Humana* 27 (57): 13–24. https://doi.org/10.1590/1980-85852503880005702.

Sandu, D. 2005. "Emerging Transnational Migration from Romanian Villages." *Current Sociology* 53 (4): 555–82.

Sayad, Abdelmayek. 2010. *La Doble Ausencia. De Las Ilusiones Del Emigrado a Los Padecimientos Del Inmigrado*. Barcelona: Anthropos Editorial.

van Schendel, Willem, and Itty Abraham. 2005. *Illicit Flows and Criminal Things: States, Borders, and the Other Side of Globalization*. Bloomington: Indiana University Press.

Scott, James C. 1985. *Weapons of the Weak. Everyday Forms of Peasant Resistance*. New Haven and London: Yale University Press.

Serrano, Jose Maria, and Maria Dolores Gadea. 2005. "What Causes the Hidden Economy in Spain?" *Applied Financial Economics Letters* 1 (3): 143–50. https://doi.org/10.1080/17446540500079885.

Sheller, Mimi, and John Urry. 2006. "The New Mobilities Paradigm." *Environment and Planning A: Economy and Space* 38 (2): 207–26. https://doi.org/10.1068/a37268.

Spyer, Patricia. 1988. *Border Fetishisms: Material Objects in Unstable Spaces*. New York: Routledge.

Stan, Sabina. 2012. "Neither Commodities nor Gifts: Post-Socialist Informal Exchanges in the Romanian Healthcare System." *Journal of the Royal Anthropological Institute* 18 (1): 65–82. https://doi.org/10.1111/j.1467-9655.2011.01731.x.

Stoica, Augustin. 2012. "Old Habits Die Hard? An Exploratory Analysis of Communist-Era Social Ties in Post-Communist Romania." *European Journal of Science and Theology* 8 (June): 171–93.

Swidler, Ann. 1986. "Culture in Action: Symbols and Strategies." *American Sociological Review* 51 (2): 273. https://doi.org/10.2307/2095521.

Tilly, Charles. 2007. "Trust Networks in Transnational Migration." *Sociological Forum* 22 (1): 3–24. https://doi.org/10.1111/j.1573-7861.2006.00002.x.

Turaeva, Rano. 2014. "Mobile Entrepreneurs in Post-Soviet Central Asia." *Communist and Post-Communist Studies* 47 (1): 105–14. https://doi.org/10.1016/j.postcomstud.2014.02.001.

Urinboyev, Rustamjon. 2016. "Migration and Transnational Informality in Post-Soviet Societies: Ethnographic Study of 'Po Rukam' Experiences of Uzbek Migrant Workers in Moscow." In *Migrant Workers in Russia: Global Challenges of the Shadow Economy in Societal Transformation*, edited by A.-L. Heusala and K. Aitamurto, 70–93. London: Routledge.

Urinboyev, Rustamjon, and Abel Polese. 2016. "Informality Currencies: A Tale of Misha, His Brigada and Informal Practices Among Uzbek Labour Migrants in Russia." *Journal of Contemporary Central and Eastern Europe* 24 (3): 191–206. https://doi.org/10.1080/0965156X.2016.1261215.

Vacchiano, Mattia, Joel Martí, Lidia Yepes-Cayuela, and Joan Miquel Verd. 2018. "Las Redes Personales En La Inserción Laboral Juvenil En Tiempos de Crisis. Un Análisis En Barcelona / Personal Networks in Job Insertion among Young Adults in Times of Crisis: An Analysis in Barcelona." *Revista Española de Investigaciones Sociológicas* 161: 121–40. https://doi.org/10.5477/cis/reis.161.121.

Verdery, Katherine. 2009. "The Rural Contribution to Emigration in 1990s Romania." *Sociologie Românească* 7 (3): 21–36.

Viruela, Rafael. 2013. "Entre Dos Crisis Económicas . Estrategias de Los Rumanos En El Mercado de Trabajo Español." *Política y Sociedad* 50 (3): 981–1008.

Viruela, Rafael, and Silvia Marcu. 2015. "Estrategias de Los Immigrantes Del Este En España Durante La Crisis Económica." *Migraciones* 38: 35–59. mig.i38.y2015.002.

Werbner, Pnina. 2013. "Migration and Transnational Studies: Between Simultaneity and Rupture." In *A Companion to Diaspora and Transnationalism*, edited by A. Quayson and G. Daswani, 106–24. Oxford: Blackwell.

Williams, Colin C., and Olga Onoshenko. 2014. "The Diverse Livelihood Practices of Healthcare Workers in Ukraine." In *The Informal Post-Socialist Economy: Embedded Practices and Livelihoods*, 21–34. London and New York: Routledge.

Wimmer, Andreas, and Nina Glick Schiller. 2002. "Methodological Nationalism and Beyond: Nation-State Building, Migration and the Social Sciences." *Global Networks* 2 (4): 301–34. https://doi.org/10.1111/1471-0374.00043.

Yalcin-Heckmann, Lale. 2014. "Informal Economy Writ Large and Small: From Azerbaijani Herb Traders to Moscow Shop Owners." In *The Informal Post-Socialist Economy: Embedded Practices and Livelihoods*, edited by Jeremy Morris and Abel Polese, 165–85. London and New York: Routledge.

Zerilli, Filippo M. 2005. "Corruption, Property Restitution, and Romanianness." In *Corruption: Anthropological Perspectives*, edited by C. Shore and D. Haller, 83–102. London and Ann Arbor: Pluto Press.

Restaurant Backyards, Food Stores, and Temples: Invisibility, Informal Labour Practices, and Migrant Networks in the Suburbs of Warsaw

Karolina Bielenin-Lenczowska and Helena Patzer

Introduction: Ethnography at Home, Ethnography Close to Home

Vignette 1

We are walking through a suburban village along a busy motorway, still surrounded by cabbage fields on the fringes. Concrete family houses in all shades of grey line the streets, along which you can notice many small shops and business establishments. In the middle of all this, a golden sikh flag is hanging high on a pole. Along the main street brightly colored signs in a decorative Hindi-like font invite customers to come in and taste the "Orient". Not too distant from them, we pass a signboard, which

K. Bielenin-Lenczowska · H. Patzer (✉)
University of Warsaw, Warsaw, Poland
e-mail: h.patzer@uw.edu.pl

K. Bielenin-Lenczowska
Federal University of Santa Catarina, Florianópolis, Brazil

A. Polese (ed.), *Informality, Labour Mobility and Precariousness*, International Political Economy Series,
https://doi.org/10.1007/978-3-030-82499-0_3

reads "Turkish barber, best in town" in Turkish, Polish, and English. Emerging diversity on the outskirts of Warsaw.

The chapter discusses the visible shift to diversity or even superdiversity in the character of a suburban neighbourhood in the outskirts of Warsaw (cf. Vertovec 2007), illustrated by the vignette above, and tries to uncover what it means for the different migrant groups in terms of access to the labour market, as well as the formal and informal practices they engage in. By doing this we challenge the understanding of informality, adding the context of migrant workers in Central-Eastern Europe. Our particular aim is showing how widely accepted informal practices in the labour market in Poland do not work for migrants, and so have to be thought anew and adapted to the condition of mobile precarious workers.

The case study is based on observations in a suburban municipality of Warsaw inhabited by a considerable, and increasing, number of migrants from countries such as Vietnam, Ukraine, Belarus, India, and Turkey.[1] As many other suburban areas, it is semi-rural, with a significant number of people involved in agriculture and at the same time multiple small businesses, both of which attract workers that, in turn, increase the demand for other services and goods. The high number of individual family houses, which all need cleaners, babysitters, gardeners, and construction workers, is an opportunity for migrants (Kindler 2011; Kindler and Szulecka 2013). The municipality is a unique case, as since the 1990s it has attracted a sizeable migrant population and it lies close to the biggest Asian trading centre in Poland (Hüwelmeier 2013). Nowadays, the place is experiencing a significant increase in the number of migrants living in the municipality, especially those coming from India and other South-Asian countries, as well as from Ukraine and Belarus (Piłat 2013). The situation is very dynamic, under the influence of the changing migration policy (Matkowska 2012; Duszczyk 2014), the shifting labour market, and the possibilities arising for migrants. It is through informal channels of communication that future migrants learn about the opportunity to go to Poland—an Eastern-European country, which is an EU member, and so offers access to the wider European labour market. Poland thus emerges as an attractive immigration destination, said to be relatively

[1] The official data of the Municipality Office from 2017 states that there are 1064 foreigners on temporary residence and 353 foreigners on permanent residence in the studied area. According to the data collected by the Institute of Public Affairs, one in seven inhabitants is a foreigner (Piłat 2013: 138).

cheap and less strict in terms of paperwork (Andrejuk and Oleksiyenko 2018: 82–83; see also: Kindler 2014). This attraction, however, connects with being pushed into the informal economy or migrant economic niches, which makes migrants more prone to abuse and discrimination, and leaves them non-integrated into the official labour market.

Migrants mostly settle in the cheaper districts of the city, among them in the suburb described in this chapter. Although for many of the new inhabitants the suburbia is mainly a sleeping place, one can observe that migrants domesticate and reorganize them in many different ways, leaving traces of their presence and becoming visible in the public space (Wise 2010). Among various migrant-led initiatives, numerous small migrant enterprises, such as ethnic restaurants and oriental food stores, have appeared in many places in Poland. What is worth highlighting, in the studied neighbourhood it is mostly Indians, Turks, and Vietnamese who establish businesses, which serve both the migrant and non-migrant communities. Migrant children attend the local kindergartens and primary school, posing a challenge to the institutions and making them rethink their everyday practices. There is no overarching state policy on diversity in educational institutions. To add to the diversity of the neighbourhood, there are multiple places of worship used by migrant communities. "Gurudwara Singh Sabha", the only Sikh temple in Central Europe, has been present in the municipality since 2009[2] and is well-attended by the Indian community. Alongside the temple, an unmarked Muslim prayer house can be found in the village. The prayer house is used mostly by the Turkish community, however, Muslims from other countries also come for the Friday *namaz*. Both of these places of worship are housed in private houses rented from the local community, and do not bear almost any visible markers of being a temple. The only sign outside the Sikh temple is a golden flag and an inscription on the gate. What is more, two Vietnamese Buddhist temples, a Muslim prayer room, and a Hindu and a Hare Krishna temple operate in the nearby villages.

We started our fieldwork by walking around the neighbourhood and noting the signs of diversity. We walked into shops to buy products, we observed how people went around their daily business. Our first visit to the Sikh temple consisted mostly of being there and observing, we hardly

[2] https://www.pism.pl/files/?id_plik=18701 (accessed 05.03.2019).

exchanged any words with the men present there, yet got gesture instructions from them what to do—tie a kerchief around our heads, take off our shoes, go up to the temple, and then come down to eat the shared meal. The instructions connected to eating were especially important—how to roll the *naan*, how to take the sauce with it, how to ask for more food, how to thank and show you had enough. Our first longer conversations also happened in the Sikh temple, where we learned about the history of the place and the precarious lives of Indian migrants in Poland. Staying for days at a time in the neighbourhood and coming everyday to the temple, we gained the trust of people. Some of the Sikhs invited us to their houses, others shared long conversations with us in the courtyard of the temple, while sipping the warm spiced *chai*. After these first encounters, we started to expand the network of interlocutors to other migrants, not only from India, but also from Vietnam, Turkey, and Ukraine, visiting them at their places of worship and work: restaurants, shops, and temples. We interacted shortly with many migrants and collected individual life stories.

The research was part of the project "Migrants and refugees domesticating space. The anthropology of mobility in the vicinity of Warsaw" (2017–2019),[3] and consisted of weekly stays and shorter visits in the field, accompanied by online ethnography. Our research participants were mostly Indians who migrated to Poland in the years 2014–2017. We also talked to individuals who have lived in Poland for 10–30 years and created the diversity of the studied neighbourhood, among them Indians (with a significant number of Sikhs), Turks, Ukrainians, Vietnamese, and Pakistanis. We have done participant observation in the community and talked at length to 30 migrants, while interacting with many others at the different spaces we observed. Among our interlocutors, there were more men than women, which is characteristic especially of Indian (Bąbiak and Gmaj 2015) and Turkish (Stryjewski 2012) migration to Poland in the recent years. These were both employees and owners of local businesses, students, and people waiting to receive their temporary residence permits.

[3] The research was done with a group of students from the Institute of Ethnology and Cultural Anthropology, University of Warsaw. We would like to thank the students participating in the group—Magda Bodzan, Julia Bogusz, Martyna Engeset-Pograniczna, Michalina Kobla, Gustaw Kowalczyk, Blanka Pawlak, Konstanty Ramotowski, Julianna Rutkowska, Daria Wiśniewska, Zuzanna Zając—for sharing fieldwork with us, and for coming up with both pressing questions and very accurate answers. The research formed the basis for BA theses, and these are available at the archive of the Institute.

Most of the interviews were conducted at migrant sites of gathering and ethnic restaurants, shops, or other businesses, which made the conversations often serendipitous. It turned out to be the best solution, because many migrants lived in crowded rooms and did not feel comfortable inviting us to their living quarters. Another factor was the long working hours in the shops and restaurants and the fact that some of our interviewees did not live in the same municipality, but journeyed to work by bus or car. The language of the research was mostly English and Polish, depending on the participants' and our knowledge of a common language. Some interactions with the newly arrived non-English-speaking Indians were non-verbal, as we lacked a common language. Yet, we participated in their everyday practices, such as cooking meals in the temple, or leisurely walking through the village.

This chapter brings a phenomenon which is close to home under the ethnographic eye and the scrutiny of anthropological analysis. We decided to study the changes in the urban environment caused by incoming migration and the way migrants adapt to their new place of life. It was difficult to leave the field, as it became partly entangled with our daily lives in Warsaw: we passed our interlocutors in the city centre, while they went shopping or were delivering food through Uber Eats, we visited the same migrant restaurants and shops, we attended migrant events. We have also become possible contacts within the new and not always understandable world of bureaucracy, permits, and border crossings. We formed some ties to the migrant communities, and became part of their extended informal networks. Even if some of our research partners did not plan to stay in Poland, the possibility of having a local contact person and possible guide was important to them. We thus call our ethnography an "ethnography close to home", a cousin of the "anthropology at home" genre, yet involving a venture into the transnational social fields which the migrant communities function in (Levitt 2001). The analysis of both the material dimensions of the city space observed while walking through the neighbourhood (Wise 2010) and the ethnographic encounters enables us to understand the consequences of the visibility and invisibility of the various groups. It also allows us to observe formal and informal communication practices and recognize the limited opportunities available to migrants in the new environment. Through unobtrusive observation and the analysis of the different languages present in the signage in public space, we can understand the labour specialization of the neighbourhood and the ways the labour market opens up for migrant workers. The lack of opportunities results in migrants being pushed into the informal labour sector or migrant economic niches, notwithstanding whether they are willing to accept it or not.

This chapter contributes to research on migrant economies and informal practices and networks in the post-socialist context. While such research is well-established globally, it is still not well represented in Central-Eastern Europe (Šimić Banović et al. 2020: 95), which is increasingly becoming a preferred migrant destination. What guides us in the analysis of the observed phenomena is the concept of "informality". The term, although widely used, still needs to be discussed and redefined, especially showing how it works differently in local contexts. Using the concept to describe migrant communities and practices is important, as it shows the work of informality on the margins of state belonging, where people are not included in the cultural intimacy of the local communities. In these areas, state institutions try to impose its power over the non-citizens. This field of research allows for a discussion of trust and confidence in the state and institutions versus informal networks.

We follow scholars who analyse informal practices in Central-Eastern Europe in order to see the "ways of getting things done" (Ledeneva 2018: 1) in post-socialist realities: the historically rooted phenomenon of doing things through "acquaintances" or "connections" (Giordano and Hayoz 2013; Morris and Polese 2015; Šimić Banović et al. 2020) and the ways of arranging things not fully in accordance with the written rules and regulations, sometimes on the verge of illegality. Furthermore, we use the concept of *załatwianie* (Pieprzyca 2018), the local practice of informal arrangements, to explain the exclusion of migrants from the Polish labour market. We understand informality as an activity neither marginal nor infrequent (Giordano and Hayoz 2013: 30), and even central to local culture and community life (Ledeneva 2018: 36). At the same time, it is difficult to define, as it is an umbrella term for different social phenomena referring to "unwritten rules and hidden practices" (ibid.: 1), being a form of "rational choice" (Giordano and Hayoz 2013: 14) in the studied society.

In this chapter, we try to provide a more complete picture of a diverse neighbourhood, which has recently emerged in the suburbs of Warsaw, and to understand the way it functions, both formally and informally. Thus, we focus on the variety of mobility-related informal practices migrants are involved in: communication, economic, and labour practices. First, we look at informal advertisements about migrant services and work offers for migrants: where these can be found and the language in which they are written, in order to understand the regimes of visibility and invisibility of the different groups. Secondly, we observe informal spaces in migrant restaurants, shops, and bars, which serve as venues of gatherings and informal economic activities. Next, we look at informal migrant networks and their role in settling in the new place: obtaining information, finding accommodation, and employment (Brunarska 2014; Harney 2014). Furthermore, we analyse informal practices and different forms of *załatwianie*, which form the basis of migrant functioning in Poland, remaining for the most part hidden.

Visibility and Invisibility

The moment in which we did our research was very particular: the number of migrants in Poland, especially from South Asia, was on the rise, increasing quite rapidly in the years 2017–2018.[4] The exact numbers are hard to determine, as the national statistics do not include all incoming foreigners. The highest influx of migrants came from India, and the overall number of Indians was estimated to be 15,000–22,000 people.[5] In the described municipality, one would see individuals or groups of Indians (among them many Sikhs), mostly men, walking the streets, doing shopping, or just taking a stroll. At the time we did research, there were queues of people in front of the Sikh temple during the Sunday celebrations. This was a significant change compared to what the community has witnessed before. The rising number of migrants became a sudden business opportunity, with Indian or "oriental" shops mushrooming in the centre of Warsaw and in the suburbs. The municipality we observed also witnessed a two-fold increase in Indian shops: from two to four in the span of three

[4] https://migracje.gov.pl/statystyki/zakres/polska/typ/dokumenty/widok/mapa/rok/2016/rok2/2019/kraj/IN/?x=0.4011&y=0.5291&level=0.5 (accessed: 26.04.2019).

[5] According to data from Indian Ministry of Foreign Affairs, in 2018 there were around 22,000 citizens of India in Poland (source: http://mea.gov.in/images/attach/NRIs-and-PIOs_1.pdf#page=5). They obtained 8362 work permits (source: https://psz.praca.gov.pl/-/8180234-informacje-o-zatrudnieniu-cudzoziemcow-w-polsce). According to the platform "Indian Community of Poland" there are about 15,000 Indians in Poland (https://www.indiancommunityofpoland.org/english).

months. However, the migration situation is very dynamic, and not only one of the new shops has closed, but also a large part of the newly arrived migrants have already gone, not having obtained temporary residence in Poland, or grabbing a better opportunity elsewhere.

Migrants face numerous restrictions when entering the Polish labour market. Apart from linguistic exclusion, foreigners experience also other discriminatory behaviour from present or future employers. There are only a number of jobs readily available to them. Recently, migrants in Warsaw and the surrounding suburbs have started working as Uber drivers, mirroring the global trend. This job is taken mostly by Ukrainians, Georgians, and migrants from other post-Soviet countries, as basic knowledge of Polish, or at least some understanding of the language, is often required for the job. Since February 2017, when Uber Eats entered Poland,[6] a visible change could also be observed in the streets of major Polish cities, with a growing number of migrants working as Uber delivery personnel. This has become a niche for Indians, Bangladeshis, and Pakistanis, especially those who have recently arrived in Warsaw and those who study at Polish universities. This was also the case for many of our interlocutors. While the emergence of Uber is not unique solely to Poland, it has created a new niche for migrant workers in a country where migrants not speaking Polish are virtually excluded from the formal labour market.[7]

Until recently, the jobs most readily accessible to migrants in Poland were domestic and care work (Kindler 2011), regardless of their education and professional skills. Ukrainian women often work in these two sectors, while men find jobs in agriculture or at construction sites. Numerous of them work without any contract, enlarging the grey zone economics. Vietnamese migrants work mostly in trade in Wólka Kosowska and run other businesses, such as restaurants and hotels (Szymańska-Matusiewicz 2016: 12), and they admit it is difficult for them to find work beyond this niche. Migrants also face a number of irregularities in their workplaces: long work hours, frequent overtime, working for the family without remuneration, informal work without a contract. Migrant business owners

[6] http://www.dlahandlu.pl/detal-hurt/wiadomosci/ubereats-zadebiutowal-w-polsce,58652.html (accessed: 08.04.2019), http://wyborcza.pl/7,156282,24482414,szara-strefa-uber-eats.html (accessed: 18.11.2019).

[7] https://www.uber.com/pl/pl/about/diversity/immigrants-at-uber/ (accessed: 18.11.2019).

also acknowledge that they experience a number of difficulties in running their businesses, such as violent or unruly behaviour, and a negative attitude towards the customers. Kebab bars in Poland even faced an outright boycott in 2017, after violence erupted in one establishment in Northern Poland.[8]

Generally, institutions and service providers in Poland have not yet adapted linguistically to the emerging more diverse society.[9] The analysed municipality is no exception: all the offices work in Polish and demand that migrants come with their own translators,[10] sometimes provided by NGOs, should they not understand Polish. Also, public services, such as healthcare and schooling, are provided in Polish, with little linguistic flexibility. Overall, the municipality does not offer many top-down signs in a language other than Polish. What is important, there are groups which are linguistically almost invisible in the written signs, while they hold a strong presence in the neighbourhood they inhabit (Bielenin-Lenczowska 2020). We understand visibility as being present in local community by way of language, written signs, occupied spaces, and established businesses.[11]

One of such visible and well-recognized groups in the municipality is the Vietnamese community, comprising three generations. Their migration to Poland dates back to 1955, to the agreement between the Polish and the Vietnamese governments, making them the largest Asian population in Central-Eastern Europe. This has continued after 1989, in the form of economic migration, with a peak in border crossings in the years 2007 and 2011 (Szymańska-Matusiewicz 2016). The Vietnamese are seen

[8] See: https://natemat.pl/221361,brutalny-atak-na-tureckich-wlascicieli-baru-z-kebabem-maczety-i-siekiery-poszly-w-ruch. Another article about the events: https://www.newsweek.pl/polska/spoleczenstwo/zamieszki-w-elku-po-zabojstwie-21-latka-zatrzymano-kilkanascie-osob/x69zf3w (accessed: 7.05.2019).

[9] An exception are ethnic and national minorities which make up 20% of inhabitants of a given municipality. In this case, they can require the administration to use their language as an "auxiliary" official language http://prawo.sejm.gov.pl/isap.nsf/download.xsp/WDU20050170141/O/D20050141.pdf (accessed: 01.03.2019).

[10] Only very recently information in English, Vietnamese and Hindi appeared in the local clinic: "Dear Patients! Patients who do not speak Polish or English are asked to come with their own translator" (noted in February 2019 by our colleague, anthropologist Tatiana Busygina-Wojtas).

[11] This is somewhat similar to how Aneta Piekut describes Warsaw's visible migrants: "foreigners of any race of nationality who are regularly encountered by local residents in their day-to-day affairs" (Piekut 2012: 190).

by their Polish neighbours in the community as a hard-working and well-off group with an elaborate culture, taking part in the local community life. The main indicator of their involvement in the municipality according to local inhabitants is taking part in local celebrations and co-sponsoring them, for example giving money for decorating the municipal Christmas tree, which is placed in front of the town hall every year. Those involved in supporting this common effort were praised, while those who did not give money for it, were treated with reserve and distrust, if not hostility. Another important sign of recognizing the community's presence, is underlining that the Vietnamese are involved in supporting intercultural communication by providing a Vietnamese-Polish translator and cultural assistant for the local elementary school. They are also active in civic and political life. An example of this was the presence of a Vietnamese candidate in the elections to the higher level local administration. Posters of this candidate were placed in the windows of one of the Vietnamese restaurants. From the guy who brought the posters to the restaurant we learned that the candidate is well known among the Vietnamese community as a long-term resident in Poland.

The second group, which is recognized in the municipality, are Turkish migrants of the first and second generation. They have their own informal spaces, invisible to those who do not know them: a prayer house placed in a rented house and an informal café with billiards and a smoking space. They are also involved in more formal and visible enterprises: kebab shops, bakeries, food stores, a barber salon, and retail trade. They are seen by local inhabitants as successful businessmen, yet not to be trusted; loud and somehow too visible, not wanting to integrate with the Polish society. What was raised against them in the municipality was the fact that they did not support the common cause of the Christmas tree, although they had the money to do so. That, together with the recent rise of islamophobia in Poland (Buchowski 2016; Bobako 2017), is the reason why they might want to remain invisible, i.e. in practising their religion. They have a prayer house, yet it is impossible to be found by someone walking through the neighbourhood: hidden in a narrow street, with no plaque or religious symbol on the outside, it looks more like a storehouse or big residential building. If you are lucky, you might see men gathering there every Friday around noon, but they tend not to make a big crowd outside. The prayer house is visited by Muslims of various origins, however, there is an apparent transnational connection to Turkey, traceable through the incoming imams and the sponsors of the decor of the prayer room itself.

Migrants from India and Pakistan have become one of the most visible groups in the municipal space; they stand out mostly due to the homogeneity of the majority population in Poland.[12] The story of Indian migration to Poland dates back to the 1980s, with a surge in numbers during the 1990s boom in textile trade. Nowadays, the community has shifted more into the ethnic food sector, establishing Indian restaurants and food stores around Poland (Kugiel and Pędziwiatr 2014: 18). As we observed, some Indian migrants are also involved in informal businesses, such as beauty parlours dedicated to Indian clients, advertised through social networks, or hand-made posters.

Lastly, Ukrainians, the biggest migrant group in Poland, are also present in the studied neighbourhood (cf. Kindler and Szulecka 2013). If we look at the public and private signage in the streets of the researched municipality, Ukrainian and Russian languages are almost non-present. The only exception we found are bottom-up advertisements for work found on sign posts and bulletin boards, as well as advertisements for money transfer services or mobile services, sometimes written in Ukrainian or Russian.[13] Among the advertisements we found on the signboards were those directly addressed to Ukrainians. Some offers were written in Ukrainian, while others named Ukrainians as preferred employees, announcing different kinds of jobs as a salesperson in stores, on construction sites, in storages, and in production. The official and unofficial signs reflect the place allocated to the migrant group within the wider community, the formal and informal labour practices they engage in, and communication strategies of minority groups (Ben-Rafael et al. 2006). Invisible migrants engage in invisible practices, as informal practices are for the most part hidden and kept out of sight (Ledeneva 2018).

[12] http://warszawa.wyborcza.pl/warszawa/7,54420,22307909,nie-przyjaznimy-sie-ale-sie-lubimy-czyli-multi-kulti-w-raszynie.html (accessed: 05.03.2019).

[13] Recently Ukrainian has been introduced as a language option to choose in the ticket vending machines in Warsaw public transport, and as well as to some ATMs.

Informal Migrant Networks and "Getting Things Done"

Vignette 2

Sunny Sunday afternoon at the Sikh temple: the courtyard is full of people, talking in different languages and drinking *chai*. Some have been here since the morning, some arrive late and join the celebrations only for a few minutes. The long line leads to the food court where volunteers serve the free meal to all visitors as a service to the community (*seva*). There is constant movement, chatter, and commotion/activity.

After taking part in the religious rituals, we also partake in the meal. We chat with our friends from the community, asking about their week, how they spend time, how they help in the temple. After all the celebrations finish, we start out with a young couple, Chitra and Samrath, who have arrived in Poland some six months earlier. We already met them a couple of times and learned their migration story and how they adapt to life in Poland. They have told us about their dream to leave India to study (Chitra), and their struggles to get a visa. Finally, they boarded a plane to Poland, not going for the dreamt university scholarship, but taking the opportunity which came their way.

After arriving in Poland, they lived in a city laying some 30 km outside Warsaw. They moved to the studied municipality on purpose, after learning there is a Sikh temple there. Renting a room in a house with other Indians, not far from the temple, they took the chance to visit the gurudwara every day and do their share of helping. The temple has become for them not only a religious centre, but also a source of valuable contacts and information. They met other Indians there, who told them about work and housing opportunities. The frequent visit can be compared to leafing through the ads' pages of a newspaper—each visit an opportunity to learn about an official or unofficial job offer, or accommodation.

That day, when walking Chitra and Samrath home from the temple, we learn about their plans of waking up early the next day and going to work on some loading or unloading of trucks in the area. When we ask about the details, they hesitate, looking around to see if someone is listening. They tell us in hushed voices that this is not regular work, just a one-day job, a chance to earn something.

On another occasion, I (Helena) stay longer at the temple with them, and meet another young Indian couple. We venture into a conversation, and then I find myself going with them to see the place of the newly met people. As it turns out, they have extra space and could sublet it to Chitra and Samrath. The visit is a venture into the world of a young immigrant couple trying to settle in Poland, with their difficulties, being deceived by greedy owners, taking care of a small child without the support of the family.

Functioning as a migrant in Poland relies on learning how to move in the new environment: obtaining legal papers, securing a place to stay, and—most importantly—getting a job. Research has shown that migrants have very limited access to the labour market in Poland, as preference is given to non-foreign applicants (Bieniecki and Frelak 2007). Finding a job in Poland often means knowing the right people: they inform about the job opening, recommend the future employee, or push the application through. This has been described in anthropological literature as *załatwianie*: "a Polish euphemism describing a range of informal behaviours carried out mostly to obtain benefits by avoiding the use of arduous legal activities or formal norms (...). In a rough translation, the phrase '*załatwić* something' means 'to get something done' or 'to get something done in an easier way'" (Pieprzyca 2018: 89).[14]

"Shadow economy", especially working without the contract and not reporting all income to the state, is widespread and socially accepted in Poland. Many Poles engage in informal labour practices, and migrant workers only follow this pattern. Some scholars argue that this is mainly determined by the high gross cost of labour for the employer (Kicinger and Kloc-Nowak 2008: 207–208), however, we would also see it as the workings of "cultural intimacy": "the recognition of those aspects of cultural identity that are considered a source of external embarrassment, but nevertheless provide insiders with their assurance of common sociality" (Herzfeld 2005: 3). This would explain the fact that it is very difficult for an outsider to get into this kind of sociality, which is connected with

[14] *Załatwianie* is often connected with *kombinowanie*, a wider term used by Nicolette Makovicky in her work among the Polish highlanders. As she writes, "*kombinowanie* is used to describe both a type of activity – an ad hoc employment of practical knowledge to deal with any number of everyday problems – and an entrepreneurial mentality that seeks opportunities for accessing income, employment, and education through unconventional means" (2018: 1). Moreover, *kombinowanie* is usually related to illegal or half-legal activities (Galbraith 2003: 6).

załatwianie and informal work. As argued in a study of Vietnamese and Ukrainian women in Poland, the "main motivations for engaging in undeclared work are, to some extent, the same for Polish nationals as they are for foreigners" (Kindler and Szulecka 2013: 658). Although undeclared work and unregistered entrepreneurship have also not been strictly controlled by the state until recently (ibid.: 568), for migrants it is much more riskier than for Polish nationals. Without knowing the local labour code and without institutional support, migrants are at risk of getting a travel ban to the EU or even deportation, while also being prone to discrimination and exploitation by employers (Kicinger and Kloc-Nowak 2008: 200–201). On the other hand sometimes, as has been argued by Marta Kindler and Monika Szulecka "informal ties are also helpful in accessing lawful employment by natives. Experiences gained in informal professional relationships among Polish nationals may also result in a move to the formal sphere of the economy" (2013: 663).

Migrants do not usually have access to local Polish networks, so they cannot follow the practice of *załatwianie*. At the same time they use very similar strategies to "get things done", relying on personal and family networks (Granovetter 1973; Šimić Banović et al., in this volume). As we observed, they either seek employment through home-country and diasporic networks, or use the services of job agencies. Knowing and trusting someone in the new place is crucial: both migrants and their employers use their network of friends to recommend someone for work (Kindler and Szulecka 2013; Kindler 2011: 13; Brunarska 2014: 166). In the observed migrant businesses, the practice of employing co-nationals is widespread: in Vietnamese restaurants, Turkish shops, and other establishments, as well as Indian shops (Piekut 2012: 190).

Getting access to the job market can thus also happen through establishing personal contacts with Polish citizens in the form of friendships and romantic relationships. This was the case for a number of our interlocutors: two Turkish men working in food shops and an Indian businessman we talked to all had Polish wives who helped them establish themselves in the new place, while an Indian couple declared that they found work through Polish acquaintances. The observed pattern has been described extensively in works on the integration of migrants into the Polish labour market (Andrejuk and Oleksiyenko 2018: 88–89), as well as studies from other places. Another way to find work is taking part in informal job markets, in which migrants are recruited by employers seeking employees. Migrants wait at a certain spot—identified by the local inhabitants—to be picked up by people seeking day labourers (Bieniecki and Frelak 2007; Brunarska 2014). As many of the migrants do not speak the official language, or speak it with an accent, they are subject to linguistic exclusion in the labour market. This has been highlighted by Hartmut Esser, who argues that: "languages and accents can act as symbols of belonging or foreignness and give rise to differentiation and discrimination. Inequalities in terms of access to education, income, central institutions, societal recognition and social contact are significantly, although not exclusively, determined by linguistic competence in the relevant national language" (2006: i). That is why migrants engage in informal economic practices and function in the economic niches available to them (cf. Guarnizo 2003; Sosnowska 2016). What is also worth noting is that for the groups practically excluded from Polish-language communication, such as first-generation Indians, Vietnamese, or Turks, information about jobs is mostly obtained through the ethnic and religious networks functioning in the new place of residence. Among smaller groups, without formal institutions or meeting places, information can also circulate in diasporic networks, via another country, with the use of social networks and mobile technology. The places of gathering, such as temples, shops, restaurants, and other ethnic businesses thus become a hotspot of activity and information exchange. For migrants from India, the Sikh temple and Indian shops are a meeting point, where they learn how to find a job, secure accommodation, deal with offices, and overcome language problems. Migrants from Vietnam meet in Vietnamese restaurants and the huge trading centres nearby. Turks meet in similar spaces, such as Turkish shops, prayer rooms, or the informal Turkish café functioning in the neighbourhood. Informal networks are thus the basis of migrant life in Poland, allowing them to legalize their stay, find work, build a safe environment for themselves, and obtain basic practical information (Chacko 2003; Lin 2011; Zhou 2004).

Ethnic Niches and Ethnic Capital

Vignette 3

When we arrived in the studied neighbourhood for the first time, we noticed a new shop was about to open. There were announcements in English on the doors saying "coming soon...", while the shop windows were covered with Polish grocery stores advertising leaflets. The next time we visited it was already open and we were able to talk to the owner. He turned out to be a young Pakistani guy, who came to Poland on the suggestion of a friend that this is a good place to do business. He was working in the shop with his friend from Punjab, whom he called his "brother". The shop bore the name, which was a hybrid of an English and Polish name, stating that they wanted to attract both the local and foreign customers, even if the owner and the seller did not speak Polish. The shop was quite spacious inside, with a multitude of Indian products, but also many Polish brands.

On one of the occasions we visited the shop, the owner asked one of our students to help him in bringing a load of goods from Germany. He could not arrange the transport, because in all the shipping companies he contacted only Polish was spoken. On another occasion, we were asked to help in arranging the transaction to buy a pet rabbit on a Polish sales website. Again, the problem was communication in Polish. However, our interlocutor did not plan to learn Polish as he did not see his future in Poland. He was disappointed with Poland and the lack of success of his business. He was already envisioning moving somewhere else in Europe, if the shop does not bring enough profit. Indeed, next year when we visited the neighbourhood, the shop had already disappeared.

Ethnic entrepreneurship has been widely studied since the 1970s, especially in the US, because of the increasing number of migrant businesses and the role they played in the adaptation to life in the new country (Portes and Rumbaut 1990; Waldinger 1993; Portes et al. 2002). However, there is still not much research done on this topic in the post-socialist context (see: Šimić Banović et al., in this volume). Migrants seizing business opportunities in Poland, establishing new shops, and expanding the ethnic niches available to them: this is connected with the development of ethnic entrepreneurship, which adds to the diversity of the studied area. When establishing such enterprises, migrants rely mostly on co-ethnics, both as a source of workers, and as a support group (Zhou 2004: 1048). The decision of how to orient the business is closely connected with the migration strategy of individual migrants: if and how long they want to stay in the host country, if they see their future there, or they plan to move elsewhere. Below we will look in detail at two branches of migrant businesses, which were important in the studied neighbourhood: the food economy and the beauty sector.

One of the most recognized areas of migrant activity is the ethnic food economy (Kershen 2002), i.e. ethnic restaurants and cafes, as well as food stores. The shops and restaurants function as ethnic niches and are established first for the needs of the migrant community, and secondly, for those who might take a liking to oriental or otherwise "exotic" food (Baranova and Fedorova 2019: 22; Zhou 2004). Using the "oriental" or "exotic" label by the shop or restaurant owners is a way of utilizing ethnic capital. Following Zhou and Lin, we define ethnic capital as "the interplay of financial capital, human capital, and social capital within an identifiable ethnic community", which is often a way to social mobility and community transformation (2005: 261). Ethnic capital becomes of fundamental value and is being highlighted, both when establishing businesses for one's own community and when seeking a more global or local reach (ibid.). When running such enterprises, migrants rely mostly on co-ethnics, both as a source of workers, and as a support group (Zhou 2004: 1048). In Poland the area of the ethnic food economy is dominated by a few groups, namely the Vietnamese, Indians, and Turks (Stryjewski 2012). This goes in line with the observations by other researchers that some ethnic groups, e.g. Korean, Cuban, Indian, and Chinese, have been seen in the literature as more entrepreneurial than others (Zhou 2004: 1041; Valdez 2002). Vietnamese restaurants and bars have since the 1990s occupied an important place in the post-socialist urban landscape, including Poland (Hüwelmeier 2013). They have become, together with kebab booths, one of the most accessible and cheap places to eat for the overworked urban dwellers. Interestingly, there has also been a rising interest in Vietnamese cuisine among the middle-class (Patzer et al. 2015).

In the analysed municipality there are two Vietnamese restaurants. They are designed for Polish customers, but also serve the needs of their own ethnic group. The offered menu is in Polish. The space for customers functions in Polish, whereas the space accessible only for the staff (kitchen, storage rooms) functions in Vietnamese. The whole restaurant is organized in such a way that it invokes the image of the Orient, with elements easily identifiable as Asian, Vietnamese, or otherwise "exotic" (i.e. huge Vietnamese fans on the walls, exotic flowers, framed pictures from Vietnam and Vietnamese landscape wallpaper). In many Vietnamese eateries in Poland Vietnamese TV stations and Vietnamese music are audible signs of Otherness and add to the thorough feeling of being in an "exotic" place; this is also the case in the observed municipality. One of the businesses we observed had an inner space for Vietnamese customers, with an outdoor and indoor space for organizing parties in the backyard, surrounded with party banners. The banners form a kind of fence around the courtyard, creating an internal, more intimate and less visible, space for the community.

Just beside the restaurant, a nondescript club can be found. With no sign board to attract clients, and a half-derelict look, it is not a welcoming place for those who are not in the know. Through the huge shop-floor windows you can see coffee tables and billiard. If you enter the club in the afternoon, you will see groups of Turkish men sitting at small tables, smoking, drinking tea, and playing cards. Coffee and snacks can be ordered, but there is no set menu to choose from. At other times, the place seems deserted. It belongs to the Turkish community, and from talking to the men frequenting the establishment, we learned that, it is so crowded in the afternoons, when the men working in the nearby retail centre finish their work. These two places are good examples of informal and intimate spaces migrant communities create. In their research about the linguistic landscape of St. Petersburg, Vlada Baranova and Kapitolina Fedorova develop the concept of "hidden diversity". They argue that linguistic landscape analysis sheds light on the power relations between different ethnic groups in the urban public space: "Importantly, there is very strong disparity between open, public, and closed, inner spaces in terms of language use. Migrants' languages can be used inside these cafés, but not outside: on the surface they are advertised through Russian [the dominant language], but when customers get inside, they can be served in other languages, and in some cases this other language can even dominate" (Baranova and Fedorova 2019: 23). Such spaces are sites of gathering, exchanging information, and establishing networks. In other words, these are spaces where migrant forms of *załatwianie* take place.

Another group present in the ethnic food economy are the Indian and Turkish migrants, who run food stores. Indian shops in Poland are relatively expensive, and mostly aimed at customers from India and other Asian countries, and Poles interested in Indian culture or "oriental" food. In the studied neighbourhood, they were advertised as Indian or Asian. Products sold in the shops are listed on the signboards, indicating which are authentic, coming "directly" from India, i.e. masalas, spices, basmati rice. The signs also indicate which products are genuinely Polish: salt, Polish sugar, a popular brand of milk, the best-known spice producers, and a brand of rice often used in Polish households. Most of the "oriental" products are shipped to Poland from Germany or the UK, which is visible on the labels indicating the price, in euros or pounds. Running the businesses transnationally is possible because of the informal networks and connections between individual migrants living in different countries. These transnational or diasporic connections between businesses and people allow ethnic economies to function.

The Turkish shops in the studied neighbourhood present a slightly different picture. One is a grocery store, and the other a Turkish bakery. They were both run by Turkish men who have been living in Poland for a long time and they both employ Turkish salesmen and bakers. We talked to the owner of the shop and an employee of the bakery. They both have a long migration history, 15 and 7 years in Poland, respectively, have wives from Poland, and graduated from Turkish universities. The shops bear signs of Turkishness in the form of flags, decorations, and the reference to the term "oriental". The products in the store are mostly imported from Turkey, "the taste of Turkey", and these include spices, preserves, coffee and tea, sweets, but also dairy products. The bakery sells sweet and savory pastries, such as baklava or börek.

Other visibly "Oriental" immigrant-run eating places are kebab bars. They are designed for quick cheap meals and are especially popular among men, and interestingly related to imaginaries of masculinity (cf. Engeset-Pograniczna 2018). Kebab bars are also a case of an ethnic business mostly identified with Turkey, but run by migrants of different origins, e.g. Bangladeshi, Turkish, Egyptian, Lebanese, or even Indian. At the time of our research in the observed neighbourhood, there were three kebab bars located along the main road. The connection to Turkey and the Middle East is seen in the interior decorations (mosaics, photos, tableware on display), and sometimes found in the names of the eateries. We observed a tendency to employ co-ethnics in these businesses, however, this was not a strict rule: one of the bars employed mostly Bangladeshis, while another had Indian employees.

Another ethnic niche we want to focus on here is the beauty sector. Many beauty parlours serve mostly the different migrant groups present in the area. Among them are the numerous Vietnamese-run nail salons in the nearby retail centre: with all signs written in Vietnamese, Vietnamese music, and Vietnamese cosmetics, they cater to the needs of Vietnamese community. One can also find migrants offering informal beauty services: during our research, one such parlour was being advertised through a handwritten and xeroxed poster in English, while others offered their services to their compatriots through the functioning networks. The advertisement we saw lists the services offered, opening hours, telephone number, and location, which is the same street where the Sikh temple is located. It specifically states that they work also on Sundays, which is the day when the Indian community in Warsaw and its vicinity gathers for prayers. We also met young Indian women who offered make-up services at the clients' homes: an informal way of earning money. Word spread through migrant networks, helping those without employment and connections get extra income. This kind of work is characteristic for women and accepted by their spouses, however, it is put under surveillance by the community. We used their services once, and the visit was preceded by a short "scouting" trip by one of the women accompanied by her husband. A different example of a migrant-run beauty salon is the "Turkish barber", which is situated in the main street of the municipality and advertised in three languages (Polish, English, and Turkish). The business is well-known throughout Warsaw, and boasts that it trims ambassadors, as well as businessmen, and Turks from all around Poland. There is a huge TV screen in the establishment, which—at the time of our visit—was showing Erdogan's visit to Poland. The barber thus became a place of political discussions, not only beautification. We attentively note such moments of everyday interactions, as they confirm the strong transnational connections of the ethnic entrepreneurs (Portes et al. 2002). Men coming to the establishment are attended to by a Turkish barber who specializes in men's haircut and beard trimming. The saloon also employs a Ukrainian woman, who is the one who caters to Polish female clients, and the only one who speaks fluent Polish and is able to settle appointments by phone. In our conversation with her, she complained about her intermediary role and the need to be constantly available: she lives very close and is sometimes forced to run to the saloon in her slippers to answer the phone.

It is not the only instance of inter-group cooperation which we observed: Ukrainian women and men often work as sellers, waiters, or other frontline personnel in the business establishments owned by other migrants who are less fluent in the Polish language. Even though most Ukrainians speak Polish, they are easily recognized as foreigners by their accent. At the same time, they are often seen as more "familiar" than other foreigners: they are said to be closer linguistically and culturally, and so both less demanding as workers, and more acceptable. Still, many of them find it difficult to overcome the ethnic niches they are ascribed to, regardless of their education. There are few Ukrainian-owned businesses in the researched area, and at the time of the research, there were no Ukrainian food stores or restaurants. This is probably connected with the low interest in Ukrainian cuisine among Poles compared to the more "exotic" Indian or Turkish food, new to the Polish palate (Głuchowski et al. 2017: 124).[15] One of the characteristics of the ethnic economy might be the presence of informal labour practices (long work hours, frequent overtime, working for the family without remuneration, informal work without a contract), which are caused by the growing need for these services, the exclusion of migrants from the mainstream labour market, as well as their vulnerability in a new place.

Conclusions

In the last few years, suburban neighbourhoods in Poland visibly changed their character, in relation to the increasing number of migrants who live and work there. In this chapter, we looked at this emerging diversity and tried to understand what it means for migrants, their networks, and possible futures. We looked at places frequented and inhabited by migrants in a suburban municipality of Warsaw: public institutions, shared public spaces, as well as ethnic restaurants and shops with their backyards and kitchens, places of worship, which become places of gathering. We analysed institutions and service providers to see if the migrant groups are somehow accounted for and addressed. As we have shown, there are migrant groups which are linguistically invisible in the written signs, and not given any official representation, while they hold a strong presence in the neighbourhood.

What we argue is that the exclusion of migrants in the linguistic sphere further results in them being pushed into the informal labour sector and dependence on their own networks. We have further shown that informality plays an important role in finding one's place in the new country, especially in the labour market. As we observed, migrants use informal networks for finding employment and accommodation. These networks are formed in the inner, often hidden spaces of gathering, such as ethnic shops, restaurants, and places of worship serving their communities. These places are crucial for exchanging information, strengthening

[15] The situation changed after we finished the research—since late 2019 there has been a rise in the number of eating places in Raszyn, among them a Ukrainian Hutsul restaurant. This will probably however change again, due to the situation with the COVID-19 epidemic.

transnational connections, and building significant reciprocal relations of care (Patzer 2018a, b). Migrants, not being able to access local networks, cannot follow the locally grounded strategy of *załatwianie*. Thus, they "get things done" through personal connections and established migrant networks. As a result, they are pushed into economic niches, which often do not correspond with their qualifications, education, and professional skills. A consequence of this is also that migrants function within the informal economy, often working irregular and precarious jobs, without contracts, and without stability.

The constantly changing migrant strategies of moving and settling, especially in the context of global migration regimes and the volatile economic situation, definitely need further investigation. 'Ethnographic revisits' at the previously researched sites are indispensable to account for such changes. They can bring new issues to light and help to establish a more nuanced view on migrant informality and its consequences over time.

Bibliography

Andrejuk Katarzyna, Oleksiyenko Olena (2018) 'Experience of Migrant Self-Employment as "Economisation of Ethnicity"', *Indian and Chinese Entrepreneurs in Poland. Studia migracyjne - Przegląd Polonijny* 3(169): 77–96.

Bąbiak Iwona, Gmaj Katarzyna (eds.) (2015) *Namaste Polsko! Sytuacja i potrzeby imigrantów z Indii w Polsce*, Warszawa: Wydawnictwo Naukowe Scholar.

Baranova Vlada, Fedorova Kapitolina (2019) '"Invisible Minorities" and "Hidden Diversity" in Saint-Petersburg's Linguistic Landscape', *Language and Communication* 68: 17–27.

Ben-Rafael Elizer, Shohamy Elana, Amara Muhammad Hasan, Trumper-Hecht Nira (2006) 'Linguistic Landscape as Symbolic Construction of the Public Space: The Case of Israel', *International Journal of Multilingualism* 3(1): 7–30.

Bielenin-Lenczowska Karolina (2020) 'A paisagem sócio-linguística: a política, a diversidade e a migração no espaço público', *Forum Lingustico* 17(4): 5275–5291.

Bieniecki Mirosław, Frelak Justyna (2007) 'Migranci ekonomiczni w Polsce-Praktyka funkcjonowania', in: Klaus W. (ed) *Migranci na polskim rynku pracy: Rzeczywistość, problemy, wyzwania*, Warszawa: Stowarzyszenie Interwencji Prawnej, pp. 47–60.

Bobako Monika (2017) Islamofobia jako technologia *władzy: studium z antropologii politycznej*, Kraków: Universitas.

Brunarska Zuzanna (2014) 'Ukraińscy migranci zarobkowi w Polsce – dlaczego tak trudno ich policzyć?', *Studia BAS* 4(40): 155–174.

Buchowski Michał (2016) 'Making Anthropology Matter in the Heyday of Islamophobia and the "Refugee Crisis": The Case of Poland', *Český lid: etnologický časopis* 103(1): 51–67.

Chacko Elizabeth (2003) 'Ethiopian Ethos and the Making of Ethnic Places in the Washington Metropolitan Area', *Journal of Cultural Geography* 20(2): 21–42.

Duszczyk Maciej (2014) 'Ewolucja polskiej polityki migracyjnej w zakresie migracji zarobkowych po 1 maja 2004 r', *Studia BAS* 4(40): 37–54.

Engeset-Pograniczna Martyna (2018) 'Kebab nasz powszedni', Magazyn Kontakt. Available online: http://magazynkontakt.pl/kebab-nasz-powszedni.html.

Esser Hartmut (2006) 'Migration, language, and integration', *AKI Research Review 4*. Berlin: Programme on Intercultural Conflicts and Societal Integration (AKI), Social Science Research Center.

Galbraith Marysia (2003) '"We Just Want to Live Normally": Intersecting Discourses of Public, Private, Poland, and the West', *Journal of the Society for the Anthropology of Europe* 3: 2–13.

Giordano Christian, Hayoz Nicolas (eds.) (2013) *Structures, Political Cultures and Social Practices*, Peter Lang.

Głuchowski Artur, Rasińska Ewa, Czarniecka-Skubina Ewa (2017) 'Rynek usług gastronomicznych w Polsce na przykładzie Warszawy', *Handel wewnętrzny* 4 (369), vol. 2: 118–133.

Granovetter Mark (1973) 'The Strength of Weak Ties', *American Journal of Sociology* 78(6): 1360–1380.

Guarnizo Louis Eduardo (2003) 'The Economics of Transnational Living', *International Migration Review* 37(3): 666–699.

Harney Nicholas D. (2014) 'Rumour, Migrants, and the Informal Economies of Naples, Italy', *International Journal of Sociology and Social Policy* 26(9/10): 374–384.

Herzfeld Michael (2005) *Cultural Intimacy: Social Poetics in the Nation-State* (2nd ed.). New York and London: Routledge.

Hüwelmeier Gertrud (2013) 'Postsocialist Bazaars: Diversity, Solidarity, and Conflict in the Marketplace', *Laboratorium* 5(1): 52–72.

Kershen Anne J. (ed.) (2002) *Food in the Migrant Experience*, London: Routledge.

Kicinger Anna, Kloc-Nowak Weronika (2008) 'Combating the Illegal Employment of Foreigners in the Enlarged European Union: Polish Country Report', in: Kupiszewski M., Mattila H. (eds) *Addressing the Irregular Employment of*

Immigrants in the European Union: Between Sanctions and Rights, Budapest: International Organization for Migration, pp. 195–250.

Kindler Marta (2011) *A 'Risky' Business? Ukrainian Migrant Women in Warsaw's Domestic Work Sector*, Amsterdam: Amsterdam University Press.

Kindler Marta, Szulecka Monika (2013) 'The Economic Integration of Ukrainian and Vietnamese Migrant Women in the Polish Labour Market', *Journal of Ethnic and Migration Studies* 39(4): 649–671.

Kugiel Patryk, Pędziwiatr Konrad (2014) 'The Indian Diaspora and Poland-India Relations', Raport. The Polish Institute of International Affairs, November 2014.

Ledeneva Alena (ed.) (2018) *The Global Encyclopaedia of Informality, Volume 1: Towards Understanding of Social and Cultural Complexity*, London, UCL Press.

Levitt Peggy (2001) *The Transnational Villagers*, Berkeley and Los Angeles: University of California Press.

Lin Jan (2011) *The Power of Urban Ethnic Places: Cultural Heritage and Community Life*, New York: Routledge.

Makovicky Nicolette (2018) 'Kombinowanie: Agency, Informality, and the Poetics of Self in Highland Poland', *Journal of the Royal Anthropological Institute* (N.S.) 24(3): 493–511.

Matkowska Marzena (2012) 'Imigranci na polskim rynku pracy', *Studia i Prace WNEiZ* 25: 77–90.

Morris Jeremy, Polese Abel (2015) *Informal Economies in Post-Socialist Spaces. Practices, Institutions and Networks*, Palgrave Macmillan.

Patzer Helena (2018a) 'Unpacking the Balikbayan Box: Long-Distance Care Through Feeding and Food-Consumption in the Philippines', *Studia Socjologiczne* 4: 131–148.

Patzer Helena (2018b) 'Meeting the Pre-textual: Intersubjective Knowledge of Long-Distance Care in the Philippines', in: Tomasz Rakowski, Helena Patzer (eds) *Pre-textual Ethnographies: Challenging the Phenomenological Level of Anthropological Knowledge-Making*, Canon Pyon: Sean Kingston Publishing.

Patzer Helena, Góralska Magdalena, Winkowska Małgorzata (2015) 'The Stadium as a Witness. A Story of a Changing Monument', *View: Theories and Practices of Visual Culture* 9, Warsaw-Moscow: Monuments of Transition, pp. 1–36. http://pismowidok.org/index.php/one/article/view/271.

Piekut Aneta (2012) 'Visible and Invisible Ethnic "Others" in Warsaw: Spaces of Encounter and Places of Exclusion', in: Grubbauer M., Kusiak J. (eds) *Chasing Warsaw: Socio-Material Dynamics of Urban Change Since 1990*, Frankfurt and New York: Campus Verlag, pp. 189–212.

Pieprzyca Paulina (2018) 'Załatwianie (Poland)', in: Ledeneva A. (ed.) *The Global Encyclopaedia of Informality, Volume 1: Towards Understanding of Social and Cultural Complexity*, London: UCL Press, pp. 89–91.

Piłat Anna (2013) 'Obcokrajowcy mieszkający w gminie Raszyn - wzory integracji i relacje ze społecznością lokalną', in: Piłat A., Segeš Frelak J., Wysieńska K. (eds) *Mała Azja w Polsce. Plany i strategie imigrantów z Azji i Bliskiego Wschodu w Polsce*, Warszawa: Instytut Spraw Publicznych, pp. 133–157.

Portes Alejandro, Guarnizo Luis Eduardo, Haller William J. (2002) 'Transnational Entrepreneurs: An Alternative Form of Immigrant Economic Adaptation', *American Sociological Review* 67(2): 278–298.

Portes Alejandro, Rumbaut Ruben G. (1990) *Immigrant America: A Portrait*, Berkeley and Los Angeles, University of California: University of California Press.

Šimić Banović Ružica, Škokić Vlatka, Alpeza Mirela (2020) 'Informal Networks Among Immigrant Entrepreneurs: Case of Croatia', in this volume.

Sosnowska Anna (2016) *Polski Greenpoint a Nowy Jork. Gentryfikacja, stosunki etniczne i imigrancki rynek pracy na przełomie XX i XXI wieku*, Warszawa: Wydawnictwo Naukowe Scholar.

Stryjewski Radosław (2012) *Integracja społeczna i gospodarcza imigrantów wyznania muzułmańskiego w Polsce. Raport z badań*, Warszawa: Fundacja Ocalenie.

Szymańska-Matusiewicz Grażyna (2016) 'Migration and Cultural Flows Between Vietnam and Poland', *Asian and Pacific Migration Journal* 25(3): 275-295.

Valdez Zulema (2002) Beyond Ethnic Entrepreneurship: Ethnicity and the Economy in Enterprise, 'The Center for Comparative Immigration Studies Working Papers' 7: 1–63.

Vertovec Steven (2007) 'Super-Diversity and Its Implications', *Ethnic and Racial Studies* 30(6): 1024–1054.

Waldinger Roger (1993) 'The Ethnic Enclave Debate Revisited', *International Journal of Urban and Regional Research* 17: 444–452.

Wise Amanda (2010) 'Sensuous Multiculturalism: Emotional Landscapes of Inter-ethnic Living in Australian Suburbia', *Journal of Ethnic and Migration Studies* 36(6): 917–937.

Zhou Min (2004) 'Revisiting Ethnic Entrepreneurship: Convergencies, Controversies, and Conceptual Advancements', *International Migration Review* 38(3): 1040–1074.

Zhou Min, Lin Mingang (2005) 'Community Transformation and the Formation of Ethnic Capital: The Case of Immigrant Chinese Communities in the United States', *Journal of Chinese Overseas* 1(2): 260–284.

Informal Networks Among Immigrant Entrepreneurs: Case of Croatia

Ružica Šimić Banović, Vlatka Škokić, and Mirela Alpeza

Introduction

There are various definitions and descriptions of the ethnic entrepreneurship (Volery 2007). In its most comprehensive scope, it can be defined as 'a set of connections and regular patterns of interaction among people sharing common national background or migration experiences' (Waldinger et al. 1990, p. 3). 'Immigrant entrepreneurs' is often used as an alternative term, but it only includes those individuals who have immigrated recently and over the past few decades and excludes those members

This research was partly implemented within the research project 'Institutional support to the immigrant entrepreneurs in Croatia: socioeconomic and legal aspects' funded by the University of Zagreb (short-term research grant in 2020).

R. Šimić Banović (✉)
Faculty of Law, University of Zagreb, Zagreb, Croatia
e-mail: ruzica.simic@pravo.hr

V. Škokić
Faculty of Economics, Business and Tourism, University of Split, Split, Croatia

A. Polese (ed.), *Informality, Labour Mobility and Precariousness*, International Political Economy Series,
https://doi.org/10.1007/978-3-030-82499-0_4

of ethnic minority groups who have been settled into the country for several centuries (Volery 2007). Therefore, it is more appropriate to use the term 'immigrants' 'when speaking specifically of the early stages in the process of ethnic entrepreneurship, that is, when an ethnic group is new in a host society and its members can clearly be considered as "immigrants"' (Volery 2007, p. 31).

In the last decade, ethnic entrepreneurs have been recognised as a significant element of economic growth and social cohesion (Mostajer Haghighi and Lynch 2012; Rath and Swagerman 2011). A significant body of research has demonstrated that social or personal networks are crucial in the formation of entrepreneurial ventures by immigrants (Light and Bonacich 1988; Sequeira and Rasheed 2006). Since they are more likely to enter self-employment (Kauffman Foundation 2009), they use extensively personal ties to look for, e.g. funding and employees. In the later stages of the business lifecycle, ethnic networks may become irrelevant and other factors, such as human capital, gain importance (Yoon 1991). However, despite the growing research on immigrant entrepreneurship and their networking practices, there exist a gap in our understanding.

Firstly, a vast majority of the studies observe entrepreneurs who immigrate to wealthy western economies with already strong ethnic communities. For instance, Hispanics, Chinese, and Koreans in North America are extensively researched, where studies mainly investigated different characteristics of immigrants (Aliaga-Isla and Rialp 2013). Immigrant entrepreneurship research in Europe has not been studied extensively and exiting research focuses on different ethnic groups in Netherlands and Germany. Countries such as UK, Denmark, Italy, and Sweden are sporadically represented. Former socialist economies of South-Eastern Europe (SEE) are almost completely excluded from the recent research on ethnic entrepreneurship. This is possibly due to a limited amount of immigrant entrepreneurs in SEE, what seems to change dramatically over the next few years. In the most recent overview of immigrant entrepreneurship research, Dabić et al. (2020), in the most recent overview of immigrant entrepreneurship research, similarly argue that researchers have to orientate towards the emerging markets within a southern hemisphere as these

M. Alpeza
Faculty of Economics, J.J. Strossmayer University of Osijek, Osijek, Croatia

countries have limited history of entrepreneurship and constraining institutional context, 'which researchers could capitalize on for advancing context-specific immigrant entrepreneurship research (p. 34)'. Exclusion of a diverse range of settings in immigrant entrepreneurship research is rather problematic as a growing number of scholars argue that, given the socially embedded nature of entrepreneurship, national and international differences in entrepreneurial networking are significant (see for instance Chand and Ghorbani 2011; Drakopoulou-Dodd and Patra 2002; Klyver and Foley 2012; Staber and Aldrich 1995). There exists an increasing evidence from the CEE economies that unique forms of personal ties develop over time (see for instance Horak and Klein 2016; Hsu 2005; Ledeneva 2008; Skokic et al. 2019; Šimić Banović 2019) and that weak formal institutions in the transition countries may be partly substituted by informal networks that can support entrepreneurship and overcome the inadequacy of the institutional environment (Aidis et al. 2008). Namely, considering that key features of the informal practices in post-socialist societies are deemed to be omnipresence and resilience, their high degree of relevance is not surprising (Polese 2015; Steenberg 2016).

Secondly, studies which examined networks among immigrant entrepreneurs, did not investigate how differences in characteristics among a different network types provide access to different resources. Also, very often the studies would assume that family networks based on kin ties and personal or social networks based on ties with other ethnic members or other community members, are a single concept (Zimmer and Aldrich 1987). However, it is important to move beyond ethnic networks as they depend on the market strategies and entrepreneurial orientation and understand the relationships which immigrant entrepreneurs form with the wider society (see Kloosterman et al. 1999). Our study shows a vital role played by the expat community and their trustworthy local partners in surmounting the issues caused by inefficient and burdensome public administration but also in addressing various private challenges.

The aim of the present study is to address this gap and to investigate networking activities among ethnic entrepreneurs in Croatia, a former socialist economy with low rates of immigrants (CBS 2019; Eurostat 2019) and an increasing emigration wave (Troskot et al. 2019). Also, we look how ethnic entrepreneurs use different network types to access different resources. Croatia has faced turbulent transitional environment and despite recent improvements, institutional framework is still not stimulating for entrepreneurs (Singer et al. 2021), with high rates of corruption (Transparency International 2019). This offers an unique opportunity to examine ethnic entrepreneurship in an environment which is characterised by complex political and economic changes and where the role of networks may be crucial for conduct of entrepreneurship (Skokic et al. 2019). In addition, research on ethnic entrepreneurship networks in Croatia is almost non-existent. Studies exploring ethnic entrepreneurs have mostly focused on migrant social entrepreneurship initiatives (Jurković and Rajković 2016; Bužinkić 2017), entrepreneurship of ethnic minorities (Kuti and Božić 2011; Lekaj et al. 2015), family businesses, and female entrepreneurship of ethnic minorities (Nreca and Gregov 2018). The closest research to the purpose of this study is done by Čapo and Kelemen (2018a, b) and Čapo (2019) on the entrepreneurial activities of foreigners in the Croatian capital and it mostly investigates the ethnological aspects.

Based on the unique set of empirical evidence, this study proposes the formal platform representing the immigrant entrepreneurs' interests and providing mentoring and training programmes, same as additional networking opportunities for them. That initiative would be built on the already developed in-group cooperation and support among immigrant entrepreneurs that is shown to be a very efficient way of compensation for the deficits of the Croatian entrepreneurial environment. Therefore, this chapter contributes to both goals of this Volume by addressing the recognised deficiencies (as outlined in the Introductory section by Polese forthcoming). First, it provides novel first-hand empirical evidence found at the intersection of informality, entrepreneurship, and migration. Second, it explores the interplay between mobility and informal practices in a post-socialist country with an emphasis on entrepreneurial networks. Hence, while focusing on post-socialism, it adds to the debate on the varieties of informality in the post-socialist context and on the 'importing' mobilities (contrary to the most literature on Eastern Europe and the former USSR exploring 'exporting' mobilities). More precisely, it contributes to the discussion on the (specific) informal governance that mostly serves as a substitute to underdeveloped formal institutions and networks, and is as such not considered to have an elitist flavour by the actors involved.

Entrepreneurial Networks: Contrasting the West and the East

Networks as the Drivers of Success: Entrepreneurial Networks in the Western Economies

In the last decade, there was a significant emphasis on entrepreneurial networks, which are considered as a key element of entrepreneurship (Hoang and Antoncic 2003; Sarasvathy 2018). Entrepreneurial activity is strongly influenced by entrepreneurial networks and their impact on the business activity is observed through different stages of business lifecycle (Zhang 2010). The entrepreneurship literature agrees there are two main different types of networks: personal/social and formal (Birley 1985). Personal networks are composed of individuals, which include acquaintances, friends, family, employees, and business friends (Jack 2005). Formal networks include relationships with formal institutions such as banks, creditors, lawyers, venture capitalists, and accountants (Das and Teng 1997). Both network types are considered to improve the likelihood of success as they provide a range of resources and support throughout the business lifecycle (Evald et al. 2006; Strobl and Kronenberg 2016).

Generic entrepreneurship literature has extensively examined different types of relations that are more effective for business development and entrepreneurs and also how different network ties are more effective for different business stages (Jack and Anderson 2002; Lechner and Dowling 2003; Strobl and Kronenberg 2016). A number of studies show that personal ties are crucial during the early stages of business formation and development (Greve and Salaff 2003; Jack 2005; Martinez and Aldrich 2011; Zhang et al. 2011). Entrepreneurs often rely on family ties to provide them with the ability to recognise the potential for entrepreneurial opportunities (Davidsson and Honig 2003). Ostgaard and Birley's (1994) study among entrepreneurs in two counties in England revealed that besides being the most important resource upon which entrepreneurs drew in the early days of a firm's development, personal networks also increased competitive advantage in the later stages of the process. For instance, personal networks were utilised to gain ideas for product development and to gain access to distribution channels.

Studies focusing on ethnic entrepreneurs demonstrate a crucial role of personal ties for business formation and development (Chand and Ghorbani 2011; Yoo 2000). Aliaga-Isla and Rialp (2012) investigate Latin Americans in Barcelona and find that the decision to set up a business is based on information provided by fellow citizens who are highly trusted by immigrants. A study by Yoo (2000) among Korean immigrants in Atlanta shows the importance of education for immigrant entrepreneurs. In their study, irrespective of possession of family ties, those immigrants with higher educational levels were more likely to start their businesses without suffering first in the low-wage and low-skilled jobs. Also, their findings demonstrate that immigrant entrepreneurs use their personal networks for different purposes, where family networks are utilised more often to gain capital whilst social networks play important role in acquiring information. In addition, a vast majority of the studies show that ethnic entrepreneurs acquired their first work experience in the family business helping their parents (Aliaga-Isla and Rialp 2012; Yoo 2000), a factor which can shape positive attitudes towards entrepreneurship. Rusinovic (2008) investigated transnational networks among first- and second-generation immigrant entrepreneurs in Netherlands. She defines transnational networks as 'contacts or associates in the home country' (p. 433). Her results demonstrate the importance of transnational networks even for the second generation of immigrants, and these ties are found to be important for the business. Transnational ties include family members and acquaintances in home countries who assist the entrepreneurs in doing business with the home country. Main motivation for becoming transnationally active relates to economic interest (to gain profit) and to assist businesses. This relates mainly to the wholesale businesses, as the import of products from the home country is often their main business activity. Similar results are outlined in a study by Kariv et al. (2009). In addition, their results show that ethnicity moderates the activities of transnational networking. For instance, Vietnamese maintain the strongest family and business ties with the home country while the Chinese and Italians were mainly involved in formal, business, and professional ethnic associations.

Mainstream entrepreneurship literature demonstrates that entrepreneurs during the business lifecycle expand their networks and search for new business contacts and engage in formal networking activities. For instance, co-operation, sales, and marketing networks gain importance in the early growth stage and it can be argued that the network transforms from strong to a calculative network (Hite and Hesterly 2001; Hsu et al. 2012; Schutjens and Stam 2003). In the later stages, entrepreneurs again modify their networks as the issues of positioning on the market, people and sales gain further importance and entrepreneurs focus their efforts on obtaining the key resources needed (Larson and Starr 1993; Strobl and Kronenberg 2016). Although organisation-to-organisation ties in these stages gain importance, research has demonstrated that some firms, such as small tourism and hospitality firms, do avoid formal networks (Hsu et al. 2012). Existing research among ethnic entrepreneurs shows that they do not avoid formal networks but have strong preference towards specific organisations. For instance, although the Korean community in the USA has a significant number of formal organisations, ethnic entrepreneurs prefer church and alumni associations and only a minority become members of business associations (Yoo 2000). However, it is not clear from the existing studies whether ethnic entrepreneurs cannot access formal organisations, are not aware of their presence or they obtain all the necessary resources, information, and support from personal ties.

From Zero to Hero? Entrepreneurial (Informal) Networks in Transition Economies

Within the former socialist economies of CEE, entrepreneurial endeavours were not seen as legitimate, 'leaving a lasting stigma on individuals pursuing entrepreneurial opportunities' (Manolova et al. 2008, p. 206). Although the communist system was not identical among the CEE countries, Ireland et al. (2008) argued that the one-party system and anti-entrepreneurial norms and beliefs were the distinct features of former socialist economies. In a majority of CEE countries, private ownership was prohibited so denying the establishment of a private sector (Rehn and Taalas 2004). Entrepreneurial responses to such turbulent environments are still relatively unexplored, although scholars argue that entrepreneurs demonstrated unique strategic responses (Manolova and Yan 2002). This unique nature is particularly demonstrated through the practice of creating informal networks (Aidis et al. 2008). Informal networks can be defined as 'alternative forms of regulation operating often outside the norms of the state. They serve as an alternative social mechanism for supporting economic interaction where formal institutions are either non-existent or ineffective. Informal networks are characterised by reciprocal exchanges of favours, they are intentionally formed and informal relations are always cultivated' (Skokic et al. 2019, p. 26). Although scholars argue that similar networking behaviour may be widespread among other CEE countries (Puffer et al. 2010), available evidence is limited to *blat* in Russia, *guanxi* in China and *veza* in Croatia. There is also growing evidence that similar practices do exist in Bosnia and Herzegovina (Koutkova 2015; Brković and Koutkova 2018), Serbia (Stanojevic and Stokanic 2018), Bulgaria (Chavdarova 2015, 2018), Macedonia (Otten 2018), and Ukraine (Polese 2013). Table 1 compares *veza* to *blat* and *guanxi*. We focus on *blat* and *guanxi* only as; (1) they are widely researched and there exist a substantial body of knowledge on these two practices; (2) their operating mechanisms are extensively researched, compared, and tested and hence represent mechanisms of social interaction that are well explicated in the literature.

Although there are significant differences between *guanxi*, *blat* and *veza*, they all draw on a web of connections to secure different favours, both in personal and business relations, exchange of favours is the main operating mechanism and are affiliated by different structures or circles of trusted people.

For instance, entrepreneurs in Russia identified *blat* as instrumental in starting up businesses, as informal networks with government officials enabled Russian entrepreneurs to overcome institutional barriers (Kets de Vries and Florent-Treacy 2003). Investigating the attributes and qualities of successful entrepreneurs in Russia, Kuznetsov et al. (2000) found that the ability to create or join existing informal networks was the most significant factor for someone to become an entrepreneur. Similarly in China, *guanxi* practice was also used to bypass inadequate formal and financial institutions (Peng 2004). For instance, a study by Hsu (2005) shows that entrepreneurs use *guanxi* to secure money for start-ups, to negotiate with tax authorities, to secure retail space, and to access potential customers. In Croatia, *veza* ties emerged during the transition period and provided the access to entrepreneurship. Therefore, it was possible to start a small firm with only limited resources but well-developed informal networks. Entrepreneurs were additionally constrained by an underdeveloped institutional framework, poor enforcement of property rights, high corruption, and by an intrusive and hostile business environment (Aidis et al. 2008; Ireland et al. 2008; Šimić Banović 2015, 2016). The conduct

Table 1 Characteristic of *veza*, *blat* and *guanxi*

Attributes	*Characteristics*		
	Veza	*Blat*	*Guanxi*
Driver for activation	Access to entrepreneurship	Maximisation of individual values (economy of shortages); access to necessities (food, jobs)	Maximisation of the value for a family; moral obligation
Origin	Transition period (1990s)	Emergence of socialism; after Russian revolution (1917)	Confucianism and ancient Chinese philosophies
Operating mechanism	Exchange of favours; one off reciprocal transaction; transaction to be completed without significant time delays	Exchange of favours; gifts and resources; unlimited exchange of favours; time lag often preferred	Exchange of favours, gifts and resources; unlimited exchange of favours; time lag often preferred
Relations	Short-term; utilitarian	Continuous and long-term; utilitarian; emotional; social	Continuous and long-term; utilitarian; emotional; social
Structure	Dyadic relationships	Dyadic relationships with extended vertical and horizontal structures, affiliated with a circle of trusted people	Dyadic relationships often embedded in or influenced by actors outside the dyad
Dynamics	Very dynamic, exchange fairly equal	Very dynamic, exchange fairly unequal	Very dynamic, exchange fairly unequal
Diversity	Diverse, but individual relations not linked in a particular network	Diverse; vertical and horizontal circles relatively closed	Diverse; circles open to expansion and can grow easily
Meaning	Positive	Negative, closely linked with corruption	Neutral to positive

Source Skokic et al. (2019, p. 34)

of business in such environments has become explicitly related to the availability of informal networks (Peng 2004; Skokic et al. 2019).

Access to informal networks happens directly or is often facilitated by a third party or intermediaries. Ledeneva (2008) in the case of *blat* explains there are two types of *blat* networks. Horizontal networks get formed by people who share similar status, known as a 'people of the circle' (svoi lyudi). Vertical networks get formed by people of different social strata interested in each other's connections and linked by kin, personal contacts, or intermediaries known as 'useful people' (nuzhnye lyudi). Similarly, entrepreneurs in Croatia establish *veza* through either direct relationship ('inner circle') or they establish *veza* to people through an extended circle of 'friends of friends'.

In the case of ethnic entrepreneurs this appears rather problematic. Ethnic entrepreneurs in CEE economies do not have large communities of ethnic entrepreneurs, as these countries were not economically attractive. Therefore, access to information and knowledge on the business environment may be limited. Without strong ethnic communities, immigrant entrepreneurs may be lacking personal ties and are not in a position to develop informal ties, which may jeopardise their entrepreneurial attempts. Although such studies have not been conducted, there is limited evidence from Skokic et al.'s (2019) study among Croatian entrepreneurs. Their longitudinal study includes the cases of returning migrants (entrepreneurs who emigrated due to political or economic reasons and returned to Croatia after the reforms in the 1990s) and shows that returning migrants suffered from weaker or non-existent *veza* ties, making it even more difficult for them to start a business. Returning migrants were not embedded into the local area and lacked social and kinship networks, which significantly reduced their access to *veza* ties.

Setting: Entrepreneurship and Migration Climate in Croatia

With 3.8 immigrants per 1000 inhabitants, Croatia is below EU-28 average and among the countries with the lowest immigrant/inhabitant ratio in the EU (Eurostat 2019). In total, in Croatia only 1.3% of the population does not have a Croatian citizenship, out of which 0.4% are EU citizens (Eurostat 2019). Less than 1% of Croatian population are third-country nationals. They have mostly immigrated because of family reunification, work, education, and humanitarian reasons. Their most common countries of origin are Bosnia and Herzegovina, Serbia, Kosovo, North Macedonia; as well as Russia, Ukraine, and USA (CPS/IHZ 2019). As an illustration of the structure of immigrants to Croatia by their citizenship, the most recent analysis (Eurostat 2019) indicates that in 2017 half of the immigrants had Croatian citizenship, only 14% were from other EU member countries, and the rest (35%) were from nonEU countries. Croatia belongs to the leading group of EU countries based on the most immigrants being the country nationals (Eurostat 2019). These are not necessarily the return migrants, but most likely the residents of countries in the region, Bosnia and Herzegovina in particular, who are dual citizenship holders. Thus, considering rather small number of immigrants, the integration policies in Croatia mainly focus on the refugees (Giljević and Lalić Novak 2018). Yet, due to necessity reasons, i.e. impassable shortage of the workforce threating tourist season and some major construction projects, there was a significant increase in the national quota for migrant workers in 2019. Consequently, there were 65,100 work permits issued in 2019. That is more than double compared to 2018, and in 2014 there were only 231 work permits (CPS/IHZ 2019).

As in the other European economies, entrepreneurship and the small and medium enterprise sector play an important role in Croatia. Thus, small and medium enterprises account for 99.7% of the total number of enterprises in Croatia, have an employment share of almost 70%, and achieve a 54% share in total exports (Alpeza et al. 2018, p. 10). Results of the GEM—Global Entrepreneurship Monitor survey, in which Croatia has been involved since 2002, provide a deeper insight into characteristics of entrepreneurial activity in Croatia (Singer et al. 2019, p. 27). According to the intensity of new business start-ups, with 9.6 new ventures per 100 adults, Croatian position is above the average of EU countries included in the GEM survey in 2018. However, entrepreneurial activity in Croatia at this early stage is characterised by a high share of new ventures started out of necessity. In 2018, on average in the EU there were 5.3 times more new ventures started because of perceived opportunity than those started out of necessity, while in Croatia there were only 1.9 times more entrepreneurs because of opportunity, which positions Croatia at the rear of EU countries involved in the GEM survey.

Furthermore, Croatia's entrepreneurial capacity is characterised by a low share of 'established' businesses (older than 42 months), indicating a low survival rate and lack of growth capacity of established entrepreneurial ventures. According to this criterion, the GEM survey in 2018 again positions Croatia at the rear of EU countries.

Entrepreneurial activity in Croatia is characterised by good technological readiness, but also by absence of new products, which is reflected in the low competitiveness of the economy as a whole (68th place in the World Economic Forum's Global Competitiveness Report ranking in 2018). In 2018, 30% of newly started entrepreneurial ventures in Croatia (compared to 13.6% in the EU) and 28.3% of 'established' businesses (compared to 7.9% in the EU) had the latest technology. Yet, at the same time, as many as 70% of newly started and more than 75% of 'established' businesses have products that are not new to anyone in the market.

Within Croatia, it is possible to discern regions where entrepreneurial activity intensifies in recent years, namely Dalmatia, Istria, Primorje, and Gorski Kotar, while lower levels of entrepreneurial activity are recorded in Lika and Banovina, and Slavonia and Baranja. With regard to the distribution of entrepreneurial activity by gender, Croatia is a significantly more 'male' country, with almost twice as many entrepreneurial ventures started by men than by women, which is the average of EU countries. Although the share of women entrepreneurs in Croatia has been slightly increasing over the 2010–2018 period, the still present gap in activity of women and men indicates the consistency of obstacles and insufficient efficiency of programmes and measures for the development and strengthening women entrepreneurship in Croatia.

Methodology

The evidence is primarily based on primary sources due to shortage of the secondary sources (as already presented in the Introductory section of this chapter). Our primary data originate from the semi-structured interviews with immigrant entrepreneurs in Croatia and representatives of institutions dealing, or supposed to be dealing, with immigrant entrepreneurship in Croatia. The representatives of the institutions were selected according to purposive sampling. Both they and initially interviewed foreign entrepreneurs served as the informants for identifying immigrant entrepreneurs. Hence, the snowball sampling procedure was employed in addition to media sources to identify the interviewees. Different initial informants and various media sources contributed to the diversity of the sample. It is important to emphasise that due to the size of the sample and availability of recent secondary sources, they played an important role in obtaining the information. The sample included a total of seven (coded as IN1, IN2, IN3, IN4, IN5, IN6, and IN7—see Table 2) institutional representatives (NGOs, think tanks, business-supporting organisations, research institutes) and thirteen entrepreneurs (coded as EU or nonEU 1–13—see Table 3). Those interviews were conducted in the period between January and November 2019 personally in Zagreb, partly via e-mail or over the phone. The interviews were carried out in

Table 2 The overview of the respondents: entrepreneurs

Code/origin	*Business description*	*Gender*	*Age range*	*Years spent in Croatia*
nonEU1	Hospitality	M	31–40	9
nonEU2	Consulting	M	31–40	10+
EU3	Small-scale production and retail	F	41–50	4
EU4	Small-scale production and retail	F	31–40	5
nonEU5	Hospitality	F	41–50	7
EU6	Hospitality	M	41–50	2
nonEU7	Hospitality	M	31–40	5
EU8	Consulting	M	31–40	4
EU9	Publishing & Consulting	M	41–50	10+
EU10	Financial services	M	51–60	8
nonEU11	IT	M	21–30	2
nonEU12	Hospitality	M	21–30	5
nonEU13	Hospitality	F	41–50	3

Table 3 The overview of the respondents: Institutional representatives

Code	*Type of institution*	*Gender*
IN1	NGO/think tank	F
IN2	Business support organisation	F
IN3	Business support organisation	M
IN4	NGO/think tank	F
IN5	NGO/think tank	F
IN6	Business support organisation	F
IN7	Research institute	F

English and lasted 45 minutes on average. Main topics of the interview were mostly chronologically sequenced around two distinctive stages of migration (Castles and Miller 2009, pp. 20–21): first, the immigration determinants and patterns; and second, the integration processes and perspectives. This necessarily leads to the analysis of interaction of macro- and micro-structures, and their intermediate mechanisms named 'meso-structures', as any migratory movement is a result of it (Castles et al. 2014, pp. 26–27). In line with the purpose of this research, the afore-stated (im)migration stages were focused on informal networks of the entrepreneurs.

Hence, the findings are contextualised since the context in migration studies today is deemed 'increasingly important as migratory phenomena become more diversified, and as the statistical grasp over migration flows and stocks becomes less secure' (King 2018, p. 39). Furthermore, even though some factors can be quantified (such as demographic and socio-economic variables, legal status, etc.), 'the complex nature of these factors as potential drivers of migration, and their interaction with each other, can only be appreciated via a more qualitative approach rooted in specific socio-economic, political, cultural and spatial settings' (King 2018, p. 39). This is in line with the notion of migration as an intrinsic part of globalisation, social transformation, and development processes rather than 'a problem to be solved' issue (Castles et al. 2014, p. 25). The overview of the Croatian migration and entrepreneurial climate is presented in the previous section.

Empirical Evidence from Croatia

Previous research has presented rather an extensive list of formal and informal barriers faced by ethnic entrepreneurs in Croatia (Table 4). It has been clearly demonstrated that native and foreign entrepreneurs in Croatia have the same set of core obstacles that they invariably stress. Their impressions are in line with the findings of the international benchmarks (like World Economic Forum 2018) and include administrative burden, unpredictable tax regulations and high tax rates, overly restrictive labour regulations, parafiscal charges, corruption, insufficient quality of the health and education system. In addition to that, both local and ethnic entrepreneurs recognise the lack of understanding for entrepreneurship and insufficient entrepreneurial spirit in the society (this is also in line with Global Entrepreneurship Monitor findings presented in Singer et al. 2019). The list keeps enlarging depending on the origin of the entrepreneur. Language barrier is present in both oral and written communication with the authorities. Moreover, the nonEU nationals are faced with undermining and costly requirements when establishing a business. That situation is mostly coupled with unclear and lengthy procedures when scaling up a business.

Table 4 Barriers faced by ethnic and native entrepreneurs in Croatia

	Key formal barriers	Key informal barriers
Non-EU, less developed countries	• lengthy asylum-seeking procedure (if applicable)	• considered to be a security threat • perceived to expect social contributions only • ignorance
Non-EU, developed countries	• a work/business permit procedure • mandatory financial capital (10x or 200x higher than for the EU / native entrepreneurs – depending on the legal form) • obligation to employ 3 Croatian citizens • ambiguous/double procedures	
EU entrepreneurs	• absence of any written instructions in English • front office employees not speaking English	• discrimination (ethnic and racial origin, LGBT, gender, age) • limited openness towards other nationalities and religions
Native entrepreneurs	• administrative burden • inefficient public administration • unpredictable tax regulations • high tax rates and parafiscal charges (including mandatory memberships) • overly strict employment regulations • labour shortages • corruption prevalence • weaknesses of the education and health system	• lack of entrepreneurial spirit • lack of understanding for entrepreneurship

Source Šimić Banović et al. (2019)

Informal practices usually indicate the deficits of the official structure of the society (Ledeneva 2008) and may serve as either substitutes or complements to the official economy (Dreher and Schneider 2010; Aligica and Tarko 2014). Furthermore, instrumentalising personal connections for 'getting things done' is rather usual and expected in Croatian society (Šimić Banović 2019). Yet, foreign entrepreneurs are still on their integration path and may not be fully aware of the extent and influence of informality in professional and private life in Croatia.

Formal Associations and Informal Networks

The (missing) satisfaction about the wider formal network/ecosystem for the ethnic entrepreneurs has already been examined in the previous section. There is currently no synchronised channel to express that to the decision- and policymakers as the association representing ethnic entrepreneurs in Croatia does not exist and, according to the key informants contacted for this research, there are no plans to establish it in the near future (IN1, IN2, IN3, IN4, IN5, IN6, IN7). There have been several initiatives aimed at training and mentoring of the (future) immigrant entrepreneurs but, apart from intrinsic networking, they do not provide advocacy or lobbying. All respondents in the recent survey on immigrant entrepreneurs in Croatia claimed not to have any information on the existence of supporting business networks (ACT Group 2019, p. 21). Yet, since all the legal subjects in Croatia are obliged to be the members of Croatian Chamber of Commerce or Croatian Chamber of Trades and Crafts,[1] the entrepreneurs are supposed to have some benefit out of it. But, similar to native entrepreneurs, the impressions on the Croatian Chamber of Commerce[2] are predominantly disappointing. This ranges from

> I've hear of them, but then I've never received any information about what they do. CCC sends a newsletter every month but it's not tailored to your needs, and I always have quick browse, but I'm not sure of their purpose. (EU8)

to more direct critique like

> 'total waste of money, time and effort' (EU9) and
> 'this is another tax'. (EU10)

Furthermore, in line with the local entrepreneurs, foreign entrepreneurs mostly do not consider Chamber's regional offices to be useful either:

> Why do I pay the Chamber of Commerce membership? I have never ever received one single email from them, let alone some locally specific advice on request or available at their website. When I opened my first shop in Istria, nobody told me that all the official information should be bilingual: in Croatian and in Italian. When we decided to open a shop in Dubrovnik, we could not find anybody to advise us on the tables in the Old town – what is allowed, where it is allowed to put them, etc. When we finally got the rules, they were changed in the next summer season. (EU3)

Most of the respondents have not heard of the Croatian Employers' Association[3] that is an independent organisation representing businesses in

[1] This also implies a mandatory membership fee.

[2] None of the respondents is a member of the Chamber of Crafts as all of them own limited liability company.

[3] Association representing businesses in Croatia, membership is paid, but voluntary (unlike Croatian Chamber of Commerce).

Croatia. In general, many foreign citizens believe that missing cultural, racial, and ethnic diversity coupled with rare public activities in English[4] impede their integration in the Croatian society. Those deficits are to some extent compensated through socialising in online and offline networks of expats (nonEU1, non EU2, EU3, nonEU5, EU8, nonEU11, nonEU12, Čapo and Kelemen 2018b, pp. 8–9). In line with that, many ethnic entrepreneurs (like EU3, EU6, nonEU7, EU8, EU9, EU10, nonEU11) consider their informal network of expats in Croatia to be their primary network for business and private purposes. For some of them, this is their only informal network that they identify and/or deem important for their private and professional life. The family members are often included in the business and are usually considered to be decisive factors in running it (nonEU2, EU3, nonEU5, EU10, nonEU13). In addition to that network, many of them (nonEU2, nonEU5, EU6, EU8, nonEU11, nonEU12) stress the importance of several reliable and competent local partners such as lawyer, accountant, key business partners, and possibly co-owners of their business. Hence, the informal network of foreign peers coupled with family members and rather small number of crucial local partners is the overview of all the ties ethnic entrepreneurs utilise to run their personal and professional life in Croatia.[5] In a nutshell:

> Is doing business as a foreigner hard? Yes. Is it impossible? No. Like anywhere else in the world, you need good and trustworthy local partners. They are hard to find here, but they do exist. (EU9)

Still, despite the trustworthiness and supportiveness of the network presented, some ethnic entrepreneurs warn of its limitations:

> My friends and acquaintances, especially the ones from the same sector, have been an immense support in obtaining the information to start my

[4] An illustration of the lack of public activities in English for the wider audience of foreigners: In November 2019, in the Facebook group *Expats in Zagreb*, there were several posts looking for 'Frozen 2' (extremely popular movie for kids) in English at any public place in Zagreb. No positive response was found among public announcements.

[5] This finding is also confirmed by series of interviews with foreign entrepreneurs featured in the Total Croatia News at https://www.total-croatia-news.com/tag/croatia-foreign-entrepreneur.

> business. But, at one point it seemed that…like… I reached some information ceiling if I rely solely on them. I need some business support institution to provide me with the unbiased information on the local market. But then again… even if I find one, I have a language barrier…And if start without proper information, it looks more like gambling to me! (nonEU11)
>
> My friends are willing to help, but they are in a way short-sighted when it comes to my company operations. Plus, even if they want to, they often don't have enough time to dedicate to my business development. (nonEU12)

Furthermore, the integration with the locals is usually quite slow, especially in cases of different racial backgrounds (see next section). That also adds to the importance of the expat networks. Closeness of the Croatian citizens towards foreigners is among the highest in the EU (EC 2018). Paradoxically, even in the regions like Dalmatia with the greatest exposures to foreigners due to tourism and employment in the seafarer industry, the openness towards immigrants is rather low (IN7).

The observed limitations of the informal networks bring back the attention to the formal networks. Yet, due to their inefficiencies it appears that informal networks compensate for the deficits of the formal networks. That kind of substitute is limited by feasibility—as pragmatically noted by the interviewed entrepreneurs. Hence, as the existing formal networks do not seem to change fast enough and adapt to the needs of ethnic entrepreneurs, an association of the ethnic entrepreneurs would be a step forward in representing their interests and enabling their further integration at the Croatian business scene.

Ethnic Communities, Transnational Personal Relations, and Reliance on 'Domestic' Informal Practices

A recent survey (ACT Group 2019) among immigrant entrepreneurs originating from the nonEU less developed or conflict-affected countries lead to somewhat unsurprising results considering aforementioned. Namely, 73% of the interviewed entrepreneurs consider irrelevant accessing local business networks in other sectors, but not of the same ethnic origin as theirs. Slightly smaller share (67%) of them considers irrelevant accessing other local business networks in the same sector, but not of the same ethnic origin as theirs. Finally, 47% of them believe that it is irrelevant for them to access business networks of people of the same ethnic origin as theirs. These data additionally confirm the need for the previously discussed platform targeting the needs of ethnic entrepreneurs in Croatia.

In addition, due to a rather small share of foreigners in Croatia, there is a small number of ethnic communities operating in a formal or informal way. For instance, there is only approximately 80 African people living in Croatia now. They founded the African society in 2013 with one of the main purposes to educate Croatian people of various backgrounds and ages about the real life in Africa:

> The love for culture, the love for existence, the love for diversity – that is the main motivation behind African society… So far we have visited about eighty schools in Croatia. And the kids ask questions like what do you eat in Africa? What is yam, what is plantain? Don't you eat tigers, lions, elephants? We have spent quite a lot of time explaining what we actually eat. (nonEU1)

Education and raising awareness on diversity is very much needed because

> people openly make very racist jokes and are convinced that they are funny… It is not only about random people in the street. We also had unpleasant situation in one Croatian ministry where, reportedly for security reasons, they did not let us into their building despite the delivery they ordered from our company. By contrast, few days after that shocking situation we had a delivery to the US Embassy in Zagreb that is known for its strict security procedure and of course, there was no problem whatsoever. (nonEU12)

In line with that, several African entrepreneurs named their feeling in the initial period in Croatia as 'being invisible despite being black' (IN7). That feeling is fully in line with previously mentioned predominant attitude of (rather homogenous and closed) Croatian population. It appears to be one of the most persistent 'invisible' barriers for the incoming migration.

The ethnic communities may serve as a platform for employment opportunities. For instance, 68% of employees in the businesses owned by entrepreneurs from the third countries are immigrants, mostly with the same ethnic background as the owner (ACT Group 2019). Still, in general, the relations in the ethnic communities are much less employed than expected at the first glance and presented in the literature (IN6).

The interviewed immigrant entrepreneurs said that they maintain their relationships with their family, friends, and acquaintances in their home country. Yet, some of them admit that happens 'scarcer and scarcer' (EU3) and for most of the entrepreneurs, their ties in their country of origin do not help with their business in Croatia as 'family back home has its own reality and can hardly be included in my business in Croatia.... yet, they are my moral support!' (nonEU12). Entrepreneurs sometimes use connections in their home country to find new business deals there (EU8, EU10), to find more affordable suppliers (EU3, EU6) or suppliers of the goods not available in Croatia (nonEU1, nonEU13). Similar to the relations within their ethnic communities in Croatia, the contacts in their home countries seem to be somewhat underexploited as well. That situation leads to the possible reliance on the informal practices developed in Croatia and often used by the Croatian citizens.

Despite their predominant reliance on informal networks coupled with lacking trust and somewhat negative experience with the formal networks, none of the interviewed entrepreneurs claimed that they had used *veza*. Many of them were not fully familiar with the concept, but when explained, they could relate it to a similar or same informal practice in their home country. Still, the ethnic entrepreneurs do not seem capable to grasp the most efficient mechanisms of the informal networks in Croatia (IN7). That is not surprising as the return migrants to Croatia experience the same issues (IN7, Skokic et al. 2019). Despite all the efforts, the missing contacts influence their businesses, i.e. increase the opportunity costs. Still...

> ..people expect things to work like they do at home, and when they don't, they get frustrated. Language, the extra bureaucracy etc. Understandable but a lot of people hide behind that... My best advice for foreigners coming to do business in Dalmatia is contained in one sentence, and if they can accept it and implement it, they will do fine, but most won't. Do not try to change Dalmatia, but expect Dalmatia to change you.... I think it is also true that entrepreneurial expectations differ a lot depending on country of origin...Romanian people, for instance, see only opportunity here, while Brits see only problems. (EU9)

As shown in the previous quote, despite the recognised disadvantages of the Croatian business climate, the entrepreneurs are also well aware of the influences connected with their background and the need for self-responsibility and adaptability to the circumstances in their new environment.

Conclusion

Despite being crucial for the ethnic entrepreneurs, the informal networks among them are still underresearched. This gap between their importance and their presence in the studies is particularly evident in Southeast European countries who have joined or are intending to join the EU and as such shall be more open to the free movement of people, labour, and goods. This study explores the case of Croatia, the 'freshest' EU member country, with long emigration tradition and somewhat small, but increasingly important, recent immigration influx. The ethnic entrepreneurs in Croatia are facing a rather long list of formal and informal barriers. Based on the interviews with immigrant entrepreneurs and institutional representatives and confirmed by secondary data, this study shows that informal networks are a key mechanism to (partly) address, diminish or speed up the process of relaxing those barriers. Hence, there are several key 'takeaways' from this study.

First, the informal networks appear to serve as substitutes to missing or insufficiently efficient institutions of the formal networks aimed at foreigners' needs, entrepreneurs in this case. Thus, the interviewed entrepreneurs in Croatia claim the groups of expatriates to be their primary platform for socialising and vital source of obtaining the information and help needed for professional and private purposes. Interestingly, when compared to the ethnic communities, the networks of foreigners originating from various countries are considered to be more active and more relied on for business purposes. The explanation for this finding may be three-fold. First, for most countries there is a rather small number of citizens originating from every single of those countries. Second, for most businesses, the access to sector-specific information is better if more people are included irrespective of their origin. Third, explanation is related to the sample: we did not specifically investigate the more numerous ethnic communities in Croatia, like Chinese with decades-long tradition or Syrian with more recent and increasing influx. In addition, due to regional specifics, we have interviewed entrepreneurs originating outside the former Yugoslav republics.

Second important finding is the 'extension' of the mentioned expat groups that the ethnic entrepreneurs consider vital for their business establishment and development, but for their private life as well. These are mostly Croatian citizens, fluent in English, who are recognised as reliable local partners. These mainly include competent accountants, co-owners, and attorneys. General population in Croatia, as demonstrated by the EU-wide surveys, is still relatively slow in integrating foreigners.

Third conclusion is also a recommendation for the way forward. Namely, in their informal networks, the ethnic entrepreneurs are used to in-group cooperation and support (as an efficient way of compensating for formal deficiencies in the business environment as well). That shall be utilised in further initiatives targeting them, particularly in mentoring and training programmes. In addition, they are keen to improve their ties with the local community, so that is another point to tackle in the initiatives of business support organisations but also in discussing and co-delivering public services aimed at foreigners in Croatia. Finally, a formal platform representing ethnic entrepreneurs in Croatia would be a robust channel to increase their visibility and address their needs in a more coherent way. Moreover, it would partly help them overcome the mechanisms of personal connections and favours in Croatia for which they, same as return migrants, do not have the local resources.

This research contributes to the literature on informal networks of ethnic entrepreneurs in post-socialist European countries, more specifically in SEE countries with similar legacy and similar EU trajectory. Despite reliance on the most recent primary data from Croatia, this study shall have broader resonance for other (post-socialist) countries sharing similar development and migration-related pattern.

Our research has several limitations. First, similar to previous research across transitional economies, significant problems were experienced with data access and the availability of reliable sources. Hence, this study did not investigate more diverse ethnic communities in Croatia. Second, although this study did embrace a significant sample of ethnic entrepreneurs, it would be valuable to extend the study across different regions in Croatia. Still, due to a very small population of ethnic entrepreneurs in Croatia, it is important to note that our sample could be deemed quite representative. Several illustrative facts are following: approximately one-third of all African entrepreneurs in Croatia (that is at the same time majority of African entrepreneurs in hospitality sector), the only Romanian entrepreneur with 100% Romanian ownership of the company and the most experienced (and informed) foreigners in consulting other Croatian and immigrant entrepreneurs were interviewed. In addition, our primary data were complemented by the most recent secondary sources. Finally, our results do not propose that a well-established formal networks or informal ties will help ethnic entrepreneurs in achieving business success. Further research in these areas is crucial for better understanding of ethnic entrepreneurs and their networking strategies.

Remark The authors are thankful to student-assistant Matea Nikolić for her research help.

References

ACT Group (2019). Social Entrepreneurship as an Enabling Environment for Migrants' Employment and Integration. Interreg Central Europe SEE ME IN, Zagreb, September 2019.

Aidis, R., Estrin, S., and Mickiewicz, T. (2008). Institutions and Entrepreneurship Development in Russia: A Comparative Perspective. *Journal of Business Venturing*, 23(6), 656–672.

Aliaga-Isla, R., and Rialp, A. (2012). How Do Information and Experience Play a Role in the Discovery of Entrepreneurial Opportunities? The Case of Latin-American Immigrants in Barcelona. *Latin American Business Review*, 13(1), 59–80.

Aliaga-Isla, R., and Rialp, A. (2013). Systematic Review of Immigrant Entrepreneurship Literature: Previous Findings and Ways Forward. *Entrepreneurship & Regional Development*, 25(9–10), 819–844.

Aligica, P. D., and Tarko, V. (2014). Crony Capitalism: Rent Seeking, Institutions and Ideology. *Kyklos*, 67(2), 156–176.

Alpeza, M., Oberman, M., and Has, M. (2018). Small and Medium Enterprises Report Croatia 2018. CEPOR—SMEs & Entrepreneurship Policy Center, Zagreb. http://www.cepor.hr/wp-content/uploads/2015/04/EN-SME-Report-2018-za-web.pdf.

Birley, S. (1985). The Role of Networks in the Entrepreneurial Process. *Journal of Business Venturing*, 1(1), 107–117.

Brković, Č., and Koutkova, K. (2018). Štela. In Ledeneva, A. (Ed.) *The Global Encyclopaedia of Informality: Understanding Social and Cultural Complexity*. Volume 1. London: UCL Press, pp. 54–58. https://doi.org/10.14324/111.9781911307907.

Bužinkić, E. (Ed.) (2017). MIGRENT—Migrant (Social) Entrepreneurship as a Tool of Socio-Economic Emancipation of Migrants. Centre for Peace Studies and Taste of Home, Zagreb. https://www.cms.hr/system/publication/pdf/92/Publikacija_Migrent.pdf.

Čapo, J. (2019). Economic Activities and Agency of "Love-Driven" International Migrants in the City of Zagreb. In Rajković Iveta, M., Kelemen, P., and Župarić-Iljić, D. (Eds.) *Contemporary Migration Trends and Flows on the Territory of Southeast Europe*. Zagreb: Filozofski fakultet, pp. 195–212. https://openbooks.ffzg.unizg.hr/index.php/FFpress/catalog/view/35/44/1824-1.

Čapo, J., and Kelemen, P. (2018a). Mixing Ethnic and Non-ethnic Economic Strategies: Migrant Entrepreneurs in Zagreb. *Narodna umjetnost*, 55(2), 29–56. https://doi.org/10.15176/vol55no202.

Čapo, J., and Kelemen, P. (2018b). The City and International Migrants. Institute of Ethnology and Folklore Research, Zagreb. http://www.citymaking.eu/wp-content/uploads/2018/09/The_City_and_International_Migrants.pdf.

Castles, S., Hein de Haas, M., and Miller, M. J. (2014). *The Age of Migration: International Population Movements in the Modern World* (5th ed.). Basingstoke: Palgrave Macmillan.

Castles, S., and Miller, M. J. (2009). *The Age of Migration: International Population Movements in the Modern World* (4th ed.). Basingstoke: Palgrave Macmillan.

CBS—Croatian Bureau of Statistics (2019). International Migration. https://www.dzs.hr/Hrv_Eng/Pokazatelji/Stanovnistvo%20-%20migracije.xlsx.

Chand, M., and Ghorbani, M. (2011). National Culture, Networks and Ethnic Entrepreneurship: A Comparison of the Indian and Chinese Immigrants in the US. *International Business Review*, 20(6), 593–606.

Chavdarova, T. (2015). Perceptions vs. Practices: Nepotism in Small Businesses in Bulgaria. In Morris, J., and Polese, A. (Eds.) *Informal Economies in Post-Socialist Spaces: Practices, Institutions and Networks*. Palgrave Macmillan UK, pp. 154–174.

Chavdarova, T. (2018). Vruzki. In Ledeneva, A. (Ed.) *The Global Encyclopaedia of Informality: Understanding Social and Cultural Complexity*. Volume 1. London: UCL Press, pp. 64–67. https://doi.org/10.14324/111.9781911307907.

CPS/IHZ—Centre for Peace Studies & Impact Hub Zagreb. (2019). Country Analysis of Self-Employment and Entrepreneurship Initiatives for TCN. BEST, Zagreb.

Dabić, M., Vlačić, B., Paul, J., Dana, L. P., Sahasranamam, S., and Glinka, B. (2020). Immigrant Entrepreneurship: A Review and Research Agenda. *Journal of Business Research*, 113, 25–38.

Das, T. K., and Teng, B. S. (1997). Sustaining Strategic Alliances: Options and Guidelines. *Journal of General Management*, 22(4), 49–64.

Davidsson, P., and Honig, B. (2003). The Role of Social and Human Capital Among Nascent Entrepreneurs. *Journal of Business Venturing*, 18(3), 301–331.

Drakopoulou-Dodd, S., and Patra, E. (2002). National Differences in Entrepreneurial Networking. *Entrepreneurship & Regional Development*, 14(2), 117–134.

Dreher, A., and Schneider, F. (2010). Corruption and the Shadow Economy: An Empirical Analysis. *Public Choice*, 144, 215–238.

EC—European Commission (2018). Special Eurobarometer 469—Integration of Immigrants in the European Union, April 2018. https://ec.europa.eu›index.cfm›ResultDoc›download›DocumentKy.

Eurostat (2019). Migration and Migrant Population Statistics, March 2019. https://ec.europa.eu/eurostat/statistics-explained/index.php/Migration_and_migrant_population_statistics#Migration_flows:_Immigration_to_the_EU_from_non-member_countries_was_2.4_million_in_2017.

Evald, M. R., Klyver, K. I. M., and Svendsen, S. G. (2006). The Changing Importance of the Strength of Ties Throughout the Entrepreneurial Process. *Journal of Enterprising Culture*, 14(1), 1–26.

Giljević, T., and Lalić Novak, G. (2018). Coordination Instruments in Croatian Integration Policy: Classification, Evaluation, and Proposals for Improvements. In Koprić, I., Lalić Novak, G., and Vukojičić Tomić, T. (Eds.) *Migrations, Diversity, Integration and Public Governance in Europe and Beyond*. Zagreb, Croatia: Institute for Public Administration, pp. 341–362.

Greve, A., and Salaff, J. W. (2003). Social Networks and Entrepreneurship. *Entrepreneurship Theory and Practice*, 28(1), 1–22.

Hite, J. M., and Hesterly, W. S. (2001). The Evolution of Firm Networks: From Emergence to Early Growth of the Firm. *Strategic Management Journal*, 22(3), 275–286.

Hoang, H., and Antoncic, B. (2003). Network-Based Research in Entrepreneurship. *Journal of Business Venturing*, 18(2), 165–187.

Horak, S., and Klein, A. (2016). Persistence of Informal Social Networks in East Asia: Evidence from South Korea. *Asia Pacific Journal of Management*, 33(3), 673–694.

Hsu, C. (2005). Capitalism Without Contracts Versus Capitalists Without Capitalism: Comparing the Influence of Chinese Guanxi and Russian Blat on Marketization. *Communist and Post-Communist Studies*, 38(3), 309–327.

Hsu, C., Liu, Z., and Huang, S. (2012). Managerial Ties in Economy Hotel Chains in China. *International Journal of Contemporary Hospitality Management*, 24(3), 477–495.

Ireland, R. D., Tihanyi, L., and Webb, J. W. (2008). A Tale of Two Politico-Economic Systems: Implications for Entrepreneurship in Central and Eastern Europe. *Entrepreneurship Theory and Practice*, 32(1), 107–130.

Jack, S. (2005). The Role, Use and Activation of Strong and Weak Network Ties: A Qualitative Analysis. *Journal of Management Studies*, 42(6), 1233–1259.

Jack, S. L., and Anderson, A. R. (2002). The Effects of Embeddedness on the Entrepreneurial Process. *Journal of Business Venturing*, 17(5), 467–487.

Jurković, R., and Rajković Iveta, M. (2016). "Okus doma": integracija azilanata protkana transnacionalnim procesima i promicanjem kulinarskih tradicija. *Studia ethnologica Croatica*, 28(1), 147–178.

Kariv, D., Menzies, T. V., Brenner, G. A., and Filion, L. J. (2009). Transnational Networking and Business Performance: Ethnic Entrepreneurs in Canada. *Entrepreneurship and Regional Development*, 21(3), 239–264.

Kauffman Foundation (2009). Kauffman Index of Entrepreneurial Activity. Retrieved 29 November 2019. http://www.kauffman.org/uploadedFiles/kiea_042709.pdf.

Kets de Vries, M. F., and Florent-Treacy, E. (2003). Roustam Tariko (A): Russian Entrepreneur. *Entrepreneurship Theory and Practice*, 27(3), 299–313.

King, R. (2018). Context-Based Qualitative Research and Multi-sited Migration Studies in Europe, In Zapata-Barrero R., and Yalaz E. (Eds.) *Qualitative Research in European Migration Studies*, IMISCOE Research Series. Cham: Springer.

Kloosterman, R., Van Der Leun, J., and Rath, J. (1999). Mixed Embeddedness: (In) Formal Economic Activities and Immigrant Businesses in the Netherlands. *International Journal of Urban and Regional Research*, 23(2), 252–266.

Klyver, K., and Foley, D. (2012). Networking and Culture in Entrepreneurship. *Entrepreneurship & Regional Development*, 24(7–8), 561–588.

Koutkova, K. (2015). The Importance of Having štela: Reproduction of Informality in the Democratization Sector in Bosnia. In Morris, J., and Polese, A. (Eds.) *Informal Economies in Post-Socialist Spaces: Practices, Institutions and Networks*. Palgrave Macmillan UK, pp. 139–153.

Kuti, S., and Božić, S. (2011). Analitičke dimenzije za istraživanje transnacionalnih aktivnosti: primjer kineskih migranata u Hrvatskoj. *Revija za sociologiju*, 41(3), 315–340. https://doi.org/10.5613/rzs.41.3.3.

Kuznetsov, A., McDonald, F., and Kuznetsova, O. (2000). Entrepreneurial Qualities: A Case from Russia. *Journal of Small Business Management*, 38(1), 101–107.

Larson, A., and Starr, J. A. (1993). A Network Model of Organization Formation. *Entrepreneurship Theory and Practice*, 17(2), 5–15.

Lechner, C., and Dowling, M. (2003). Firm Networks: External Relationships as Sources for the Growth and Competitiveness of Entrepreneurial Firms. *Entrepreneurship & Regional Development*, 15(1), 1–26.

Ledeneva, A. (2008). Blat and Guanxi: Informal Practices in Russia and China. *Comparative Studies in Society and History*, 50(1), 118–144.

Lekaj, A., Gregov, Z., and Varlandy Supek, M. (2015). Temeljne značajke manjinskog poduzetništva na primjeru poduzetništva albanske nacionalne manjine u Hrvatskoj. *Obrazovanje za poduzetništvo - E4E*, 5(2), 69–82. Retrieved from https://hrcak.srce.hr/149531.

Light, I., and Bonacich, E. (1988). *Immigrant Entrepreneurs: Koreans in Los Angeles, 1965–1982*. University of California Press.

Manolova, T. S., Eunni, R. V., and Gyoshev, B. S. (2008). Institutional Environments for Entrepreneurship: Evidence from Emerging Economies in Eastern Europe. *Entrepreneurship Theory and Practice*, 32(1), 203–218.

Manolova, T. S., and Yan, A. (2002). Institutional Constraints and Entrepreneurial Responses in a Transforming Economy: The Case of Bulgaria. *International Small Business Journal*, 20(2), 163–184.

Martinez, M. A., and Aldrich, H. E. (2011). Networking Strategies for Entrepreneurs: Balancing Cohesion and Diversity. *International Journal of Entrepreneurial Behavior & Research*, 17(1), 7–38.

Mostajer Haghighi, A., and Lynch, P. (2012). Entrepreneurship and the Social Integration of New Minorities: Iranian Hospitality Entrepreneurs in Scotland. *Tourism Review*, 67(1), 4–10.

Nreca, K., and Gregov, Z. (2018). Uloga žena u obiteljskim poduzećima albanske nacionalne manjine u Hrvatskoj. *Obrazovanje za poduzetništvo - E4E*, 8 (Special issue), 73–94. Retrieved from https://hrcak.srce.hr/196916.

Ostgaard, T. A., and Birley, S. (1994). Personal Networks and Firm Competitive Strategy—A Strategic or Coincidental Match? *Journal of Business Venturing*, 9(4), 281–305.

Otten, J. (2018). Vrski. In Ledeneva, A. (Ed.) *The Global Encyclopaedia of Informality: Understanding Social and Cultural Complexity*. Volume 1. London: UCL Press, pp. 62–64. https://doi.org/10.14324/111.9781911307907.

Peng, Y. (2004). Kinship Networks and Entrepreneurs in China's Transitional Economy. *American Journal of Sociology*, 109(5), 1045–1074.

Polese, A. (2013). *Drinking with Vova: A Ukrainian Entrepreneur Between Informality and Illegality. The Informal Post-Socialist Economy: Embedded Practices and Livelihoods*. London and New York: Routledge.

Polese, A. (2015). Informality Crusades: Why Informal Practices Are Stigmatized, Fought and Allowed in Different Contexts According to an Apparently Ununderstandable Logic. *Caucasus Social Science Review*, 25(1), 1–26

Polese, A. (forthcoming). Labour Mobility and Precariousness: Why Informality Ends Up Replacing and Supplementing the State for the Invisible and the Vulnerable. In Polese, A., Turaeva, R., and Urinboyev, R. (Eds.) *Informality and Mobility in Eastern Europe: People, Spaces, Movements*. London: Palgrave.

Puffer, S. M., McCarthy, D. J., and Boisot, M. (2010). Entrepreneurship in Russia and China: The Impact of Formal Institutional Voids. *Entrepreneurship Theory and Practice*, 34(3), 441–467.

Rath, J., and Swagerman, A. (2011). *Promoting Ethnic Entrepreneurship in European Cities*. Luxembourg: Publications Office of the European Union.

Rehn, A., and Taalas, S. (2004). 'Znakomstva I Svyazi' (Acquaintances and connections)–Blat, the Soviet Union, and Mundane Entrepreneurship. *Entrepreneurship & Regional Development*, 16(3), 235–250.

Rusinovic, K. (2008). Transnational Embeddedness: Transnational Activities and Networks Among First- and Second-Generation Immigrant Entrepreneurs in the Netherlands. *Journal of Ethnic and Migration Studies*, 34(3), 431–451.

Sarasvathy, Saras D. (2018). What Makes Entrepreneurs Entrepreneurial? Darden Case No. UVA-ENT-0065. Available at SSRN: https://ssrn.com/abstract=909038.

Schutjens, V., and Stam, E. (2003). The Evolution and Nature of Young Firm Networks: A Longitudinal Perspective. *Small Business Economics*, 21(2), 115–134.

Sequeira, J. M., and Rasheed, A. A. (2006). Start-Up and Growth of Immigrant Small Businesses: The Impact of Social and Human Capital. *Journal of Developmental Entrepreneurship*, 11(4), 357–375.

Šimić Banović, R. (2015) Cutting the Red Ribbon but Not the Red Tape: The Failure of Business Environment Reform in Croatia. *Post-Communist Economies*, 27(1), 106–128.

Šimić Banović, R. (2016). Is Culture an Underpinning or Undermining Factor in the Business Environment of the Transitional Countries? In Ateljević, J., and Trivić, J. (Eds.) *Economic Development and Entrepreneurship in Transition Economies: Issues, Obstacles and Perspectives*. Springer International Publishing, pp. 11–37.

Šimić Banović, R. (2019). Uhljeb—A Post-Socialist Homo Croaticus: A Personification of the Economy of Favours in Croatia? *Post-Communist Economies*, 31(3), 279–300.

Šimić Banović, Ružica, Alpeza, Mirela, and Brzozowski, Jan (2019). Immigrant Entrepreneurship in Croatia: Exploring Its Potential, (Rising) Barriers and Integration Patterns, September 1. Available at SSRN: https://ssrn.com/abstract=3528132.

Singer, S., Šarlija, N., Pfeifer, S., and Oberman Peterka, S. (2019) Što čini Hrvatsku (ne)poduzetničkom zemljom? - GEM Hrvatska 2018, CEPOR—SMEs and Entrepreneurhip Policy Center, Zagreb. http://www.cepor.hr/wp-content/uploads/2019/04/GEM2018zaweb.pdf.

———. (2021). Što čini Hrvatsku (ne)poduzetničkom zemljom? GEM Hrvatska 2019–2020. CEPOR, Zagreb.

Skokic, V., Lynch, P., and Morrison, A. (2019). Veza: An Informal Network of Tourism Entrepreneurs. *Annals of Tourism Research*, 77, 26–37.

Staber, U., and Aldrich, H. E. (1995). Cross-National Similarities in the Personal Networks of Small Business Owners: A Comparison of Two Regions in North America. *Canadian Journal of Sociology*, 20(4), 441–467.

Stanojevic, D., and Stokanic, D. (2018). Veza. In Ledeneva, A. (Ed.) *The Global Encyclopaedia of Informality: Understanding Social and Cultural Complexity*. Volume 1. London: UCL Press, pp. 58–62. https://doi.org/10.14324/111.9781911307907.

Steenberg, R. (2016). The Art of Not Seeing Like a State: On the Ideology of "Informality". *Journal of Contemporary Central and Eastern Europe*, 24(3), 293–306. https://doi.org/10.1080/0965156X.2016.1262229.

Strobl, A., and Kronenberg, C. (2016). Entrepreneurial Networks Across the Business Life Cycle: The Case of Alpine Hospitality Entrepreneurs. *International Journal of Contemporary Hospitality Management*, 28(6), 1177–1203.

Transparency International (2019). Corruption Perceptions Index 2018: Croatia. https://www.transparency.org/country/HRV.

Troskot, Z., Prskalo, M. E., and Šimić Banović, R. (2019). Ključne odrednice iseljavanja visokokvalificiranog stanovništva: slučaj Hrvatske s komparativnim osvrtom na nove članice EU-a. *Zbornik radova Pravnog fakulteta u Splitu*, 56(4), 877–904.

Volery, T. (2007). Ethnic Entrepreneurship: A Theoretical Framework. In Dana, L. P. (Ed.) *Handbook of Research on Ethnic Minority Entrepreneurship*. Cheltenham, UK: Edward Elgar Publishing, pp. 30–41.

Waldinger, R., Aldrich, H., and Ward R. (1990). Opportunities, Group Characteristics and Strategies. In Waldinger, R., Aldrich, H., and Ward, R. (Eds.) *Ethnic Entrepreneurs: Immigrant Business in Industrial Societies*. London: Sage, pp. 13–48.

Yoo, J. K. (2000). Utilization of Social Networks for Immigrant Entrepreneurship: A Case Study of Korean Immigrants in the Atlanta Area. *International Review of Sociology*, 10(3), 347–363.

Yoon, I. J. (1991). The Changing Significance of Ethnic and Class Resources in Immigrant Businesses: The Case of Korean Immigrant Businesses in Chicago. *International Migration Review*, 25(2), 303–332.

Zhang, J. (2010). The Problems of Using Social Networks in Entrepreneurial Resource Acquisition. *International Small Business Journal*, 28(4), 338–361.

Zhang, J., Soh, P. H., and Wong, P. K. (2011). Direct Ties, Prior Knowledge, and Entrepreneurial Resource Acquisitions in China and Singapore. *International Small Business Journal*, 29(2), 170–189.

Zimmer, C., and Aldrich, H. (1987). Resource Mobilization Through Ethnic Networks: Kinship and Friendship Ties of Shopkeepers in England. *Sociological Perspectives*, 30(4), 422–445.

World Economic Forum (2018). The Global Competitiveness Report 2018. Cologny/Geneva, Switzerland. https://www.weforum.org/reports/the-global-competitveness-report-2018.

"Performance of Illegality" Toward Migrants Living with HIV in Russia: From Social Exclusion to Deportation

Daniel Kashnitsky

The Border Spectacle

Immanuel Kant was one of the first thinkers who raised the issue of the "reception of strangers" and attitudes of the host community. In his work, "Perpetual Peace" Kant was talking about the indisputable right "of each pilgrim to be met without hostility on foreign territory" (Kant 1994). He mentioned that the shortcomings of the state system lag behind the flows of humanity moving from one country to another. The states rarely have time to respond to the challenges of migration and almost instinctively enact prohibitions. Kant gave a cosmopolitan perspective for further reflection in this area. He defined the concept of the world as a "universal state," where every citizen should have the right to asylum. One hundred years after Kant, at the end of the nineteenth century, Henry Sidgwick discussed immigration from the nationalist standpoint concluding it was the right and duty of the national state to make decisions on accepting

D. Kashnitsky (✉)
Center for Qualitative Studies, Higher School of Economics, Moscow, Russia

A. Polese (ed.), *Informality, Labour Mobility and Precariousness*, International Political Economy Series,
https://doi.org/10.1007/978-3-030-82499-0_5

migrants and refugees (Sidgwick 2015). His position formed the basis of many modern proponents of strict state control over migration.

Žižek's concept of power (1997) implies a certain dark supplement of itself—a "superstructure" so that the authorities can reproduce themselves and harness "the other." He formulates the ideological and political significance of "saving face" in relation to the political discourse. The effectiveness the policy depends on the mechanism of hypocrisy, whereas non-transparent migration policy is an intrinsic part of it. Hence, we can say that migration policy is a sensitive intersection of security and electoral practices, economic pragmatism, and global ambitions of individual states. This complexity has multiple repercussions on opportunities of inclusion and well-being of thousands of migrants (Sassen 1988; Castañeda et al. 2015). An important milestone of this discussion was timed to the appearance of the collection "Boundaries: National Autonomy and Its Limits" edited by Peter Brown and Henry Shue (1981), where the questions about the manifestations of illegality on daily lives of migrants were raised by several scholars. The American sociologist De Genova (2002) took it further and introduced the concept of the "Border Spectacle," a special exclusionary ritual, which is designed to fully demonstrate tough migration policy and effective border control. The reverse non-public side of this "border spectacle" is the growing number of undocumented or semi-documented migrants recruited to do the heavy unskilled labor, beneficial to the state and large businesses. By implementing this policy, the state maintains a conservative electoral image and at the same time solves some economic problems.

This is the essence of the "Border spectacle" where deportations, amnesty, and legal uncertainty (Reeves 2015) destroys the fragile ideological dichotomy of being either included or excluded. It outlines a longer continuum of "legality - illegality" (Dave 2014). Spontaneous police raids, practice of street check-ups, detention and deportation practices create a space for the "border spectacle" that becomes a permanent condition not only at the passport control, but all throughout the migratory experience in the host country. Borders become filtering mechanisms to stimulate an unequal exchange of values (Kearney 2004). The boundaries delimit people into those who are expelled immediately and deported and those who are included, often as "illegal" migrants.

Migrants are imposed perception of deportation that defines the horizon of their life in the receiving country. The deportability of "illegal immigrants" is the defining category for a large group of people. Often this category has no limitation. Hence, deportability is rather a disciplinary measure, often yielding enforcement (De Genova 2002). The inclusion of undocumented migrants is usually carried out through various forms of labor submission, and constantly associated with various types of civil exclusion. Borderline has become a ceremonial space where "effective state control" is performed. This is a space of exchange, separation, and sovereign exclusion (Rosas 2006; Schinkel 2009) where power is positioned above the law and arbitrariness emerges.

Maintaining the discourse about human trafficking and smuggling allows the state to remain paternalistic, applying protection only to "their" citizens, but even on some of the migrant groups, especially females, who need to be saved from the crimes of the illegal migration. Sometimes countermeasures against trafficking and other crimes unleashes hands for expelling anyone who is depicted in the media under the generalized image of an "illegal migrant," when necessary (Schenk 2018). Likewise, the policy of citizenship turns into the actual politics of differences. Unequal and discriminatory policies of citizenship produce illegality and social exclusion of migrants (Hills et al. 2002). Among various arguments in favor of closing borders is that migrants threaten the values of local culture, humanity, public safety, and hygiene. Cole (2000) argues that political systems operating in a similar way violate their own principles and self-destruct. At the same time, informality appears to be a major constituent of social policymaking in post-socialist countries, including Russia (Morris and Polese 2014) hence excluded groups of population have some extent of flexibility to develop coping strategies and navigate throughout the maze of state bureaucracy and social barriers (Dave 2014; Ooms and Kruja 2019). Informality allows to constantly negotiate social practices in the public life and sometimes redefine the borders between the individual and the regulating state (Haid 2016).

I held 15 semi-structured interviews with undocumented labor migrants who shared about their experience living in limbo in Russia as the result of their HIV status. In addition to the interviews, I analyzed 10 recorded cases from the Steps Foundation. Cases are in fact notes of case managers who accompanied migrants with HIV while they were clients of the Foundation. Among all the diversity of migration flows, we decided to focus on migrants from most prevalent origins of labor migration flows, such as Armenia, Belarus, Kyrgyzstan, Moldova, Ukraine, Uzbekistan, Tajikistan—all former republic of the Soviet Union that still keep visa-free regime with Russia. That means migrants can freely come to Russia as tourists. All their problems with legal status started when they had to confirm their residence status—that is when HIV status becomes an unsurmountable barrier for a documented status.

The interviews with the migrants took place in autumn 2020 whereas cases referred to service delivery and interactions with the migrants during 2017–2020. We consulted civil society organizations that support people living with HIV in sending countries and in Russia to recruit our study participants.

HIV Positive Migrants as the Catch 22 of the Russian Epidemic

As for the Russian Federation, there has been very little reflection about the ethics of migration and state management of migration flows. The existing arbitrary approach to migration is very seldom challenged. Public debate on migrants is predominantly discussed in the terms of economic efficiency and national security, with little consideration of rights of migrants and refugees. There is, for example, little debate about the 98% rejection rate of refugee application even though Russia signed the Refugee Convention back in 1992 (Kubal 2019). At the same time, Russia supports a large visa-free space that extends from Central Asia to the Arctic, from Moldova to the Far East. So, the exclusion mechanisms in Russia, unlike other global host regions of migration such as the European Union and the USA, tend to affect most international migrants not at the border itself—many can enter visa-free, but at further stages of integration—in the cities and towns of the host country when they meet with the migration authorities, police, and employers.

The Russian state seems to understand the pragmatic benefits of attracting international labor migrants but prevents them from raising voices to protect their rights and build a strong civil society. Therefore, migrants are held in the so-called legal uncertainty (Reeves 2015). Migrants never know exactly how many documents they need to collect to be perfect in the eyes of the state—to be eligible, to medical care, to legal work, finally, to freely move in the city. Migrants never fully realize what else the police or state officials can cling to. On the one hand, the migrant residence system is being reformed and seems to be more transparent (for example, the new work permit system introduced in 2017), but a recent statement made by the Ministry of Internal Affairs about "illegal" migrants making about two million people (about 1/5 of the total) does not indicate that authorities are close to the solution. The undocumented migrants comprise different groups of labor migrants and unrecognized refugees, who are characterized by extremely low mobility in the host country because of the fear of being caught and deported. At the same time, they have to accept an extremely low level of remuneration and they rarely complain about violations of labor rights, lack of access to social entitlements as they fear prosecution by the migration authorities. In such a situation, undocumented migrants remain a submissive labor force experiencing multiple barriers in access to health services (Demintseva and Kashnitsky 2016), education, and other social entitlements. Moreover, even fully legal migrants will be deported from Russia after two administrative violations, for example, for crossing the road in the wrong place or for slight speeding. That is, foreigners in Russia are required to keep to perfection—their career and life plan are often at stake. Specific ways to exclude migrants are country-specific. For example, in order to obtain a work permit, it is required to provide a certificate of the absence of social diseases such as HIV and tuberculosis, it is also necessary to buy private medical insurance, which, in reality, does not provide services and turns out to be the only a condition for admission.

It is also necessary to have a residence registration, which is actually very difficult to obtain as landlords often do not want to disclose that they are leasing out their apartments. There is also structural discrimination at the level of state institutions, which manifests itself in quite explicit statements by officials about the need to deport foreigners (Schenk 2018).

Failure to meet one of the multiple requirements for legalization imposed by the state, migrants find themselves in illegality with no access to secured labor and medical insurance. That creates an additional burden for the state—numerous people who become placed out of the epidemiological statistics, out of the reach of health prevention policy, vaccinations, and other important population health measures.

Exclusion of HIV-Positive International Migrants

The case of HIV-positive international migrants is a demonstration of an extreme degree of exclusion and an example of state biopolitics (Rushton 2012; Ingram 2010). When an international migrant is tested positively on HIV, they must leave Russia. However, few leave voluntarily as the state does not actually have the necessary regulation to immediately deport them. These migrants stay in Russia undocumented by default, with no access to vital antiretroviral therapy. They are afraid to contact state medical clinics and hospitals for fear of being reported to migration authorities. This limbo situation thus can last for years.

Deportation is very rarely enacted in Russia as it is costly to deport people. However, migrants are aware that the testing results are immediately communicated to the state migration service and their names are included into the blacklist at the border control. That means an international migrant if found HIV would not be able to reenter Russia for life once he or she has left Russia. Low information about HIV, stigma, and fear to deceive relatives not meeting economic expectations of family members keep many migrants with HIV locked in Russia for years. Their health is getting deteriorated in spite of the fact that free treatment is available in all main home countries of migrants due to the remaining presence of international donors who support provision of antiretroviral therapy.

Nowadays when an affordable and effective antiretroviral therapy exists in most countries of the world (UNAIDS 2017), HIV infection, remaining a chronic disease, is no longer considered an epidemiological threat. if a person living with HIV is on treatment, the viral load goes down to an untransmittable level (UNAIDS 2019a) and so these people can live and work and be healthy migrants and beneficial for the host country. However, by leaving groups of HIV-positive foreigners in Russia with no access to treatment, the state in fact exacerbates the epidemiological situation, creating all prerequisites for future development of the epidemic throughout the region of Eastern Europe and Central Asia.

Today Russia remains the only country in the European region that keeps a residence ban for migrants who were tested positive with HIV (UNAIDS 2019b). This protective measure is doubtful both from epidemiologic and human rights perspective. Russia has now the highest burden of the HIV epidemic in East Europe and Central Asia region with over 1% population being HIV-positive whereas all key migration-sending countries such as Kyrgyzstan, Uzbekistan, Tajikistan, have from five to six times lower prevalence rate of HIV as compared to Russia (UNAIDS 2019b). This means that this HIV residence ban for the migrants is no longer beneficial for Russia from both epidemiologic and national security points of view.

A case of a young man from Turkmenistan, aged 34, who has been living in Russia with HIV for eight years. He had come to Russia as a student migrant in 2010 and four years later got tested positive on HIV.

It was just in time to allow him to complete his degree but after graduation, he remained in Russia undocumented with no perspective to receive a residence visa. He now receives limited support from a local NGO. Living in illegality, he shares that he has a strong fear of leaving home during large public events in the city. For example, during the World Cup in 2018, he did not leave his apartment fear of being detained. He rarely uses public transport, afraid of document checks. He would like to go back to his home country and live with his family, but Turkmenistan provides no antiretroviral therapy to its citizens. Apparently, he has not even a shady perspective to live a healthy life in his home country for HIV positive people are not recognized as a group. Russia is his only hope after all because of the regular support he receives at the community NGOs. This situation illustrates an absurdist deadlock created by the two stated—one sending and one receiving country of migration. This illustration of a young man who has no sustainable solution to follow antiretroviral treatment and reside in legality and so, he objectifies as a puppet in a transnational border spectacle.

Another story of a student from Vietnam, aged 28, who won a scholarship to go to Russia to do study at the doctoral program in math and computer science. Half a year ago he was found HIV-positive after doing an anonymous test in a private clinic. He fears the moment when he needs to extend his student visa as he will have to pass the HIV test again. He is perplexed and does not know what he has to do. After graduation, he was planning to return home, but now he understands that he would not be able to do civil service as he planned before. The Vietnamese government has obligatory testing for infectious diseases and refuses to employ should HIV be detected. As a result, he would have to accept a lower paid job. He is hopeful that he would be able to stay in Russia and find employment in his profession. For the moment, the situation for him looks insoluble. He is also apprehensive of talking about his status with parents back home, as stigma toward people living with HIV is persistent. This case is another vivid vignette that shows how migrants who could have a normal life are locked in limbo in Russia with no realistic solution for the nearest time.

The first act of the border spectacle sets the scene for HIV positive migrants in Russia by delegitimizing their residence, depriving them from access to healthcare, and breaking off their professional trajectories, whereas in the second act we see migrants staying in fact staying in the country in illegality, some—in semi-legality with limited rights (Kubal

2013). They get deported physically very rarely as deportations mechanisms are complicated, costly, and enacted quite rarely. Hence a lot of HIV-positive migrants keep staying in the host country in various in-between statuses with varying levels of access to job market and social services. Here is another case that illustrates professional barrier.

A male migrant from Kyrgyzstan, 36 y.o. who had been tested positively for HIV one month prior to the interview shared "*I'm an experienced cook at the Jewish-cuisine restaurant in Moscow; I've been working there for 8 years and time is coming to renew my medical certificate [required for all restaurant employees in Russia]... I'm afraid with my HIV I am not going to make it; I won't pass the checkups and will no longer be able to work. What I'm going to do?*" For him, a professional migrant with a good income, his future career seems seriously jeopardized despite the fact that he will have no issue with a residence permit—Kyrgyzstan is a part of the Eurasian Economic Union, so citizens of member-countries can freely live and work all across the union. This is why this migrant is formally not deportable, however, this career is not likely to continue the way he has been developing it. The young male considers going back to his home country, where he is at least eligible for HIV care, but he apprehends his HIV status would be disclosed to his family and friends back home. At the time of the interview, he felt he was at a crossroads but without a clear plan.

"This is not a terrible diagnosis; these are inhumane laws" and the burden of stigma for female migrants.

An intersectional paradigm needs to be applied to understand multiple forms of discrimination and oppression where gender and ethnicity are considered as a part of the complex interplay of the determinants of health (Hankivsky and Christoffersen 2008; Inverardi 2018). Female migrants with HIV experience have additional fear that their status would be revealed in the community and the wider society (Castañeda et al. 2015; Bredström 2006).

A 27-year-old female from Uzbekistan arrived in Russia six years ago. She worked as a sex worker and was undocumented before she got to know she was HIV-positive. Luckily, her HIV-positive status was confirmed at a private clinic, hence the results were not reported to the authorities. Most of her clients were migrants from Central Asia. She was sure that HIV would not affect her, despite some unprotected contacts. Back then, she had no knowledge about the ways of HIV transmission.

Now she visits a healthcare NGO in Moscow and pays for her antiretroviral drugs. She has no plans to return to Uzbekistan due to fear of a possible disclosure of her HIV status and a negative reaction from family members. "*Most likely the family will simply refuse any contacts with me.*" She believes that it is easier for her to remain in Russia, even with an undocumented status, so she will continue to live and work in Russia and provide herself with HIV services from a private infection disease clinic and she will avoid risking stigma and possible disclosure in her home country.

Another female, an ethnic Russian from Kazakhstan, 41. y.o., relocated to Russia with her husband and two children. Already in Russia, they trying applying for Russian citizenship under a special resettlement program for compatriots—former citizens of Soviet Union. But they got refusal because of HIV positive status of the female migrant. She decided to divorce her husband to allow the rest of the family to re-apply while and returned to Kazakhstan. During the interview she shared: "*Can you imagine, I haven't seen my children for a year! I had to divorce my husband, to assist my family in legalization, I had to leave my children and leave the country. This is not a terrible diagnosis; these are inhumane laws.*" The future looks vague for her now.

The situation with migrants being infected with HIV has a trickle-down effect. Many of them end up returning to their home countries when their health gets worse. Because of the stigma, they often do not tell their diagnosis to their wives. The wives of migrants are at high risk of being exposed to HIV. Hence, labor migrants contribute to the spread of the HIV epidemic in their home countries (Kramer et al. 2008; Kashnitsky 2020) and in Russia throughout the whole region of Eastern Europe and Central Asia. When they finally decide to leave Russia where they have been stuck in limbo (Luo et al. 2012). This is especially widespread in the countries of Central Asia that provide up to half of international migrants to the Russian Federation (FMS 2019). This is a point of contradiction between restrictive migration policy and the semi-legality leading to disfunction from the epidemiologic perspective.

Although Russia limits access to migrants with HIV, they are largely present in the country, remaining completely invisible to the state, so we could even discuss the epidemiologic aspect of the problem in terms of spread of the HIV from Russia to family members and the larger society in migrant-sending countries. This brings us to the question that the issue of migrants with HIV is a global transnational issue and solutions should be mitigated on international level putting more focus on evidence-based interventions that have an aim to easy access to testing and treatment for all, especially, vulnerable groups, including migrants (Ooms and Kruja 2019; UNAIDS 2019b).

Conclusion

Russia is not the only country in the world that excludes migrants living with HIV from access to social protection and medical care, although Russia (and Saudi Arabia) is the largest host country in the world and the only European country with such restrictive policy toward a group of people with a chronic condition who can remain healthy migrants if they are on treatment (UNAIDS 2019b). Female migrants are particularly vulnerable because of the HIV residence ban, stigma in both host and sending countries as well as the regional spread of HIV that affects migrants' families back home, specifically, wives of male migrants who are the vast majority among migrants (Agadjanian and Zotova 2014). These life dramas usually remain in silence and out of media and policy attention because of stigma and discrimination.

Migrants living with HIV are just a vignette to demonstrate a larger issue of how undocumented migration is managed in the country (Kashnitsky 2020). Significant presence of undocumented migrants seems to serve the interests of the state capitalism in Russia as they can be exploited and cannot form political community (Reeves 2015) and protect their rights (Schenk 2018). At the global level, the rise of intolerance, often triggered by electoral waves of hatred and dehumanization of migrants is increasingly linked to the production of inequalities between native people and the migrants in terms of access to social entitlements when politicians support the identity of local population only on the grounds that they are locally belonging.

The "border spectacle" (De Genova 2002) continuously creates the image of an aggressive "migrant" who comes to seize the resources of the local population. But this spectacle is only intended to veil the exploitation of the host society. Thus, migration control is gradually becoming an immanent mechanism to exercise power, a kind of "executioner's performance" (Foucault 1979), and the question would be then, is it only for migrants alone? Indeed, there were examples when such mechanisms hit throughout the whole society. Against the background of increasing migration flows and, at the same time, the growth of right-wing populism in the world, it is important for social researchers to try to better understand what values, fears, and structural mechanisms of the reproduction of power are behind migration policies and where exactly governments create a lose–lose situation such as with international migrants' health and the uncontrolled spread of the epidemics throughout the whole region of Eastern Europe and Central Asia. Certainly, the specific circumstances of the "border spectacle" in Russia is different from other host countries, however, in general, the situation in Russia fits into the global dynamics of increasing exclusion and growing inequality (De Genova 2002) and exclusion of vulnerable groups. On the other hand, informal practices sometimes allow irregular migrants to develop coping strategies. They navigate throughout the constantly altering realm of informality and often survive in the very periphery of the host society remaining silent and suspended with no access to care and hope for a stable future.

Acknowledgements I would like to express my gratitude to Moscow-based Shagi Foundation for providing me with access to their case database of undocumented migrants with HIV who have been their clients. I also thank my supervisors, Ekaterina Demintseva from the Higher School of Economics, and Anna Bredström from Linköping University for productive discussions and thorough feedback provided to the draft of this article.

Funding This article is the product of a research project implemented as part of the Basic Research Programme at the National Research University Higher School of Economics (HSE).

Literature

Agadjanian, Victor, and Zotova, Natalia. 2014. Immigration and HIV Risks: Central Asian Female Migrants in the Russian Federation. *Demographic Review* 1(2): 85–109 (in Russian).

Bredström, A. 2006. Intersectionality: A Challenge for Feminist HIV/AIDS Research? *European Journal of Women's Studies* 13(3): 229–243.

Brown, P., Shue, H. 1981. *Boundaries, National Autonomy and Its Limits*. Maryland Studies in Public Philosophy. Totowa, NJ: Rowman and Littlefield.

Castañeda, H., Holmes, SM., Madrigal, DS., DeTrinidad Young, ME., Beyeler, N., Quesada, J. 2015. Immigration as a Social Determinant of Health. *Annual Review of Public Health* 36(1): 375–392.

Cole, P. 2000. *Philosophies of Exclusion. Liberal Political Theory and Immigration*. Edinbu rgh: Edinburgh UP.

Dave, B. 2014. Becoming "Legal" Through "Illegal" Procedures: The Precarious Status of Migrant Workers in Russia. Russian Analytical Digest. No. 159, 20.

De Genova, N. 2002. 'Migrant' 'Illegality' 'and Deportability in Everyday Life'. *Annual Review of Anthropology* 31: 419–447.

Demintseva, E., and Kashnitsky, D. 2016. Contextualizing Migrants' Strategies of Seeking Medical Care in Russia. *International Migration* 55 (2): 29–42.

Ending AIDS. Progress Towards the 90–90–90 Targets. UNAIDS 2017. Accessed 23 March 2020. http://www.unaids.org/ru/resources/documents/2017/20170720_Global_AIDS_update_2017.

Federal Migration Service. 2019. Migration Statistics in the Russian Federation. Accessed 23 March 2020. https://xn--b1aew.xn--p1ai/Deljatelnost/statistics/migracionnaya.

Foucault, M. 1979. Discipline and Punish: The Birth of the Prison. New York: Random House.

Haid, C. G. 2016. The Janus face of urban governance: State, informality and ambiguity in Berlin. *Current Sopciology* 65(2).

Hankivsky, O., Christoffersen, A. 2008. Intersectionality and the Determinants of Health: A Canadian Perspective. *Critical Public Health* 18(3): 271–283.

Hills, J., Le Grand, J., Piachaud, D. 2002. *Understanding Social Exclusion.* Oxford, UK: Oxford University Press.

Ingram, A. 2010. Biosecurity and the International Response to HIV/AIDS: Governmentality, Globalisation and Security. *Area* 42(3): 293–301.

Inverardi, G. 2018. Accessing HIV Prevention, Testing, Treatment Care and Support in Europe as a Migrant with Irregular Status in Europe: A Comparative 16-Country Legal Survey. The European HIV Legal Forum. Accessed 23 March, 2020. Available online at: https://www.hareact.eu/en/publication/accessing-hiv-prevention-testing-treatment-care-and-support-europe-migrant-irregular-0.

Kant, I. 1994. Towards Eternal Peace // Op.: in 8 vol. M., 1994. T. 7. S. 5–56.

Kashnitsky, D. 2020. The Russian HIV Residence Ban and State Control of Migration. *Journal of Public Health Policy* 41: 453–463.

Kearney, Michael. 2004. The Classifying and Value-Filtering Missions of Borders'. *Anthropological Theory* 4(2): 131–156.

Kramer, MA., van Veen MG., Op de Coul ELM., Geskus RB., Coutinho RA., van de Laar MJW., Prins M. 2008. Migrants Travelling to Their Country of Origin: A Bridge Population for HIV Transmission? *Sex Transm Infect* 84: 554–555.

Kubal, A. 2013. Conceptualizing Semi-Legality in Migration Research. *Law & Society Review* 47(3): 555–588.

Kubal, A. 2019. *Immigration and Refugee Law in Russia Socio-legal Perspectives.* Law in Context. Cambridge University Press.

Luo, J., Weine, S., Bahromov, M., Alexandra, MS., Golobof MD. 2012. Does Powerlessness Explain Elevated HIV Risk Among Tajik Labor Migrants? An Ethnographic Study. *Journal of HIV/AIDS & Social Services* 11(2): 105–124.

Morris, J., Polese, A. 2014. *The Informal Post-Socialist Economy: Embedded Practices and Livelihoods*. Routledge.

Ooms, G., Kruja, K. 2019. The Integration of the Global HIV/AIDS Response into Universal Health Coverage: Desirable, Perhaps Possible, but far from Easy. *Globalization and Health* 15: 41.

Reeves, M. 2015. Clean Fake: Authenticating Documents and Persons in Migrant Moscow. *American Ethnologist* 40(3, Winter 2015): 508–524.

Rosas, G. 2006. The Managed Violences of the Borderlands: Treacherous Geographies, Policeability, and the Politics of Race. *Latino Studies* 4(4): 401–418.

Rushton, S. 2012. The Global deBate over HIV-Related Travel Restrictions: Framing and Policy Change. *Global Public Health* 7(2): 159-S175.

Sassen, S. 1988. *The Mobility of Labor and Capital: A Study in International Investment and Labor Flow*. New York: Cambridge Univ. Press

Schenk, C. 2018. *Why Control Immigration?: Strategic Uses of Migration Management* in Russia. University of Toronto Press.

Schinkel, W. 2009. Illegal Aliens and the State, or: Bare Bodies Versus the Zombie. *International Sociology* 24(6): 779–806.

Sidgwick, H. 2015. Henry Sidgwick on Restrictions to Immigration 2015. *Population and Development Review* 41(4): 709–711.

UNAIDS. 2019a. Strategy 2016–2021.

UNAIDS. 2019b. Still Not Welcome. HIV-related Travel Restrictions. UNAIDS Explainer.

Žižek, Slavoj. 1997. Multiculturalism, or, the Cultural Logic of Multinational Capitalism. *New Left Review* 225: 28–51, 2008 Violence, the New York: Picador.

Staying

Institutions and the Informal Economy: Tax Morale of Small Businesses in Armenia and Georgia

Johanna Paquin

Introduction—The Informal Economy and Institutional Incongruences

Among the republics of the former Soviet Union, Georgia has often been cited as the top reformer and prime example of the emerging economies of Eastern Europe outside of the Baltic states that are now part of the European Union (Ishiyama et al. 2018; Rekhviashvili 2018; Bertelsmann Stiftung 2018b; Rekhviashvili and Polese 2017; Chaminade and Moskovko 2015; Aliyev 2014). From the thorough eradication of corruption to reforming the business environment to induce private sector development and establishing democratic institutions, the country seems to be leaving its socialist heritage behind and to be well on its way to becoming a market economy. Its neighbour Armenia, on the other hand, seems to be stuck with its Soviet heritage in the form of a more authoritarian political system, an economy with a less conducive business

J. Paquin (✉)
Goethe University Frankfurt, Frankfurt, Germany
e-mail: paquin@soz.uni-frankfurt.de

A. Polese (ed.), *Informality, Labour Mobility and Precariousness*, International Political Economy Series,
https://doi.org/10.1007/978-3-030-82499-0_6

environment and slower private sector development. Indeed, many of the important sectors of the country's economy are still largely in the hands of a few oligarchs who survived the Soviet regime (Bertelsmann Stiftung 2018a). While Georgia has climbed up the World Bank's *Ease of Doing Business* Ranking from a 116th place in the 2006 report (the first report to rank all economies) to 7th in the world in 2020. Armenia, on the other hand, over this period of time has ranked consistently somewhere between 32nd (in 2013) and 61st (in 2011) with ups and downs and a 41st place in the most recent report of 2020. While this is a particular large discrepancy, the same pattern can be seen across numerous politico-economic indices[1] when comparing the two countries.

Interestingly, Georgia much more than Armenia is notoriously famous for its large informal economy both before and after independence from Soviet rule (Mars and Altman 1983; Buehn and Schneider 2012; Light 2014; Aliyev 2017; Schneider and Medina 2018), having been "arguably the most corrupt Soviet republic" (Light 2014, p. 322). In 1983, long before the collapse of the USSR, Mars, and Altman have written their testimony of the large scale of practices spread across all social, political, and economic realms in the country. They argue that complex interpersonal networks function as the "cultural bases" of the informal economy in Georgia (Mars and Altman 1983, p. 546). Moreover, much of the recent literature and data points to the fact that Georgia has not been able to get rid of this informal heritage to this day. Even in the wake of thorough reforms both in the public and private sector, the most recent calculations and studies show its informal economy to be one of the largest in the world (Schneider 2005; Schneider et al. 2010; Buehn and Schneider 2012; Schneider and Medina 2018). Schneider and Medina (2018) estimate the Georgian informal economy to be worth 53.07% of its overall GDP in 2015, compared to the size of 35.96% of GDP of the Armenian informal economy. Aliyev argues that "although recent institutional reforms in Georgia have succeeded in undermining the importance of informality within institutions, the reforms have so far failed to significantly decrease the population's reliance on informal economic practices" (2015, p. 65). The country, as some say, is "trapped in informality" (Kopeček 2019). This is paradoxical, as informality is all too often seen

[1] See for instance the Bertelsmann Transformation Index, Global Competitiveness Index, Corruption Perception Index, Index of Economic Freedom, Global Innovation Index, or the International Property Rights Index.

as a heritage from socialist rule and a decrease in informal practices is associated with the shift from a centrally planned to a market economy. This stems from the assumption that informal economic activity as well as other informal practices flourish in those contexts where the state is seen as something to be avoided rather than trusted, as has reportedly been the case in many of the Soviet republics. So how can we explain the much larger informal economy in Georgia despite the reforms and economic development? A possible explanation can be found in the institutional literature.

Institutions are "the rules of the game in a society, or, more formally, [...] the humanly devised constraints that shape human interaction" (North 1990, p. 1). These rules can be both formal (i.e. codified laws and regulations) and informal (i.e. socially shared, unwritten rules that express norms, values, and beliefs). Informal institutions can include codes of conduct, norms of behaviour, conventions, traditions, moral values, or religious beliefs that "have passed the test of time" (Pejovich 1999, p. 166) and "are a part of the heritage that we call culture" (North 1990, pp. 36–40). There are a number of scholars from different schools of institutionalism who claim that formal and informal institutions depend on each other to work (Williams et al. 2015; Baumol and Blinder 2008; Helmke and Levitsky 2004; Pejovich 1999; North 1990). With regard to the informal economy, the argument goes that institutional reforms can only be effective in reducing informal economic activity if they are in accordance with formal institutions. Williams et al. note that "the greater [...] the non-alignment of these formal and informal institutions, the greater is the likelihood of participation in the informal economy" (Williams et al. 2015, p. 295). This "institutional incongruence hypothesis" builds on earlier works of institutionalists from different fields and earlier versions of its underlying argument can be found for instance in North (1990), Feige (1997), Pejovich (1999, 2006a, b), and Gërxhani (2004). The hypothesis was further developed and empirically tested in a number of contexts by Colin C. Williams and colleagues (Williams et al. 2015; Williams and Vorley 2015; Williams and Franic 2016; Williams and Horodnic 2015a, b, 2016, 2017a, b, c).

This chapter applies the institutional incongruence hypothesis to the Georgian and Armenian private sector in order to explain the prevalence of widespread informal economic activity among Georgian entrepreneurs to this date. A comparative case study of the business environment in Georgia and Armenia with a special focus on the tax morale of small businesses provides the frame of the analysis. Informal economic practices seem to play a major role for these businesses in their day-to-day operations; and similar to the country-level results much more so for Georgian than for Armenian businesses. According to the World Bank's Enterprise Surveys in 2013, 18% of Armenian small businesses (5–19 employees) stated that they were competing against unregistered or informal firms, whereas in Georgia as much as 52% of businesses responded the same way. In Armenia, 11.4% of those businesses identified the practices of competitors in the informal sector as the main obstacle in running their business, compared to 16% of the Georgian respondents (World Bank 2020a). The institutional incongruence hypothesis predicts that entrepreneurs are less likely to operate informally when their own values, norms, and morals overlap with those reflected in state institutions, i.e. when there is a larger discrepancy between formal and informal institutions. The hypothesis thus predicts a larger institutional incongruence in Georgia compared to Armenia. The empirical evidence, however, presents contradictory results. The qualitative and quantitative secondary data used here points to the fact that there is a considerably larger gap between formal and informal institutions in Armenia than in Georgia. This leads to the assumption that there must be other motivations for entrepreneurs in Georgia for doing business informally than simply not agreeing with the norms and values propagated by state institutions. This comparison between the two countries, it has to be noted, ends with the latest available data measuring the size of the informal economy, which is from the year 2015. Armenia has been on a changing course since the peaceful "Velvet Revolution" of 2018, which was followed by a change in government and reforms in numerous domains. The results of these reforms remain yet to be seen—and will undoubtedly provide the basis for further institutional research.

The Informal Economy, Informal Institutions, and Tax Morale

In the social science literature of the past few decades, the concept of "informality" has been applied to a wide range of phenomena, as Polese points out in the introduction to this book.[2] The term has been used in various social, political, and economic contexts across many disciplines—from informal governance and politics (Christiansen and Neuhold 2013a, b; Adriaenssens and Hendrickx 2012; Ledeneva 2013; Hale 2011), informal business practices and networking (Lomnitz 1988; Ledeneva 1998, 2006; Aliyev 2014), informal payments and corruption (Polese 2014; Polese et al. 2017; Baez-Camargo and Ledeneva 2017; Horodnic and Williams 2018), informal housing (Abrams 1964; Mangin 1967; Turner 1977; Drakakis-Smith 1981; Kombe and Kreibich 2000; Bredenoord and van Lindert 2010), informal land tenure (Bukhari 1982; Al Sayyad 1993; Payne 2002), or informal institutions (Böröcz 2000; Lauth 2000; Helmke and Levitsky 2004; Gel'Man 2004; Grzymala-Busse 2010; Williams and Franic 2016; Voigt 2016). Two of those dimensions of informality will be particularly relevant for this analysis: informal institutions as part of the independent variable, and the informal economy as the dependent variable. Hence, I will refer to *informal institutions* as all institutionalized informal practices—i.e. those that function as commonly shared rules in a society (Helmke and Levitsky 2004). The *informal economy* (i.e. shadow economy) is defined here as "all income-earning activities that are not regulated by the state in social environments where similar activities are regulated" (Castells and Portes 1991, p. 12). While it is helpful for the analysis to clearly differentiate the dimensions of informality—or "flavours of informality" to use the terminology from the introduction to this volume—in practice they are often interrelated or intertwined. In the course of this chapter, I will refer to *informality* only where more than one of those dimensions can be found.

The study of institutions as entities that shape social, political, and economic behaviours has a long tradition in the social sciences. There is a general consensus that institutions are one of the key factors that influence economic development, and many even argue that they are the main reason for differences in economic development across countries (North et al. 2013; Acemoglu and Robinson 2012; Olson 2000). One of the channels through which institutions create growth is through creating

[2] See Boanada-Fuchs and Boanada Fuchs 2018 for a detailed "taxonomic understanding of informality". The authors analyse existing literature and find seven "thematic dimensions of informality", namely an economic, legal, technical, organizational, political, social, and a cultural dimension. See also the *Global Encyclopedia of Informality*, edited by Alena Ledeneva, for a collection of informal practices around the world.

favourable conditions for businesses. In developing and transition countries, in particular, small businesses make up much of the private sector. Entrepreneurship has been shown to be an engine of economic growth (Wright and Stigliani 2013; Wright and Marlow 2012; Acs et al. 2008) and plays an important role for transition countries to move from centrally planned to market economies (Puffer et al. 2010; Djankov et al. 2006). A country's institutional environment made up of formal and informal rules determines its entrepreneurial capacity (Williams and Vorley 2015; Valdez and Richardson 2013; Acs et al. 2008; Frederking 2004). Through a wide range of mechanisms such as enforcement, incentives, and precepts, formal institutions establish the boundaries of entrepreneurial activities (Webb et al. 2009, p. 495). Weak or misfunctioning institutions can become detrimental to entrepreneurial culture and hamper growth (Baumol 1990). But not only the formal institutions determine how business is done. Wade-Benzoni et al. (2001) argue that informal institutions provide the cultural basis for legitimating entrepreneurship, while formal institutions provide the regulatory frame. The implication all too often drawn from this realization is that developing countries only need to adopt the right institutional framework in order to embark on the road to economic development (World Bank 2000c). However, time and again over the past few decades, attempts to validate this logic have not quite proven successful. Correspondingly, there is another emerging consensus among institutionalists about the fact that formal institutions work very differently in different contexts and that merely transplanting institutions that work well in the so-called "Western" or "industrialized" countries does not necessarily lead to the desired results (North et al. 2011; Williamson 2009; Scott 1998; De Soto 2000). One possible explanation that has received increased attention across the different streams of institutional theory is that formal institutions, their effectiveness, and their transformation are critically dependent on the informal institutions in a given context. Douglass North notes that "[c]hanging merely the formal rules will produce the desired results only when the informal norms are complementary to that rule change" (1997, p. 16). Pejovich (1999) argues that the discrepancy between formal and informal rules changes behaviours by altering transaction costs.[3] This implies that institutional

[3] For detailed overviews of the "New Institutionalisms" in the Social Sciences, see for example Koelble et al. (1995), Hall and Taylor (1996), Brinton and Nee (1998), Peters (2005), Nee (2005), Lowndes and Roberts (2013).

reforms can only be successful if this embeddedness of formal institutions in informal rules, norms, and practices are adequately taken into account (Nee 2005). Large discrepancies between formal and informal institutions, or in other words between the actors' and the states' moralities, will lead to less compliance with formal rules. In the words of Feige: "when formal and informal institutions clash, non-compliant behaviours proliferate" (1997, p. 22). This "institutional incongruence" results from the difference in "what is defined as legitimate by formal and informal institutions" (Webb et al. 2009, p. 495).[4] Van Schendel and Abraham (2005) distinguish informal practices (not limited to but including informal institutions) according to whether they adhere to the same norms as those of the states. There are those practices that are generally accepted (licit informality) and those that are frowned upon (illicit informality) by society. Both kinds of informal practices can be legal or illegal in the eyes of the state. A similar distinction is made by Polese and Morris (2015), who argue that informality exists "in spite of" a state—where informality exists in spheres that are regulated by the state—or "beyond the state" in those contexts that the state is not (yet) regulating. Polese (2015), distinguishes between informal practices that harm the society directly or indirectly and points out that direct violation of rules comes mostly from illicit practices, but at the same time the widely accepted informal practices might do indirect harm. Williams et al. (2015) put this in relation to the size of the informal sector and argue that the informal economy will be largely absent if there is a symmetry between formal and informal institutions. On the other hand, if an asymmetry exists between them the informal economy will be larger (2015, pp. 297–298). Webb et al. argue that growth-oriented entrepreneurs even "proactively seek to take advantage of institutional incongruence and weak enforcement of the formal institution's laws and regulations" in order to recognize and exploit opportunities in the informal economy (2009, p. 497).

It is important to add here that the relationship between formal and informal rules is a reciprocal one. Not only do formal institutions depend on informal institutions to be effective, but they can in turn contribute

[4] Legitimacy is defined here as "a generalized perception or assumption that the actions of an entity are desirable, proper, or appropriate within some socially constructed system of norms, values, beliefs, and definitions" (Suchman 1995, p. 574).

to changes in the values, norms, and informal rules within a society.[5] The following analysis will leave this out as the interactions between formal and informal institutions are likely to be of importance over the course of a certain time period. This chapter attempts merely to uncover the discrepancy between formal and informal institutions at a given point in time. The ways in which institutional dynamics have contributed to the institutional framework currently in place are of lesser importance for this argument.

Taxes are not only one key component of the business environment that is affected by these dynamics between formal and informal institutions; they are also crucially linked to the informal economy (Torosyan and Filer 2012). In fact, the informal economy can be defined in terms of taxes, when it is seen as "socially legitimate activity which is formally legal in every respect except that it is not declared to the authorities for tax, social security and/or labour law purposes" (Williams et al. 2015, p. 296).[6] Studies have found low levels of tax morale to be associated with higher occurrences of various informal practices (Horodnic and Williams 2019), such as activity in the shadow economy (Halla 2012; Torgler and Schneider 2007; Torgler, Schaffner, and Macintyre 2007), tax evasion (Brink and Porcano 2016; Dell'Anno 2009; Torgler 2004a), participation in undeclared work (Windebank and Horodnic 2017) and underreporting salaries (Williams and Horodnic 2017b).[7] In transition countries in particular, "frequent changes in the tax system, combined with high tax levels, unpredictable behaviour of state officials in applying tax regulations, and inadequate access to external capital" lead entrepreneurs to employ evasion strategies, reducing profits, and tax payments to protect the capital base of their businesses (Smallbone and Welter 2009, p. 19). The impact of taxes on the level of entrepreneurial activity has been demonstrated to be both complex and even paradoxical (Verheul et al. 2002; Bowen and Clercq 2008). As Salimath and Cullen argue, "on the one hand, high tax rates reduce the return on entrepreneurship, while on the other hand self-employment may offer greater opportunities to

[5] See Helmke and Levitsky (2004) for a typology of informal institutions and their interactions with formal institutions. Other works systematizing these relationships include Lauth (2000), Grzymala-Busse (2010), and Voigt (2019).

[6] A distinction is made here between the "informal" and the "illegal" economy.

[7] For a detailed review on the topic see Horodnic (2018) and Horodnic and Williams (2019).

evade or avoid tax liabilities" (Salimath and Cullen 2010, p. 364). This means that tax morale—defined commonly as "the intrinsic motivation to pay taxes" (Torgler and Schneider 2007)—is an expression of an incongruence between formal and informal institutions. The wider the gap between formal and informal institutions, the less the likelihood of businesses to be willing to pay taxes. This argument has also been made by Horodnic, who states that "an asymmetry in a society between the laws and regulations of its formal institutions and the socially shared unwritten norms of its informal institutions will result in low tax morale" (2018, p. 869). When we look at informal economic practices of businesses—in the terminology used here that is income-generating practices hidden from taxation—they are illegal from the eyes of the state. More specifically, the activities themselves are legal but not declaring them to the tax authorities is not. From a society's (or business's) point of view we can continue to assume that hiding these activities from taxation would be more "licit" in those contexts with diverging formal and informal institutions, meaning cheating on taxes would be more acceptable. In other words, if the entrepreneurs' values and norms are consistent with those reflected in the formal institutions, the tax morale (and correspondingly tax compliance) will be high. On the other hand, if there is an incongruence between those formal and informal institutions, entrepreneurs will be less likely to declare all their income to the tax authorities.

Tax Morale of Small Businesses in Georgia and Armenia

Is there lower tax morale and thus a larger institutional gap that can explain the stronger reliance on informal economic practices in Georgia compared to Armenia? The analysis in this chapter will provide some insights. For a long time, tax compliance in economics was explained by the neoclassical approach of Allingham and Sandmo (1972) who described it as a utility-maximizing function of the rational taxpayers behaviour dependent on the utility ratio of their monetary gains from tax evasion versus expected penalty and the probability of being detected. Later research, however, has shown that the model fails to explain variations in tax compliance and the observed levels of tax compliance are often higher than the model would predict (Torgler, Schaffner, and Macintyre 2007; Leonardo 2011). Thus, in search for the "non-pecuniary aspects of the tax compliance decision", *tax morale* (sometimes also *tax* morality) became a key term in this field of research from the 1990s (Horodnic 2018, p. 869; see also Torgler 2007, 2011).

Tax morale is not easy to measure. The *World Values Survey (WVS)* includes a question that aims at capturing people's attitudes towards paying taxes. In this question, people are asked whether cheating on taxes is justifiable and to respond on a scale from 1 to 10 on which 1 means "never justifiable" and 10 means "always justifiable". In Georgia, 78% of respondents answered "never justifiable" and the mean value across all respondents was 1.45. In Armenia, on the other hand, only 68% found cheating on taxes never to be justifiable, with a mean of 2.08 across all answers (Inglehart et al. 2014).[8] While this question regarded in isolation may give the observer a general idea of how "licit" tax evasion is across a society, it is doubtful that just the answers to this question reflect the issue in its totality. With a sensitive issue like this, it is likely that respondents would choose an answer that they feel is expected of them, especially if they really do not declare their taxes all too accurately. Therefore, the analysis takes into consideration a variety of the factors that have been associated with tax morale to paint a more complete picture.

Tax morale is influenced by factors that concern the formal institutional framework, such as trust in public authorities, perceived government quality and fairness, tax rates, deterrence measures, government expenditure, and corruption. However, the literature has also identified numerous influences stemming from informal rules. Examples include cultural factors, such as patriotism, religious values, sense of community, trust in other taxpayers, and the perception of tax fraud.[9] With regard to formal institutions, trust in the formal institutional framework and public authorities is a key influencing factor for tax morale (OECD 2019). A vast amount of studies have found a high trust in public institutions to be associated with high tax morale (Alm and Torgler 2006; Feld and Torgler 2007; Daude et al. 2012; Filippin et al. 2013; Andriani 2015).[10] Trust in public institutions just by itself can be seen as an institutional incongruence because it expresses the diverging norms of the state and individuals.

[8] Since the latest available year for which there is an estimate of the size of the informal economy in Schneiders and Medinas paper is 2015, I will use data from the same time period for the analysis.

[9] For a detailed literature review on tax morale and its influencing factors see Horodnic (2018).

[10] For a thorough review, again, see Horodnic (2018).

The data shows that while the levels of trust in public institutions reported in Armenia and Georgia in large-scale surveys are not too different, the percentage of people reporting a distrust in formal institutions in Armenia is considerably higher across various formal institutions. Figure 1 shows the discrepancy between the two countries. These results stem from the regional survey "Caucasus Barometer" in 2015 that is conducted regularly by the Caucasus Research Resource Center (CRRC). In the survey, respondents are asked the question "Please tell me how much do you trust or distrust Armenia's / Georgia's ...?" The interviewees can answer the question on a scale from 1 ("fully trust") to 5 ("fully distrust"). Figure 1 shows the cumulated percentage of people who answered that they "fully distrust" or "rather distrust" the respective institutions.

Being asked whether the courts in the country favour some citizens or treat all equally, Georgians replied with 58% and 22%, respectively; of the Armenian respondents 74% thought courts were biased, and only 11% found them to be fair (The Caucasus Research Resource Centers 2015). When asked about the fairness of the most recent general elections, 28% of Georgian respondents stated that they perceived them as conducted "completely fairly" as opposed to 10% of Armenians who replied the same way. 15% and 41%, respectively said that the last elections were conducted "not at all fairly" (ibid.). Closely related to the trust in institutions is the perception of corruption. It has been found by a number of studies that higher levels of corruption are also associated with lower levels of tax

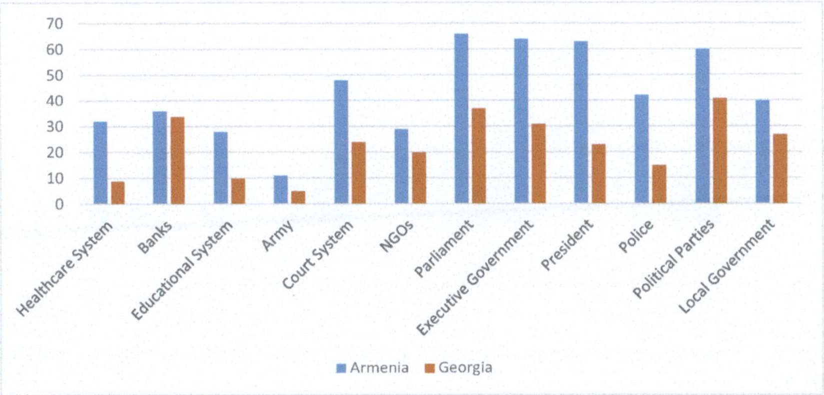

Fig. 1 Institutional distrust in Georgia and Armenia (*Source* CRRC 2015)

morale (e.g. Alasfour et al. 2016; Gerstenbluth et al. 2012; Torgler 2006; Torgler et al. 2008; Williams and Krasniqi 2017; Williams and Martinez Perez 2014). Again, this deviance is clearly an expression of an institutional incongruence. In Georgia, corruption has been the number one issue that was tackled by the reforms led by president Saakashvili in the aftermath of the Rose Revolution of 2003. The campaign was a remarkable success. In Armenia, on the other hand, corruption seems to be far more widespread today. Moreover, there is a large discrepancy between the perception of corruption and the actual experience. The following graph from the EBRD's report of the *Life in Transition Survey* (LiTS) shows that in Armenia the perceived corruption is much higher than the actually experienced corruption. While 34% of respondents have affirmed to the statement that "people like themselves usually or always have to make unofficial payments or gifts", only 11% of the respondents have confirmed that "they or a member of their household have made unofficial payments or gifts in the past 12 months" (EBRD 2016). Together with Moldova, Armenia is the only state in the transition region with such a large discrepancy between these two values (Fig. 2).

Among the small businesses in the World Bank Enterprise Surveys in 2013 in Armenia, 17.1% stated corruption to be a major constraint and 6% even saw it as the biggest obstacle in operating their business. In Georgia, only 4% of small businesses see corruption as a major, and as little as 0.3% as their biggest obstacle (World Bank 2020a). 10.6% of small firms in Armenia reported to have received at least one bribe payment request

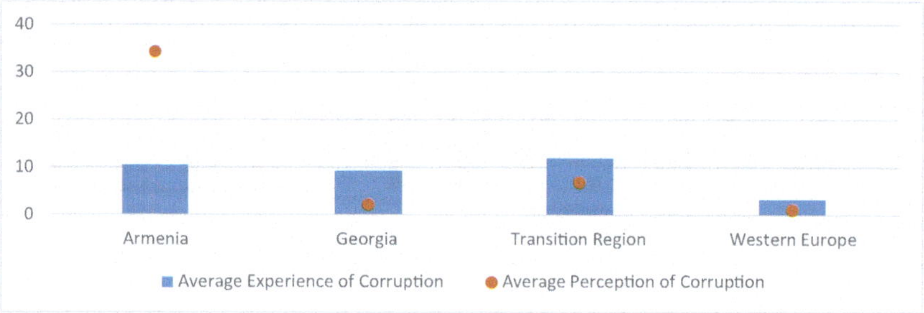

Fig. 2 Experience versus perception of corruption in 2016 (*Source* EBRD 2016)

in the last 12 months, in contrast to 1.5% of Georgian firms who claimed this to be the case. The frequency of those requests was also much higher in Armenia, with the respondents saying that on average, 9.3% of public transactions required them to make a gift or informal payment, compared to a mere 1% in Georgia (ibid.). 36.1% of the Armenian businesses were expected to give gifts to get a construction permit (7.4% in Georgia) and 5% said they were expected to do so in order to "get things done" (1.7% in Georgia). Furthermore, 9.3% of Armenian small firms reported that they were expected to give gifts in meetings with tax officials—in stark contrast to the small firms in Georgia of which none were expected to do so (ibid.). Higher perceptions of government fairness have also been correlated with higher levels of tax morale (Frey 2003; Cummings et al. 2005, 2009; Torgler and Schaffner 2007; Torgler et al. 2008; Alm and Gomez 2008; Alasfour et al. 2016; Cyan et al. 2016). In Armenia, the general public feels treated less fairly by the government. To the question "Under the present government in /country/, do you completely agree, somewhat agree, somewhat disagree, or completely disagree that people like yourself are treated fairly by the government?" an overall of 74% of Armenians replied with completely or somewhat disagree. In comparison, these answers were only chosen by an aggregated 47% of Georgian respondents (The Caucasus Research Resource Centers 2017). In a survey of small and medium enterprises (SMEs) in Armenia, only 14% found the statement "The state administration is fair, impartial and uncorrupted" to be true (Mikaelian 2017). But not only the perceived fairness of the government and tax system, also the tax rates themselves are a determinant of tax morale with higher tax rates leading to lower tax morale (Torgler 2002, 2003a; Torgler, Schneider, and Schaltegger 2007; Lago-Peñas and Lago-Peñas 2010; Alasfour et al. 2016; Williams and Martinez Perez 2014; Molero and Pujol 2011; Williams and Krasniqi 2017). In Armenia, tax rates are unfavourable for small and medium enterprises, leading to many businesses hiding part of their economic activity from the tax authorities (Bertelsmann Stiftung 2020). This reflects the answers given by small businesses in the Enterprise Surveys, in which 36.8% find tax rates to be a major constraint for their activities, followed by 23.7% who identify the tax administration as a major constraint. Out of 15 possible options that were listed, 30.6% of small businesses picked tax rates as their biggest obstacle, compared to 18.2% who said the tax administration was the biggest challenge they faced (World Bank 2020a). While the government has made efforts of simplifying the system in the meantime, the main problems in the tax system still persist today (Bertelsmann Stiftung 2020).

Tax morale has also been shown to be connected to government quality and spending effectiveness (Alasfour et al. 2016; Alm and Gomez 2008; Barone and Mocetti 2011; Frey and Torgler 2007; Molero and Pujol 2011; Sipos 2015; Vythelingum et al. 2017). In Georgia, there is high satisfaction with public services. More than 80% of respondents of the LiTS III claimed satisfaction with electricity, pipeline gas, and telephone lines, slightly less are satisfied with the provision of water and heating, and only around 30% are satisfied with roads and postal services (EBRD 2016). The satisfaction of Armenian's with public services in their country was slightly lower overall with only telephone lines with more than 80% satisfaction, and roads just below 30% (ibid.).

While there is not a single indicator that describes the perceived quality and spending effectiveness of the government, the results of the SME survey carried out by the Caucasus Institute paint a clear picture of the attitudes of small businesses towards the government in Armenia (Fig. 3).

Unfortunately, we do not have similar evidence for Georgia, although the same study was originally planned to be carried out there as well (Mikaelian 2017). However, we can add some findings from other sources that hint towards this indicator. Table 1 lists the experience of small firms in Armenia and Georgia dealing with government regulations that are likely to affect their views on the subject. All these factors put together lead to the assumption that government effectiveness seen from a small business perspective should be higher in Georgia than Armenia.

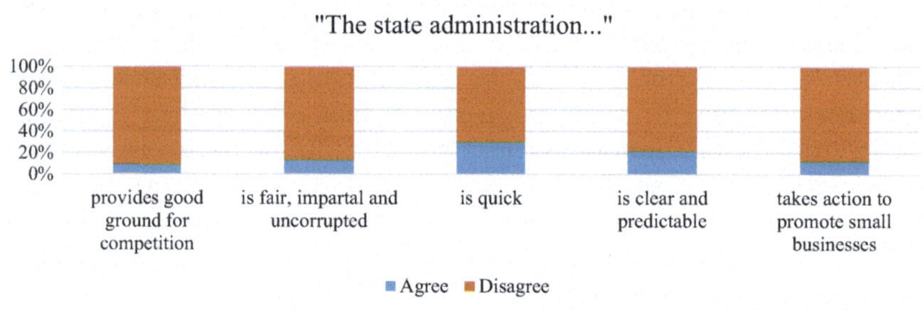

Fig. 3 State perception of SMEs in Armenia (*Source* Mikaelian 2017, p. 34)

Table 1 Attitudes towards the government of small businesses (5–19 employees)

	Armenia	*Georgia*
Senior management time spent dealing with the requirements of government regulation (%)	10.7	1.1
Days to obtain an operating license	15.5	–
Days to obtain a construction-related permit	54.6	35.4
Per cent of firms identifying business licensing and permits as a major constraint	5.2	0

Source World Bank (2020), Enterprise Surveys from 2013

Among the influencing factors relating to the formal institutional framework is also the level of democracy. More democratic countries, according to the literature, display higher levels of tax morale (Feld and Frey 2004, 2006; Lago-Peñas and Lago-Peñas 2010; Frey and Torgler 2007; Martins and Gomes 2014; Sá et al. 2015; Torgler 2003a, b, c; Williams and Krasniqi 2017). Again, Georgia clearly scores higher on that point. Across several indices that set out to measure democracy, Georgia ranks higher than Armenia. On the Polity2 Index of the Center for Systemic Peace, Georgia scores 7 on a scale from 1 to 10 (fully democratic), compared to the Armenian score of 5 (Marshall and Elzinga-Marshall 2017). The Freedom House Index (FHI) categorized Armenia as a "semi-consolidated authoritarian regime", and Georgia as a "transitional or hybrid regime" in 2015 (and this is unchanged to date), meaning that Georgia is significantly more democratic than Armenia (Freedom House 2015). But not only the actual level of democracy plays a role. Studies show that both the actual level of democracy and the taxpayers' attitudes towards democracy determine tax morale. People who identify fiscal redistribution to be essential (i.e. governments should tax the rich to enable support for the poor) show higher tax morale. (OECD 2019). Being asked how important it is for a good citizen to support those who are worse off, 75% of Georgians find it to be "extremely important" but only 47% of Armenians responded the same way. (Inglehart et al. 2014). Those who perceive democracy to be the best system of government for their country tend to think that cheating on taxes is unjustifiable (OECD

2019). In the EBRD's *Life in Transition Survey* from 2016, 53% of Georgians stated that democracy is "preferable to any other form of political system" (EBRD 2016). In Armenia, on the other hand, 66% of respondents claimed their support for the democracy (ibid.). The same question with the same possible answers asked in the Caucasus Barometer Survey in 2015 had 52% of Georgian respondents reply the same way, but only 48% of Armenians (The Caucasus Research Resource Centers 2015).

Furthermore, individuals who believe that they are living in a meritocratic society have significantly higher tax morale (OECD 2019). Being asked about whether certain basic democratic institutions exist in their home country, the number of respondents who confirm this is significantly lower than the average of the transition region across all eight institutions asked[11] (EBRD 2016). The Caucasus Barometer 2017 found that Georgians perceive their country more as a democracy than Armenians do. Specifically, 31% of Georgians believe Georgia to be a full democracy or a democracy with minor problems and only 16% of Armenians think the same about their country. On the other side of the spectrum, 57% of Georgians state that Georgia is not a democracy, or a democracy with major problems, as opposed to 70% of Armenian respondents who chose either of these answers (The Caucasus Research Resource Centers 2015).

Finally, there are some structural conditions related to the formal institutional framework that can affect tax morale. Those are listed in Table 2 with data from the World Bank from 2015 (World Bank 2020b).

Several of these influencing factors related to formal institutions were described here not only in absolute terms, but expressed as the perception of taxpayers. The deviances already constitute institutional congruences, showing the ways in which individual and state moralities diverge. There are, however, a number of other informal institutions that can affect tax morale without being directly linked to formal institutions (Horodnic 2018). One of those informal institutions is religion, which is positively associated with tax morale (Alm and Torgler 2006; Andriani 2015; Bilgin 2014; Daude et al. 2012; Feld and Torgler 2007; Trüdinger and Hildebrandt 2012; Vythelingum et al. 2017; Feld et al. 2008; Frey and Torgler 2007; Sá et al. 2015; Torgler 2003a, d, 2005, 2006,

[11] The following institutions were asked about: free elections, freedom of speech, peace and stability, law and order, gender equality, fair justice system, independent press, and political opposition.

Table 2 Structural conditions and their effects on tax morale

Indicator	*Armenia*	*Georgia*	*Correlation (±)*	*Sources*
GDP (Current US$)	11.527 billion	14.95 billion	+	Doerrenberg and Peichl (2010), Gerstenbluth et al. (2012), Williams and Krasniqi (2017), Williams and Martinez Perez (2014)
Inflation (GDP deflator, in annual %)	1.214	5.833	–	Doerrenberg and Peichl (2010)
Income inequality (Gini coefficient)	32.4	36.5	–	Gerstenbluth et al. (2012), Trüdinger and Hildebrandt (2012), Williams and Krasniqi (2017), Williams and Martinez Perez (2014)

Source World Bank (2020b)

2012; Torgler, Schaffner, and Macintyre 2007, Torgler, Schneider, and Schaltegger 2007, 2009) and there seems to be no difference in tax morale between different religions (Alasfour et al. 2016; Kanniainen and Pääkkönen 2009; Torgler 2003d). The World Value Surveys in Armenia 2011 and Georgia 2014 show a difference in religiosity between the two countries. 84.9% of Georgians find religion to be "very important" in life, 12% find it "rather important". In Armenia, the numbers are 57.5% and 31.8%, respectively. 21% of Georgian respondents are active or inactive members of a church or religious organization, compared to 2.7% in Armenia. Furthermore, 97.1% of Georgians see themselves as religious persons, compared to 88.5% of Armenians (Inglehart et al. 2014). National pride is also positively correlated with tax morale (Bilgin 2014; MacGregor and Wilkinson 2012; Martinez-Vazquez and Torgler 2009; Ristovska et al. 2013; Trüdinger and Hildebrandt 2012; Kondelaj et al. 2016; Leonardo and Martinez-Vazquez 2016; Martins and Gomes 2014; Torgler 2003b, d, 2004b; Torgler and Schneider 2004). In the WVS Wave 6, 76.2% of Georgians said they were "very proud" of their own nationality, followed by 19.9% who are "quite proud". Of the Armenian respondents, 71.4% and 22.3% chose those answers respectively (Inglehart et al. 2014). Furthermore, a stronger sense of community and civic participation is associated with higher tax morale (Çevik 2016; Russo 2013; Trüdinger and Hildebrandt 2012). In Armenia, 67.8% of respondents "strongly agreed" with the statement "I see myself as a member of my local community", whereas 29.4% "agreed". In Georgia, 66.1% strongly agreed and 28.5% agreed (Inglehart et al. 2014).

In terms of trust, not only the trust in institutions (vertical trust) but also the trust in individuals (horizontal trust) is related to tax morale. The higher one's trust in the other taxpayers, it has been shown, the higher the likelihood to be honest with one's own taxes (Horodnic 2018; Çevik 2016; Kondelaj et al. 2016; Martins and Gomes 2014; Trüdinger and Hildebrandt 2012; Torgler et al. 2008). Horizontal trust is significantly higher in Georgia than Armenia. The 2015 round of the Caucasus Barometer Survey included the question "Generally speaking, would you say that most people can be trusted, or that you can't be too careful in dealing with people?" with answers on a scale from 1 ("you can't be too careful") to 5 ("most people can be trusted"). Of the Armenian respondents, only 5% stated that most people can be trusted, and 33% responded that you can't be too careful. Georgians, on the other hand, are 8% trusting, and only 16% find that you can't be too careful (The Caucasus Research Resource Centers 2015). Scholars seem to agree that civic participation is strong in Georgia compared to its neighbours (Ishiyama et al. 2018; Broers 2005; Nodia 2005). The data on this is, however, ambivalent. In the 2013 round of the Caucasus Barometer survey, Armenians were much more likely to report being engaged in volunteer work than Georgians (31% versus 19%). The Freedom House's Nations in Transit measure of civil society, which "assesses the growth of nongovernmental organizations (NGOs), their organizational capacity and financial sustainability, and the legal and political environment in which they function; the development of free trade unions; and interest group participation in the policy process" measures civil society in the two countries at the same level in 2013. On the other hand, when Ishiyama et al. calculate from the World Value Surveys that Georgians are much more likely to report being members of civil society organizations, such as church or religious organizations, environmental, professional, charitable/humanitarian, and other organizations. In total, 18% of Georgian respondents report these active memberships, compared to 9% of Armenians (Ishiyama et al. 2018). The stronger civic participation in Georgia can be attributed to two factors: (1) a relatively weak state that does not have control over important natural resources and thus cannot suppress or co-opt civil society groups (Ishiyama et al. 2018) and (2) Georgia's identification with Europe since independence with civil society playing an important role for cultivating the ties (Ishiyama et al. 2018; Broers 2005).

Overall, the analysis presented here paints a clear picture—the institutional incongruence that affects tax morale is likely to be much smaller in Georgia than Armenia. The analysis shows that an incongruence between formal and informal institutions cannot explain the persisting high informal economy in Georgia compared to Armenia. The institutional incongruence hypothesis put forward by Williams et al. predicted a large discrepancy between formal and informal institutions in Georgia as an explanation for the high shares of informal activity. The analysis shows, however, that at least if this discrepancy is expressed in terms of tax morale, the hypothesis does not hold true. The formal and informal institutions that are correlated with tax morale almost exclusively predict a higher tax morale for Georgia than for Armenia. Important formal institutional factors, such as the trustworthiness of government and public services, the level of democracy, and institutional quality are also higher in Georgia across the board. The same is true for informal factors influencing tax morale, such as horizontal trust, national pride, perception of democracy, and a sense of community. The norms and values prescribed by the state overlap with individual norms and values, and this clearly more in Georgia than Armenia. Tax evasion and the under-reporting of taxes should therefore less licit in Georgia and correspondingly, activities in the informal economy less frequent.

Discussion and Implications for Further Research

While Armenia seems to be a case where institutional incongruences leads to lower tax morale and higher informal economic activity, Georgia has that high informality despite high congruence of formal and informal institutions and despite thorough market reforms. The institutional incongruence hypothesis does not explain the difference in the size of the informal economy between Georgia and Armenia. How can these results be interpreted?

First of all, we cannot rule out the possibility that there are problems with the definition and measurement of the variables included in this analysis. Definitions in the theoretical literature do not always match those applied in the collection of data, and some variables could only be measured through proxies. We also cannot make out the exact size of the effects of the single variables on tax morale. It is possible that some effects are larger than others. Furthermore, there might be other factors beyond the social institutions that are responsible for tax morale on an individual level. Some studies have shown that there are some intrinsic motivations to pay or to evade taxes that are not necessarily connected to the shared rules in a society (Torgler 2001, 2003c; Horodnic 2018)—some might simply be "predisposed not to evade" and others may enjoy evading taxes and see it as a game against the state (Kirchler et al. 2007).

However, while these factors are very important to keep in mind when interpreting the results, they are still not very likely to be solely responsible for the large discrepancy between the two countries that we have seen. The sheer amount of studies that have shown correlations between the institutional factors and tax morale that provided the basis for this argument lead to the conclusion that Georgia is in fact somewhat of an outlier. The size of the informal economy in Georgia does not seem to be correlated to the level of tax morale, as is the case in many other countries. Since the data does not show a significant institutional gap here, other mechanisms have to be in place that explains the higher level of informal economic activity. One possible explanation might be that informality is simply differently connotated in Georgia. Since most of the data evaluated here is from surveys and thus reflects subjective perceptions rather than objective facts, it could also be the case that informality is perceived differently in Georgia than in Armenia. It could be that certain forms of informal behaviours and informal practices such as the non-declaration of revenues or hiring workers informally are not actually perceived as "cheating on taxes" as it was put in the World Value Survey.

Incorporating these findings into the framework of informality put forward in the introduction to this volume it can be argued that the potpourri of "flavours of informality" might simply be different in the two countries. To use the typology from the introduction of this volume, it could be assumed that in Armenia, the elitist flavour of informality is more present combined with the informal payments and corruption flavour, as the data shows. These two flavours of informality create an institutional incongruence, where individual values and norms drift apart from those promoted by the state. The third flavour of informality that we see, the shadow economy, is therefore a direct result of the other two flavours. Informal economic activity of small businesses in Armenia can thus be seen as a strategy to circumvent the state that is perceived as being elitist and corrupt. In Georgia, on the other hand, informality seems to exist mostly in terms of the shadow economy flavour. The state is not something that many entrepreneurs seek to avoid, there is a general satisfaction with public services, the government, and the tax system, and the framework conditions for businesses are good. Hence, informal economic practices are most likely not seen as "evasion strategies" and thus are not perceived as something illegal, or even illicit because they are so deeply rooted in the socio-economic structures and culture. Furthermore, the fact that this flavour of informality is likely to have only indirect harmful effects (see Introduction to this book) might make it even more licit.

While this remains a hypothesis for now, this explanation matches some other findings on informality in Georgia. Lela Rekhviashvili, for instance, found that petty traders used informal practices in the wake of the neoliberal institutional reforms to compensate newly emerging market-related risks (Rekhviashvili 2018). The neoinstitutionalist literature, she argues, "assigns a narrow instrumental meaning to informal practices and institutions, and dismisses the possibility that reproduction of these practices is important for non-economic reasons, primarily for providing continuity for culturally specific ways of social integration, for asserting pre-existing (versus instrumental) social bonds and mutuality" (Rekhviashvili 2017, pp. 19–20). For further research, this implies that more qualitative research is needed in order to find out how these different perceptions, uses, and flavours of informality play out in reality. Ideally, ethnographic research with the same cohort will give further insights on the intrinsic motivations of actors to apply different forms of informal economic practices. Furthermore, the analysis needs to drawback to the interactions between formal and informal institutions (and their reforms) over the course of time. The comparison of newer data shows that some of the trends have reversed with the change of government in Armenia in the aftermath of the 2018 revolution. Trust in state institutions is now even higher than in Georgia (The Caucasus Research Center 2020). Tracing the ways in which these changes in the formal and informal institutional spheres do or do not affect informality might provide important insights for further theory-building.

References

Abrams, Charles (1964): Housing in the Modern World. London: Faber and Faber.

Acemoglu, Daron; Robinson, James A. (2012): Why Nations Fail. The Origins of Power, Prosperity and Poverty. New York: Crown Publishers.

Acs, Zoltan J.; Desai, Sameeksha; Hessels, Jolanda (2008): Entrepreneurship, Economic Development and Institutions. In Small Business Economics 31 (3), pp. 219–234. https://doi.org/10.1007/s11187-008-9135-9.

Adriaenssens, Stef; Hendrickx, Jef (2012): Institutions, Trust and Relations: A Comparative Analysis Explaining Informal Economic Activities (HUB Research Papers, 2012/33).

Al Sayyad, Nezar (1993): Informal Housing in a Comparative Perspective: On Squatting, Culture, and Development in a Latin American and a Middle Eastern Context. In Review of Urban & Regional Development Studies 5 (1), pp. 3–18. https://doi.org/10.1111/j.1467-940X.1993.tb00120.x.

Alasfour, Fadi; Samy, Martin; Bampton, Roberta (2016): The Determinants of Tax Morale and Tax Compliance: Evidence from Jordan: Emerald Group Publishing Limited (Advances in Taxation). Available online https://doi.org/10.1108/S1058-749720160000023005.

Aliyev, Huseyn (2014): The Effects of the Saakashvili Era Reforms on Informal Practices in the Republic of Georgia. In Studies of Transition States and Societies 6 (1), pp. 19–33.

Aliyev, Huseyn (2015): Institutional Transformation and Informality in Azerbaijan and Georgia. In Jeremy Morris, Abel Polese (Eds.): Informal Economies in Post-Socialist Spaces. Practices, Institutions and Networks. Basingstoke: Palgrave Macmillan, pp. 51–69.

Aliyev, Huseyn (2017): When Informal Institutions Change. Institutional Reforms and Informal Practices in the Former Soviet Union. Ann Arbor: University of Michigan Press.

Allingham, Michael G.; Sandmo, Agnar (1972): Income Tax Evasion: A Theoretical Analysis. Journal of Public Economics, 1(3–4), 323–338. https://doi.org/10.1016/0047-2727(72)90010-2.

Alm, James; Gomez, Juan Luis (2008): Social Capital and Tax Morale in Spain. In Economic Analysis and Policy 38 (1), pp. 73–87. https://doi.org/10.1016/s0313-5926(08)50007-5.

Alm, James; Torgler, Benno (2006): Culture Differences and Tax Morale in the United States and in Europe. In Journal of Economic Psychology 27 (2), pp. 224–246. https://doi.org/10.1016/j.joep.2005.09.002.

Andriani, Luca (2015): Tax Morale and Prosocial Behaviour: Evidence from a Palestinian Survey. In Cambridge Journal of Economics 40 (3), pp. 821–841. https://doi.org/10.1093/cje/bev019.

Baez-Camargo, Claudia; Ledeneva, Alena (2017): Where Does Informality Stop and Corruption Begin? InFormal Governance and the Public/Private Crossover in Mexico, Russia and Tanzania. In The Slavonic and East European Review 95 (1), p. 49. https://doi.org/10.5699/slaveasteurorev2.95.1.0049.

Barone, Guglielmo; Mocetti, Sauro (2011): Tax Morale and Public Spending Inefficiency. In International Tax and Public Finance 18 (6), pp. 724–749. https://doi.org/10.1007/s10797-011-9174-z.

Baumol, William J. (1990): Entrepreneurship: Productive, Unproductive, and Destructive. In Journal of Political Economy 98 (5), pp. 893–921.

Baumol, William J.; Blinder, Alan S. (2008): Macroeconomics. Principles and Policy. Mason, Ohio: South-Western.

Bertelsmann Stiftung (2018a): BTI 2018. Country Report—Armenia. Edited by Bertelsmann Stiftung. Gütersloh. Available online at https://www.bti-project.org/fileadmin/files/BTI/Downloads/Reports/2018/pdf/BTI_2018_Armenia.pdf, checked on 1/17/2019.

Bertelsmann Stiftung (2018b): BTI 2018. Country Report—Georgia. Edited by Bertelsmann Stiftung. Gütersloh.

Bertelsmann Stiftung (2020): BTI 2020. Country Report—Armenia. Edited by Bertelsmann Stiftung. Gütersloh.

Bilgin, Cevat (2014): Determinants of Tax Morale in Spain and Turkey: an Empirical Analysis. In European Journal of Government and Economics 3 (1), p. 60. https://doi.org/10.17979/ejge.2014.3.1.4297.

Boanada-Fuchs, Anthony; Boanada Fuchs, Vanessa (2018): Towards a Taxonomic Understanding of Informality. In International Development Planning Review 40 (4), pp. 397–420. https://doi.org/10.3828/idpr.2018.23.

Böröcz, József (2000): Informality Rules. In East European Politics and Societies 14 (2), pp. 348–380. https://doi.org/10.1177/0888325400014002006.

Bowen, Harry P.; Clercq, Dirk de (2008): Institutional Context and the Allocation of Entrepreneurial Effort. In Journal of International Business Studies 39 (4), pp. 747–767. https://doi.org/10.1057/palgrave.jibs.8400343.

Bredenoord, Jan; van Lindert, Paul (2010): Pro-Poor Housing Policies: Rethinking the Potential of Assisted Self-Help Housing. In Habitat International 34 (3), pp. 278–287. https://doi.org/10.1016/j.habitatint.2009.12.001.

Brink, William D.; Porcano, Thomas M. (2016): The Impact of Culture and Economic Structure on Tax Morale and Tax Evasion: A Country-Level Analysis Using SEM. In John Hasseldine (Ed.): Advances in Taxation, vol. 23: Emerald Group Publishing Limited (Advances in Taxation), pp. 87–123.

Brinton, Mary C.; Nee, Victor (1998): The New Institutionalism in Sociology. New York: Russell Sage Foundation. Available online at http://search.ebscohost.com/login.aspx?direct=true&scope=site&db=nlebk&AN=1066625.

Broers, Laurence (2005): After the 'Revolution': Civil Society and the Challenges of Consolidating Democracy in Georgia. In Central Asian Survey 24 (3), pp. 333–350. https://doi.org/10.1080/02634930500310444.

Buehn, Andreas; Schneider, Friedrich (2012): Shadow Economies Around the World: Novel Insights, Accepted Knowledge, and New Estimates. In International Tax and Public Finance 19 (1), pp. 139–171. https://doi.org/10.1007/s10797-011-9187-7.

Bukhari, M. Saleem (1982): Squatting and the Use of Islamic law. A Case Study of Land Occupation in Madinah Munawara, Saudi Arabia. In Habitat International 6 (5–6), pp. 555–563. https://doi.org/10.1016/0197-3975(82)90022-4.

Castells, Manuel; Portes, Alejandro (1991): World Underneath: The Origins, Dynamics, and Effects of the Informal Economy. In Alejandro Portes, Manuel Castells (Eds.): The Informal Economy. Studies in Advanced and Less Developed Countries. 2. pr. Baltimore, MD: Johns Hopkins University Press, pp. 11–37.

Çevik, S. (2016): Tax Morale and Tax Compliance in Socio-political Context. In M. Aydin, S. S. Tan (Eds.): Political Economy of Taxation. Istanbul: IJOPEC Publication, pp. 37–57.

Chaminade, Cristina; Moskovko, Maria (2015): Radical Institutional Change: Enabling the Transformation of Georgia's Innovation System. In Cornell University, INSEAD, WIPO (Eds.): The Global Innovation Index 2015. Effective Innovation Policies for Development. Geneva: World Intellectual Property Organization, pp. 113–119.

Christiansen, Thomas; Neuhold, Christine (2013a): Informal Politics in the EU. In Journal of Common Market Studies 51 (6), pp. 1196–1206. https://doi.org/10.1111/jcms.12068.

Christiansen, Thomas; Neuhold, Christine (Eds.) (2013b): International Handbook on Informal Governance. Cheltenham, UK: Elgar.

Cummings, Ronald C.; Martinez-Vazquez, Jorge; McKee, Michael; Torgler, Benno (2005): Effects of Culture on Tax Compliance: A Cross Check of Experimental and Survey Evidence. In SSRN Electronic Journal. https://doi.org/10.2139/ssrn.661921.

Cummings, Ronald G.; Martinez-Vazquez, Jorge; McKee, Michael; Torgler, Benno (2009): Tax Morale Affects Tax Compliance: Evidence from Surveys and an Artefactual Field Experiment. In Journal of Economic Behavior & Organization 70 (3), pp. 447–457. https://doi.org/10.1016/j.jebo.2008.02.010.

Cyan, Musharraf R.; Koumpias, Antonios M.; Martinez-Vazquez, Jorge (2016): The Determinants of Tax Morale in Pakistan. In Journal of Asian Economics 47, pp. 23–34. https://doi.org/10.1016/j.asieco.2016.09.002.

Daude, C.; Gutiérrez, H.; Melguizo, Á. (2012): What Drives Tax Morale? Edited by OECD Development Centre. Paris (Working Paper, No. 35).
De Soto, Hernando (2000): The Mystery of Capital. Why Capitalism Triumphs in the West and Fails Everywhere Else. London: Black swan.
Dell'Anno, Roberto (2009): Tax Evasion, Tax Morale and Policy Maker's Effectiveness. In The Journal of Socio-Economics 38 (6), pp. 988–997. https://doi.org/10.1016/j.socec.2009.06.005.
Djankov, Simeon; Qian, Yingyi; Roland, Gérard; Zhuravskaya, Ekaterina (2006): Entrepreneurship in China and Russia Compared. In Journal of the European Economic Association 4 (2–3), pp. 352–365. https://doi.org/10.1162/jeea.2006.4.2-3.352.
Doerrenberg, P.; Peichl, A. (2010): Progressive Taxation and Tax Morale. Edited by The Institute for the Study of Labor (IZA). Bonn (IZA Discussion Paper, No. 5378).
Drakakis-Smith, D. W. (1981): Urbanisation, Housing, and the Development Process. London: Taylor & Francis.
EBRD (2016): Life in Transition. A Decade of Measuring Transition, checked on 1/30/2019.
Feige, Edgar L. (1997): Underground Activity and Institutional Change: Productive, Protective, and Predatory Behavior in Transition Economies. In Joan M. Nelson, Charles Tilly, Lee Walker (Eds.): Transforming Post-Communist Political Economies. Washington, DC: National Academies Press, pp. 21–34.
Feld, L. P.; Frey, Bruno S. (2004): Illegal, Immoral, Fattening or What? How Deterrence and Responsive Regulation Shape Tax Morale. Edited by Faculty of Business Administration and Economics, Department of Economics, Philipps-Universität. Marburg (Marburg Working Papers on Economics, 2004-26).
Feld, L. P.; Torgler, Benno (2007): Tax Morale After the Reunification of Germany: Results from a Quasi-Natural Experiment. Edited by CESifo Group. Munich (CESifo Working Paper, No. 1921).
Feld, L. P.; Torgler, Benno; Dong, B. (2008): Coming Closer? Tax Morale, Deterrence and Social Learning After German Unification. Edited by Center for Research in Economics, Management and the Arts. Basel (Working Paper, No. 2008-9).
Feld, Lars P.; Frey, Bruno S. (2006): Tax Evasion in Switzerland: The Roles of Deterrence and Tax Morale. In SSRN Electronic Journal. https://doi.org/10.2139/ssrn.900351.
Filippin, Antonio; Fiorio, Carlo V.; Viviano, Eliana (2013): The Effect of Tax Enforcement on Tax Morale. In European Journal of Political Economy 32, pp. 320–331. https://doi.org/10.1016/j.ejpoleco.2013.09.005.

Frederking, Lauretta Conklin (2004): A Cross-National Study of Culture, Organization and Entrepreneurship in Three Neighbourhoods. In Entrepreneurship & Regional Development 16 (3), pp. 197–215. https://doi.org/10.1080/0898562042000197126.

Freedom House (2015): Nations in Transit 2015. Democracy on the Defensive in Europe and Eurasia. Edited by Freedom House.

Frey, Bruno S. (2003): Deterrence and Tax Morale in the European Union. In European Review 11 (3), pp. 385–406. https://doi.org/10.1017/s1062798703000346.

Frey, Bruno S.; Torgler, Benno (2007): Tax Morale and Conditional Cooperation. In Journal of Comparative Economics 35 (1), pp. 136–159. https://doi.org/10.1016/j.jce.2006.10.006.

Gel'Man, Vladimir (2004): The Unrule of Law in the Making: The Politics of Informal Institution Building in Russia. In Europe-Asia Studies 56 (7), pp. 1021–1040. https://doi.org/10.1080/1465342042000294347.

Gerstenbluth, M.; Melgar, N.; Pagano, J. P.; Rossi, M. (2012): How Do Inequality Affect Tax Morale in Latin America and Caribbean. In Revista de Economia del Rosario 15 (2), pp. 123–135.

Gërxhani, Klarita (2004): The Informal Sector in Developed and Less Developed Countries: A Literature Survey. In Public Choice 120 (3/4), pp. 267–300. https://doi.org/10.1023/B:PUCH.0000044287.88147.5E.

Grzymala-Busse, Anna (2010): The Best Laid Plans: The Impact of Informal Rules on Formal Institutions in Transitional Regimes. In Studies in Comparative International Development 45 (3), pp. 311–333. https://doi.org/10.1007/s12116-010-9071-y.

Hale, Henry E. (2011): Formal Constitutions in Informal Politics: Institutions and Democratization in Post-Soviet Eurasia. In World Politics 63 (04), pp. 581–617. https://doi.org/10.1017/S0043887111000189.

Hall, Peter A.; Taylor, Rosemary C. R. (1996): Political Science and the Three New Institutionalisms. In Political Studies 44 (5), pp. 936–957. https://doi.org/10.1111/j.1467-9248.1996.tb00343.x.

Halla, Martin (2012): Tax Morale and Compliance Behavior: First Evidence on a Causal Link. In The B.E. Journal of Economic Analysis and Policy 12 (1), pp. 1–12.

Helmke, Gretchen; Levitsky, Steven (2004): Informal Institutions and Comparative Politics: A Research Agenda. In Perspectives on Politics 2 (4), pp. 725–740.

Horodnic, Adrian V.; Williams, Colin C. (2018): Informal Payments by Patients for Health Services: Prevalence and Determinants. In The Service Industries Journal 38 (11–12), pp. 841–855. https://doi.org/10.1080/02642069.2018.1450870.

Horodnic, Ioana A. (2018): Tax Morale and Institutional Theory: A Systematic Review. In International Journal of Sociology and Social Policy 38 (9/10), pp. 868–886.

Horodnic, Ioana A.; Williams, Colin C. (2019): Tackling Undeclared Work in the European Union. In SSRN Electronic Journal. https://doi.org/10.2139/ssrn.3437406.

Inglehart, R.; Haerpfer, C.; Moreno, A.; Welzel, C.; Kizilova, K.; Diez-Medrano, J. et al. (Eds.) (2014): World Values Survey: Round Six—Country-Pooled Datafile Version. JD Systems Institute. Madrid. Available online at https://www.worldvaluessurvey.org/WVSDocumentationWV6.jsp, checked on 11/29/2020.

Ishiyama, John; Mezvrishvili, Lia; Zhgenti, Nina (2018): An Oasis of Democracy in an Authoritarian Sea? Civil Society, Social, and Institutional Trust in Georgia. In Communist and Post-Communist Studies 51 (1), pp. 19–26. https://doi.org/10.1016/j.postcomstud.2018.01.005.

Kanniainen, Vesa; Pääkkönen, Jenni (2009): Do the Catholic and Protestant Countries Differ by Their Tax Morale? In Empirica 37 (3), pp. 271–290. https://doi.org/10.1007/s10663-009-9108-5.

Kirchler, E.; Muehlbacher, S.; Kastlunger, B.; Wahl, I. (2007): Why Pay Taxes? A Review of Tax Compliance Decision. Andrew Young School of Policy Studies, Georgia State University. Atlanta, GA (International Studies Program Working Paper, No. 07-30).

Koelble, Thomas A.; Cook, Karen Schweers; Levi, Margaret; Granovetter, Mark; Swedberg, Richard; March, James G. et al. (1995): The New Institutionalism in Political Science and Sociology. In Comparative Politics 27 (2), pp. 231–243. https://doi.org/10.2307/422167.

Kombe, Wilbard J.; Kreibich, Volker (2000): Reconciling Informal and Formal Land Management: An Agenda for Improving Tenure Security and Urban Governance in Poor Countries. In Habitat International 24 (2), pp. 231–240. https://doi.org/10.1016/S0197-3975(99)00041-7.

Kondelaj, M. H.; Sameti, M.; Amiri, H.; Moayedfar, R. (2016): Analyzing Determinants of Tax Morale Based on Social Psychology Theory: Case Study of Iran. In Iranian Economic Review 20 (4), pp. 581–598.

Kopeček, Vincenc (2019): Trapped in Informality? A Study of Informal Politics in Georgia's Javakheti. In Caucasus Survey 7 (1), pp. 60–78. https://doi.org/10.1080/23761199.2018.1480191.

Lago-Peñas, Ignacio; Lago-Peñas, Santiago (2010): The Determinants of Tax Morale in Comparative Perspective: Evidence from European Countries. In European Journal of Political Economy 26 (4), pp. 441–453. https://doi.org/10.1016/j.ejpoleco.2010.06.003.

Lauth, Hans-Joachim (2000): Informal Institutions and Democracy. In Democratization 7 (4), pp. 21–50. https://doi.org/10.1080/13510340008403683.
Ledeneva, Alena (1998): Russia's Economy of Favors: Blat, Networking and Informal Exchange.
Ledeneva, Alena (2006): How Russia Really Works. The Informal Practices That Shaped Post-Soviet Politics and Business. Ithaca, NY: Cornell University Press (Culture and Society After Socialism).
Ledeneva, Alena (2013): Can Russia Modernise? Sistema, Power Networks and Informal Governance. Cambridge: Cambridge University Press. Available online at http://site.ebrary.com/lib/alltitles/docDetail.action?docID=10659324.
Leonardo, G. (2011): Politics and Tax Morale. The Role of Trust, Values, and Beliefs, in Shaping Individual Attitudes Towards Tax Compliance. Dissertation Thesis. Georgia Institute of Technology, Georgia State University, Atlanta.
Leonardo, G.; Martinez-Vazquez, J. (2016): Politicians, Bureaucrats, and Tax Morale: What Shapes Tax Compliance Attitudes? Atlanta, GA (International Studies Program Working Paper, No. 16-08).
Light, Matthew (2014): Police Reforms in the Republic of Georgia: The Convergence of Domestic and Foreign Policy in an Anti-Corruption Drive. In Policing and Society 24 (3), pp. 318–345. https://doi.org/10.1080/10439463.2013.784289.
Lomnitz, Larissa (1988): Informal Exchange Networks in Formal Systems: A Theoretical Model. In American Anthropologist (90), pp. 42–55.
Lowndes, Vivien; Roberts, Mark (2013): Why Institutions Matter. The New Institutionalism in Political Science. Basingstoke: Palgrave Macmillan (Political Analysis).
MacGregor, Jason; Wilkinson, Brett (2012): The Effect of Economic Patriotism on Tax Morale and Attitudes Toward Tax Compliance: Emerald Group Publishing Limited (Advances in Taxation). Available online at https://doi.org/10.1108/S1058-7497(2012)0000020009.
Mangin, W. (1967): Latin American Squatter Settlements: A Problem and a Solution. In Latin American Research Review 2, pp. 67–98.
Mars, Gerald; Altman, Yochanan (1983): The Cultural Bases of Soviet Georgia's Second Economy. In Soviet Studies 35 (4), pp. 546–560.
Marshall, Monty G; Elzinga-Marshall, Gabrielle C. (2017): Global Report 2017. Conflict, Governance, and State Fragility. Hg. v. Center for Systemic Peace. Vienna: VA.
Martinez-Vazquez, Jorge; Torgler, Benno (2009): The Evolution of Tax Morale in Modern Spain. In Journal of Economic Issues 43 (1), pp. 1–28. https://doi.org/10.2753/jei0021-3624430101.

Martins, A.; Gomes, C. (2014): Tax Morale, Occupation and Income Level: An Analysis of Portuguese Taxpayers. In Journal of Economics, Business and Management 2 (2), pp. 112–116.

Mikaelian, Hrant (2017): Self-Employment, Micro and Small Business in Armenia. Emergence of Entrepreneurship. Edited by Caucasus Institute. Yerevan.

Molero, Juan Carlos; Pujol, Francesc (2011): Walking Inside the Potential Tax Evader's Mind: Tax Morale Does Matter. In Journal of Business Ethics 105 (2), pp. 151–162. https://doi.org/10.1007/s10551-011-0955-1.

Nee, Victor (2005): The New Institutionalism in Economics and Sociology. In Neil J. Smelser, Richard Swedberg (Eds.): The Handbook of Economic Sociology. 2nd ed. Princeton, NJ, New York: Princeton University Press; Russell Sage Foundation, pp. 49–74.

Nodia, Ghia (2005): Civil Society Development in Georgia: Achievements and Challenges. Policy Paper. Edited by Caucasus Institute for Peace, Democracy, and Development. Tbilisi.

North, Douglass C. (1990): Institutions, Institutional Change and Economic Performance. New York: Cambridge University Press (Political Economy of Institutions and Decisions).

North, Douglass C. (1997): Understanding Economic Change. In Joan M. Nelson, Charles Tilly, Lee Walker (Eds.): Transforming Post-Communist Political Economies. Washington, DC: National Academies Press, pp. 13–18.

North, Douglass C.; Wallis, John Joseph; Webb, Steven B.; Weingast, Barry R. (2011): Limited Access Orders: Rethinking the Problems of Development and Violence. Stanford University. Available online at https://thepolicypractice.com/library/limited-access-orders-rethinking-the-problems-of-development-and-violence/, checked on 1/21/2019.

North, Douglass C.; Wallis, John Joseph; Weingast, Barry R. (2013): Violence and Social Orders. A Conceptual Framework for Interpreting Recorded Human History. Cambridge: Cambridge University Press.

OECD (2019): Tax Morale. What Drives People and Businesses to Pay Tax? Paris: OECD Publishing.

Olson, Mancur (2000): Big Bills Left on the Sidewalk: Why Some Nations Are Rich, and Others Poor. In Mancur Olson, Satu Kähköhnen (Eds.): A Not-So-Dismal Science: Oxford University Press, pp. 37–60.

Payne, G. K. (2002): Land, Rights and Innovation: Improving Tenure Security for the Urban Poor. London: ITDG Publishing.

Pejovich, Svetozar (1999): The Effects of the Interaction of Formal and Informal Institutions on Social Stability and Economic Development. In Journal of Markets and Morality 2 (2), pp. 164–181.

Pejovich, Svetozar (2006a): The Effects of the Interaction of Formal and Informal Institutions on Social Stability and Economic Development. In K.

C. Roy, Jörn Sideras (Eds.): Institutions, Globalisation and Empowerment. Cheltenham, U.K., Northampton, MA: Edward Elgar, pp. 56–74.
Pejovich, Svetozar (2006b): The Uneven Results of Institutional Changes in Central and Eastern Europe: The Role of Culture. In Social Philosophy and Policy 23 (1), pp. 231–254. https://doi.org/10.1017/S0265052506060109.
Peters, Brainard Guy (2005): Institutional Theory in Political Science. The "New Institutionalism". 2nd ed. London: Continuum.
Polese, Abel (2014): Informal Payments in Ukrainian Hospitals: On the Boundary Between Informal Payments, Gifts, and Bribes. In Anthropological Forum 24 (4), pp. 381–395. https://doi.org/10.1080/00664677.2014.953445.
Polese, Abel (2015): Informality Crusades: Why Informal Practices Are Stigmatized, Fought and Allowed in Different Contexts According to an Apparently Ununderstandable Logic. In Caucasus Social Science Review 2 (1), pp. 1–26.
Polese, Abel; Morris, Jeremy (2015): My Name Is Legion. The Resilience and Endurance of Informality Beyond, or In spite of, the State. In Jeremy Morris, Abel Polese (Eds.): Informal Economies in Post-Socialist Spaces. Practices, Institutions and Networks. Basingstoke: Palgrave Macmillan, pp. 1–21.
Polese, Abel; Williams, Colin C.; Horodnic, Ioana A.; Bejaković, Predrag (Eds.) (2017): The Informal Economy in Global Perspective. Varieties of Governance. New York: Palgrave Macmillan (International Political Economy Series).
Puffer, Sheila M.; McCarthy, Daniel J.; Boisot, Max (2010): Entrepreneurship in Russia and China: The Impact of Formal Institutional Voids. In Entrepreneurship Theory and Practice 34 (3), pp. 441–467. https://doi.org/10.1111/j.1540-6520.2009.00353.x.
Rekhviashvili, Lela (2017): Why Read Informality in a Substantivist Manner? On the Embeddedness of the Soviet Second Economy. In Abel Polese, Colin C. Williams, Ioana A. Horodnic, Predrag Bejaković (Eds.): The Informal Economy in Global Perspective. Varieties of Governance. New York: Palgrave Macmillan (International Political Economy Series), pp. 15–36.
Rekhviashvili, Lela (2018): Counterbalancing Marketization Informally: Georgia's New-Institutionalist Reform and Its Discontents. In Abel Polese, Lela Rekhviashvili, Borbála Kovács, Jeremy Morris (Eds.): Post-socialist Informalities. Power, Agency and the Construction of Extra-legalities from Bosnia to China. London, New York: Routledge, pp. 75–92.
Rekhviashvili, Lela; Polese, Abel (2017): Liberalism and Shadow Interventionism in Post-Revolutionary Georgia (2003–2012). In Caucasus Survey 5 (1), pp. 27–50. https://doi.org/10.1080/23761199.2017.1283471.
Ristovska, Maja; Mojsoska-Blazevski, Nicica; Nikolov, Miso (2013): An Alternative View to the Tax Evasion: The Effect of Tax Morale on Paying Taxes

in Macedonia and EU Countries. In Serbian Journal of Management 8 (2), pp. 169–183. https://doi.org/10.5937/sjm8-3776.

Russo, Francesco F. (2013): Tax Morale and Tax Evasion Reports. In Economics Letters 121 (1), pp. 110–114. https://doi.org/10.1016/j.econlet.2013.07.004.

Sá, C.; Martins, A.; Gomes, C. (2015): Tax Morale Determinants in Portugal. In European Scientific Journal August 2015 Special Edition, pp. 236–254. Available online at http://www.eujournal.org/index.php/esj/issue/view/198, checked on 12/23/2020.

Salimath, Manjula S.; Cullen, John B. (2010): Formal and Informal Institutional Effects on Entrepreneurship: A Synthesis of Nation-Level Research. In International Journal of Organizational Analysis 18 (3), pp. 358–385. https://doi.org/10.1108/19348831011062175.

Schneider, Friedrich (2005): Shadow Economies Around the World: What Do We Really Know? In European Journal of Political Economy 21 (3), pp. 598–642. https://doi.org/10.1016/j.ejpoleco.2004.10.002.

Schneider, Friedrich; Buehn, Andreas; Montenegro, Claudio E. (2010): New Estimates for the Shadow Economies All Over the World. In International Economic Journal 24 (4), pp. 443–461. https://doi.org/10.1080/10168737.2010.525974.

Schneider, Friedrich; Medina, Leandro (2018): Shadow Economies Around the World: What Did We Learn Over the Last 20 Years? International Monetary Fund.

Scott, James C. (1998): Seeing Like a State. How Certain Schemes to Improve the Human Condition Have Failed. New Haven: Yale University Press.

Sipos, Agnes (2015): Determining Factors of Tax-Morale with Special Emphasis on the Tax Revenues of Local Self-governments. In Procedia Economics and Finance 30, pp. 758–767. https://doi.org/10.1016/s2212-5671(15)01325-8.

Smallbone, David; Welter, Friederike (Eds.) (2009): Entrepreneurship and Small Business Development in the Former Post-Socialist Economies. London: Routledge (Routledge Studies in Small Business, 14).

Suchman, Mark C. (1995): Managing Legitimacy: Strategic and Institutional Approaches. In The Academy of Management Review 20 (3), pp. 571–610. https://doi.org/10.2307/258788.

The Caucasus Research Resource Centers (2015): Caucasus Barometer 2015 Regional Dataset. Available online at https://caucasusbarometer.org/en/cb2015/codebook/, checked on 12/29/2020.

The Caucasus Research Resource Centers (2017): Caucasus Barometer 2017 Regional Dataset. Available online at https://caucasusbarometer.org/en/cb2017/codebook/, checked on 12/29/2020.

The Caucasus Research Resource Centers (2020): Caucasus Barometer 2020 Regional Dataset. Available online at https://caucasusbarometer.org/en/cb2017/codebook/, checked on 12/29/2020.
Torgler, Benno (2001): What Do We Know About Tax Morale and Tax Compliance? In International Review of Economics and Business (RISEC) 48 (3), pp. 395–419.
Torgler, Benno (2002): Speaking to Theorists and Searching for Facts: Tax Morale and Tax Compliance in Experiments. In Journal of Economic Surveys 16 (5), pp. 657–683. https://doi.org/10.1111/1467-6419.00185.
Torgler, Benno (2003a): Tax Morale and Institutions. In SSRN Electronic Journal. https://doi.org/10.2139/ssrn.663686.
Torgler, Benno (2003b): Tax Morale in Transition Countries. In Post-Communist Economies 15 (3), pp. 357–381. https://doi.org/10.1080/1463137032000139052.
Torgler, Benno (2003c): Tax Morale, Rule-Governed Behaviour and Trust. In Constitutional Political Economy 14 (2), pp. 119–140. https://doi.org/10.1023/a:1023643622283.
Torgler, Benno (2003d): To Evade Taxes or Not to Evade: That Is the Question. In The Journal of Socio-Economics 32 (3), pp. 283–302. https://doi.org/10.1016/s1053-5357(03)00040-4.
Torgler, Benno (2004a): Cross-Culture Comparison of Tax Morale and Tax Compliance: Evidence from Costa Rica and Switzerland. In International Journal of Comparative Sociology 45 (1–2), pp. 17–43. https://doi.org/10.1177/0020715204048309.
Torgler, Benno (2004b): Tax Morale in Asian Countries. In Journal of Asian Economics 15 (2), pp. 237–266. https://doi.org/10.1016/j.asieco.2004.02.001.
Torgler, Benno (2005): A Knight Without a Sword? The Effects of Audit Courts on Tax Morale. In Journal of Institutional and Theoretical Economics 161 (4), p. 735. https://doi.org/10.1628/093245605775075988.
Torgler, Benno (2006): The Importance of Faith: Tax Morale and Religiosity. In Journal of Economic Behavior & Organization 61 (1), pp. 81–109. https://doi.org/10.1016/j.jebo.2004.10.007.
Torgler, Benno (2007): Tax Compliance and Tax Morale: Edward Elgar Publishing. Available online at https://doi.org/10.4337/9781847207203.
Torgler, Benno (2011): Tax Morale and Compliance: Review of Evidence and Case Studies for Europe: The World Bank (Policy Research Working Papers). Available online at https://doi.org/10.1596/1813-9450-5922.
Torgler, Benno (2012): Tax morale, Eastern Europe and European enlargement. In Communist and Post-Communist Studies 45 (1-2), pp. 11–25. https://doi.org/10.1016/j.postcomstud.2012.02.005.

Torgler, Benno; Demir, Ihsan C.; Macintyre, Alison; Schaffner, Markus (2008): Causes and Consequences of Tax Morale: An Empirical Investigation. In Economic Analysis and Policy 38 (2), pp. 313–339. https://doi.org/10.1016/s0313-5926(08)50023-3.

Torgler, Benno; Schaffner, M. (2007): Causes and Consequences of Tax Morale: An Empirical Investigation. Edited by Center for Research in Economics, Management and the Arts. Basel (Working Paper, No. 2007-11).

Torgler, Benno; Schaffner, Markus; Macintyre, Alison (2007): Tax Compliance, Tax Morale and Governance Quality. Edited by Center for Research in Economics, Management and the Arts (CREMA) (CREMA Working Paper Series, 2007-17).

Torgler, Benno; Schneider, Friedrich (2004): Does Culture Influence Tax Morale? Evidence from Different European Countries. Center for Research in Economics, Management and the Arts. Basel (Working Paper, No. 2004-17).

Torgler, Benno; Schneider, Friedrich (2007): What Shapes Attitudes Toward Paying Taxes? Evidence from Multicultural European Countries. In Social Science Quarterly 88 (2), pp. 443–470. https://doi.org/10.1111/j.1540-6237.2007.00466.x.

Torgler, Benno; Schneider, Friedrich; Schaltegger, Christoph A. (2007): With or Against the People? The Impact of a Bottom-Up Approach on Tax Morale and the Shadow Economy. Edited by QUT—School of Economics and Finance. Brisbane (Working/Discussion Paper, No. 211).

Torgler, Benno; Schneider, Friedrich; Schaltegger, Christoph A. (2009): Local Autonomy, Tax Morale, and the Shadow Economy. In Public Choice 144 (1-2), pp. 293–321. https://doi.org/10.1007/s11127-009-9520-1.

Torosyan, Karine; Filer, Randall (2012): Tax Reform in Georgia and the Size of the Shadow Economy. Edited by Institute for the Study of Labor (IZA). Bonn (IZA Discussion Paper, No. 6912).

Trüdinger, Eva-Maria; Hildebrandt, Achim (2012): Causes and Contexts of Tax Morale: Rational Considerations, Community Orientations, and Communist Rule. In International Political Science Review 34 (2), pp. 191–209. https://doi.org/10.1177/0192512112447117.

Turner, J. F. C. (1977): Housing by People: Towards Autonomy in Building Environments. New York: Pantheon Books.

Valdez, Michael E.; Richardson, James (2013): Institutional Determinants of Macro-Level Entrepreneurship. In Entrepreneurship Theory and Practice 37 (5), pp. 1149–1175. https://doi.org/10.1111/etap.12000.

van Schendel, Willem; Abraham, Itty (2005): Illicit Flows and Criminal Things. States, Borders, and the Other Side of Globalization. Bloomington: Indiana University Press. Available online at http://search.ebscohost.com/login.aspx?direct=true&scope=site&db=nlebk&db=nlabk&AN=171502.

Verheul, Ingrid; Wennekers, Sander; Audretsch, David; Thurik, Roy (2002): An Eclectic Theory of Entrepreneurship: Policies, Institutions and Culture. In Cristiano Antonelli, Bo Carlsson, Steven Klepper, Richard Langlois, J.S Metcalfe, David Mowery et al. (Eds.): Entrepreneurship: Determinants and Policy in a European-US Comparison, vol. 27. Boston, MA: Springer US (Economics of Science, Technology and Innovation), pp. 11–81.

Voigt, Stefan (2016): How to Measure Informal Institutions. In SSRN Electronic Journal. https://doi.org/10.2139/ssrn.2748214.

Vythelingum, P.; Soondram, H.; Jugurnath, B. (2017): An Assessment of Tax Morale Among Mauritian Taxpayers. In Journal of Accounting and Taxation 9 (1), pp. 1–10. https://doi.org/10.5897/jat2016.0224.

Wade-Benzoni, Kimberly A.; Hoffman, Andrew John; Thompson, Leigh; Moore, Don A.; Gillespie, James; Bazerman, Max H. (2001): Barriers to Resolution in Ideologically Based Negotiations: The Role of Values and Institutions. In Academy of Management Review 27 (1), pp. 41–57. https://doi.org/10.2139/ssrn.2944066.

Webb, Justin W.; Tihanyi, Laszlo; Ireland, R. Duane; Sirmon, David G. (2009): You Say Illegal, I Say Legitimate: Entrepreneurship in the Informal Economy. In Academy of Management Review 34 (3), pp. 492–510. https://doi.org/10.5465/amr.2009.40632826.

Williams, Colin C.; Franic, Josip (2016): Explaining Participation in the Informal Economy in Post-Socialist Societies: A Study of the Asymmetry Between Formal and Informal Institutions in Croatia. In Journal of Contemporary Central and Eastern Europe 24 (1), pp. 51–65. https://doi.org/10.1080/0965156X.2015.1118817.

Williams, Colin C.; Horodnic, Ioana A. (2015a): Evaluating the Prevalence of the Undeclared Economy in Central and Eastern Europe: An Institutional Asymmetry Perspective. In European Journal of Industrial Relations 21 (4), pp. 389–406. https://doi.org/10.1177/0143831x14568835.

Williams, Colin C.; Horodnic, Ioana A. (2015b): Tackling the Informal Economy in Southeast Europe: An Institutional Approach. In Southeast European and Black Sea Studies 15 (4), pp. 519–539. https://doi.org/10.1080/14683857.2015.1056980.

Williams, Colin C.; Horodnic, Ioana A. (2016): An Institutional Theory of the Informal Economy: Some Lessons from the United Kingdom. In International Journal of Social Economics 43 (7), pp. 722–738. https://doi.org/10.1108/ijse-12-2014-0256.

Williams, Colin C.; Horodnic, Adrian V. (2017a): Rethinking Informal Payments by Patients in Europe: An Institutional Approach. In Health Policy 121 (10), pp. 1053–1062. https://doi.org/10.1016/j.healthpol.2017.08.007.

Williams, Colin C.; Horodnic, Ioana A. (2017b): Evaluating the Illegal Employer Practice of Under-Reporting Employees' Salaries. In British Journal of Industrial Relations 55 (1), pp. 83–111. https://doi.org/10.1111/bjir.12179.

Williams, Colin C.; Horodnic, Ioana A. (2017c): Explaining the Informal Economy in Post-Communist Societies: A Study of the Asymmetry Between Formal and Informal Institutions in Romania: Springer International Publishing (The Informal Economy in Global Perspective). Available online at https://doi.org/10.1007/978-3-319-40931-3_7.

Williams, Colin C.; Horodnic, Ioana A.; Windebank, Jan (2015): Explaining Participation in the Informal Economy: An Institutional Incongruence Perspective. In International Sociology 30 (3), pp. 294–313. https://doi.org/10.1177/0268580915578745.

Williams, Colin C.; Krasniqi, Besnik (2017): Evaluating the Individual- and Country-Level Variations in Tax Morale. In Journal of Economic Studies 44 (5), pp. 816–832. https://doi.org/10.1108/jes-09-2016-0182.

Williams, Colin C.; Martinez Perez, Alvaro (2014): Explaining Cross-National Variations in Tax Morality in the European Union: An Exploratory Analysis. In Studies of Transition States and Societies 6 (1), pp. 5–18.

Williams, Nick; Vorley, Tim (2015): Institutional Asymmetry: How Formal and Informal Institutions Affect Entrepreneurship in Bulgaria. In International Small Business Journal 33 (8), pp. 840–861.

Williamson, Claudia R. (2009): Informal Institutions Rule: Institutional Arrangements and Economic Performance. In Public Choice 139 (3–4), pp. 371–387. https://doi.org/10.1007/s11127-009-9399-x.

Windebank, Jan; Horodnic, Ioana A. (2017): Explaining Participation in Undeclared Work in France: Lessons for Policy Evaluation. In International Journal of Sociology and Social Policy 37 (3/4), pp. 203–217. https://doi.org/10.1108/IJSSP-12-2015-0147.

World Bank (2020a): Enterprise Surveys Data. The World Bank. Available online at https://www.enterprisesurveys.org/en/data.

World Bank (2020b): World Bank Open Data. The World Bank. Available online at https://data.worldbank.org/, checked on 11/29/2020.

World Bank (2000c): Reforming Public Institutions and Strengthening Governance. A World Bank Strategy. Edited by The World Bank. Washington, DC.

Wright, Mike; Marlow, Susan (2012): Entrepreneurial Activity in the Venture Creation and Development Process. In International Small Business Journal 30 (2), pp. 107–114. https://doi.org/10.1177/0266242611432793.

Wright, Mike; Stigliani, Ileana (2013): Entrepreneurship and Growth. In International Small Business Journal 31 (1), pp. 3–22. https://doi.org/10.1177/0266242612467359.

Left in the 'Shadows': The Informal Moral Economy of the Russian Far East

Aimar Ventsel

In Russia, free market capitalism ended in 1928 when Stalin officially terminated the New Economic Policy, or NEP, disrupting the coexistence of private and public sectors. Economic reforms in the early Yeltsin era, i.e. at the outset of the 1990s, gave the green light to new non-state enterprises across multiple spheres of the economy and of various sizes. Notwithstanding the fact that this transition seems to have been well studied—especially at the level of small and medium-sized enterprises—we should be careful in painting generalisations. Russia comprises 85 regions, the more recent economic history of which can differ substantially. Thus, while private cafes and shops began mushrooming in Moscow already in 1987, it took them five more years to reach the Far East. Regions also vary tremendously in their economic structure, the share of big enterprises in regional GDP, importance of the public sector, or the main sources of income.

A. Ventsel (✉)
Department of Ethnology, University of Tartu, Tartu, Estonia
e-mail: punkrock@folklore.ee

A. Polese (ed.), *Informality, Labour Mobility and Precariousness*, International Political Economy Series,
https://doi.org/10.1007/978-3-030-82499-0_7

The enthusiasm with which former socialist countries embraced private entrepreneurship had political underpinnings, with the emerging entrepreneurial class having been associated with freethinking and a democratic ideology. While in some of these states these expectations can be said to have been more or less enacted, in Russia they remained largely unfulfilled. Although entrepreneurship in Russia has taken on different forms, it is more often related to the shadow economy, power struggles, and the criminal underworld, rather than democratisation (see Cartwright 2001; Humphrey 2002; Kuznetsov and Kuznetsova 2005; Tarasov et al. 2013; Ventsel 2005). This is manifested in the rare coverage of small and medium-sized enterprises (SMEs) in the media, even though a large proportion of the local population depends on such enterprises for a living. Yet the private sector continues to thrive, and despite an increasing tax burden, the everyday life of Russians hinges on the services and goods supplied by them.

Existing studies on private entrepreneurship in Eastern Europe tend to be critical of the utilitarian approach of the neoclassic economy (cf. Estrin et al. 2006: 4; Kuznetsov and Kuznetsova 2005; Hann 2003; Heady 2003; Nee and Young 1991). Indeed, studies of entrepreneurship in post-Soviet Russia, whether in anthropology (Hann 2000; Köllner 2013), economics (Berkowitz and DeJong 2001; Estrin et al. 2006), or sociology (Barsukova and Radaev 2012; Sakaeva 2012) imply that the actors saw the new market economy as an opportunity, but their strategies were more directed towards establishing social and economic security, rather than harvesting immediate profit.

One popular theoretical framework for analysing entrepreneurship in Russia is the concept of *blat*. Upon revising the existing literature on the topic, one is left with an impression that *blat* is a natural part of Russian private business. Russia is described as a 'network state' (Humphrey 2012), where political and economic networks can have a semi-feudal nature (Gambold and Miller 2003; Shlapentokh 2007). The originator of the academic concept of *blat* Alena Ledeneva argues that it was a Soviet-era strategy, a system of favours, used to gain access to scarce resources, mainly goods (Ledeneva 1998, 2006). Structurally, it is viewed as a circle of people engaged in delayed reciprocity. This circle is a closed one and based on a 'negative solidarity', or mistrust of outsiders (Radaev 1998, 2005). Further, the reciprocal group can be seen as a network of people tied with common interests, based on pragmatic calculations, and not on kin, clan, or other social or cultural ties. In general, modern *blat* is interpreted as an informal network linking entrepreneurs and political power structures (Puffer et al. 2010). Through such connections, entrepreneurs can exploit state resources or sign profitable deals with state structures, like ministries (Hsu 2005). Indeed, post-Soviet Russian history is full of corruption scandals where private enterprises receive overpaid orders from government institutions for the services they deliver to the latter, such as construction, catering, software installation, and the like.[1]

Such a definition of *blat* is characteristic of the Republic of Sakha, where it is used to indicate the corruption that links governmental and entrepreneurial networks. The expression is, however, rarely used to describe reciprocal practices between ordinary people or private enterprises. Here, *blat* is hardly the Maussian gift exchange, as some authors argue (Hsu 2005; Humphrey 2000; Ssorin-Chaikov 2000), but rather a cynical and sinister patron–client relationship.

For these reasons, *blat* is unsuited for analysis of the attitudes, strategies, and practices of the entrepreneurs examined here. While *blat* is in some sense a 'genuine Russian' phenomenon, then more suited for the purpose of this article is the 'genuine Chinese' concept of reciprocal networking—*guanxi*. In contrast to *blat*, seen largely as a reaction to the Soviet economy, *guanxi* dates back to the pre-Socialist period. Several authors acknowledge many similarities between *blat* and *guanxi* as a 'survival kit' or 'safety network' (Ledeneva 2008). Both are reciprocal, informal, and based on delayed reciprocity. However, gradual differences are important. *Guanxi*, as *blat*, is a network and practice that constantly changes its form. There are, nevertheless, certain features that characterise most variations of *guanxi*. First of all, *guanxi* is an inclusive and constantly expanding network, based on long-term relationships. At its core, *guanxi* is based on kinship ideology, linking both biological and fictive kin. This network channels not only the transmission of goods and money, but also favours. In fact, some scholars argue that the most important part of *guanxi* is indeed the 'small favours', which are sometimes exchanged in a highly ritualistic way.

The informal economy has been studied from various angles. In general, the majority of the scholarship is focused on how entrepreneurs

[1] The best source for current corruption scandals involving state officials or institutions is the web site of well-known Russian opposition figure Alexei Navalny—www.navalny.com.

avoid paying taxes, explore loopholes in the legislation, are partly a form of corruption, or at the very least are operating partially underground to engage in illicit or illegal transactions (Benton 1994; Donovan 2008). Ethnicity and ethnic networks are also a popular topic in the studies of the informal economy (Ibañez-Tirado and Marsden 2020; Meagher 2006; Turaeva 2014). This is also present in this chapter but from a different perspective. As a rule, what lies at the core of this literature is the relationship between the state and entrepreneurship, and in particular how both parties are connected and intertwined via different forms of corruption, networking and social links (Humphrey 2002; Ledeneva 1998, 2000, 2006; Oka 2015). In this chapter, I will discuss small-scale entrepreneurship in the Republic of Sakha (Yakutia), in the Russian Far East. In a setting where state institutions have lost interest in SMEs, and regard them as an economically insignificant sector, then kinship and clan ideologies, the distancing from formal institutions, and strong elements of a moral economy, become central to business activity. Ethnicity plays an important role here because trade is accompanied by the ideology of a nation's health, as I will go on to show. The kind of entrepreneurship studied here is quite exceptional. In the Republic Sakha, the state has little interest in controlling or taxing this sector, and entrepreneurs go out of their way in order to avoid economic or social ties with state structures. The logical answer of the entrepreneurs was to show a high degree of mistrust in the state, similar to research in Ukraine (Polese 2014). Sometimes during my research, I had a feeling that the SME here live on a different planet to the big companies whose link to state structures is vital to their existence. In a way it is typical for Russia, where disadvantaged groups tend to live in their own social setting, showing little interest in the 'outside world' (see Morris 2016).

I have been conducting fieldwork in the region since 2000, and have witnessed changes in small and medium-sized businesses. Until 2005 in the capital town Yakutsk, the typical post-socialist kiosk economy was very visible. Kiosks started to slowly disappear after that, but it was not until 2010 that the city government pursued an aggressive policy of removal. The formal reason for this was the rather unsuitable appearance of the kiosks in the urban environment, and the suspicion that they were engaged in money laundering. In parallel to the relicensing of small shops and kiosks, the republic introduced harsh regulations on the sale of alcohol. This was driven by the republic's policy to improve people's health. There was a heavy promotion of abstinence and support for engagement in sports and a healthy lifestyle. As a result, the shops were allowed to sell alcoholic beverages only from 14.00 to 20.00. This came as the final blow for many small shops, which earned a substantial share of their income from the late-night alcohol trade. This was also a boom time for small creative cafés, bakeries, video hire and sales, and stylish clothing shops. Many such places were run by enthusiasts, who engaged in these forms of entrepreneurship on a part-time basis. From 2012 to 2013, these interesting places began to close, replaced by rather conventional cafés, cocktail bars, swanky restaurants, travel agencies, and furniture shops. I was told by some of my acquaintances, how suddenly certain kinds of small enterprises encountered difficulties in getting credit, or experienced increased taxation or health controls. This is anecdotal evidence, but I simultaneously noticed that many smaller craftsmen gave up their leases in the city centre to move to modest workshops in the basements and yards on the outskirts of the city. Now, when I needed to repair my rucksack or boots, or clean my suit, I had to ask people where I could find these services. The places of these enterprises in the inner city were now occupied by offices, department stores, large and fancy restaurants, or stores that sold electronics or fashionable clothing.

This article draws on the extensive background knowledge of the region obtained over the last twenty years. I have had an interest in new forms of property and the local economy since first visiting the region (Ventsel 2003, 2005, 2010, 2011, 2017). Formally, I participated in a research project on Arctic urban development, and conducted research in the Republic of Sakha in July and December 2013. During the fieldwork, I recorded approximately 10 interviews with different entrepreneurs, whose enterprises were classified as small or microenterprises. In July 2014, I spent several weeks in the Verkhoiansk district (approximately 1000 km northeast from Yakutsk) where I conducted formal interviews and informal conversations with people who were engaged in more or less informal entrepreneurship as shop owners, providers of transport services, or fish, meat, and dairy traders. From 2013 onwards, I conducted research in villages near Yakutsk where I was able to carry out short interviews with people I would describe as having a 'broad profile'. These offered primarily transport and tourist services, fished and sold their catch, and preferred to do all of these undocumented. In autumn 2016, I studied SMEs north of the Arctic circle, in villages close to the coast of the Arctic Ocean. Interviews were also made with government and district administration officials, including one former Minister of the

Economy of the Republic of Sakha. Additionally, I talked to various other entrepreneurs, visited their enterprises, and observed how they operated. Common to all informants was that none of them had studied the profession they were engaged in at the time of our encounter. Most of them were ethnic Sakha, one was Armenian, and a handful were local Russians. Half of my respondents originated from a village, roughly 70 per cent were women, and all of them were in their forties or fifties. Informal interviews are well suited to study informal economic practices in a post-Socialist setting. Semiformal activities of SME are difficult to study in the Republic of Sakha as well. My experience shows that the entrepreneurs involved prefer to talk about their businesses off the record. They tend to decline any formal meetings, especially when those meetings are recorded. There is almost a paranoia when it comes to talking about entrepreneurship among the informants presented below. To gain the opportunity to meet people, I relied on the personal networks I had established over two decades. The researcher's access to the world of SMEs also needs certain cultural skills—or 'learning the glossary' (Polese 2014: 88)—that could be accumulated only after years of experience. For example, among the Sakha people there is certain unwritten etiquette about how to talk about money making. In order to hide my informants' identity, I have changed all the names. Statistics presented in this article are from the works of local economists.

I start with a short introduction of the Republic of Sakha, and the economic environment. Then I present some case studies and place them in a larger context. All the entrepreneurs I introduce had certain common features. Namely, their activity was not driven solely by economic rationale, but also moral categories. Moreover, it was very important for them to avoid contact with state institutions. One result of such an attitude was a conscious decision to operate more or less in the shadows. The chapter finishes with an analytical discussion that demonstrates how morality plays an important role in entrepreneurship.

The Republic of Sakha (Yakutia) and Its Economic Environment

The Republic of Sakha (Yakutia) is the largest territorial unit of the Russian Federation and belongs to the easternmost administrative unit, the Far North Federal District, containing nine different territorial units. The republic is a large but sparsely populated territory, covering more than three million square kilometres that is home to less than a million people, of whom 55 per cent are the titular ethnic group Sakha (or Yakut) speaking a Turkic language. Approximately 73 per cent of the population in the Republic of Sakha are urban.

The republic is famous for its diamond resources, producing 30 per cent of the world's diamonds and almost 100 per cent of Russia's diamonds. In addition to diamond mining, the Republic of Sakha has significant resources including gold, gas, oil, precious metals, coal, and timber. Natural resource extraction is the domain of big companies from the region, as is large-scale construction. The difference between the two is that while mining and oil extraction is dominated by state-owned companies, the construction sector is dominated by private companies. Recently, the republic has seen the expansion of large Russian oil extraction companies in its territory, with Surgutneftegaz and Lukoil, traditionally located in Western Siberia, being the two major examples. This drives up wages in a tiny segment of the workforce connected to such companies, and has an impact on the service and construction sector, especially in the capital of Yakutsk.

The official definition of SME in Russia remains blurred, and is constantly changing, but as of 2009, 181,514 small enterprises operated in the Russian Far North[2] (Stepanova and Nogovitsyn 2011: 49). Approximately 80 per cent of these were 'microenterprises', or enterprises that hired up to 15 employees on a permanent basis. It must be mentioned that in relation to the number of enterprises per capita, the Far North is slightly below the average Russian level. In terms of regional distribution, about 47 per cent of small enterprises are located in Primorski Krai, 19 per cent in Khabarovsk Krai, and approximately 8 per cent in the Republic of Sakha (Yakutia) (ibid.: 49–59). Small enterprises tend to have very low investment and profit levels, and 60 per cent of small enterprises in the Russian Far North are engaged in wholesale and retail services (Egorov 2006: 192; Stepanova and Nogovitsyn 2011: 65).

[2] A construction firm in Yakutsk which I studied in 2013 illustrates this ambivalence well. The enterprise had less than 15 permanent workers which makes it a 'microenterprise'. These permanent workers were mainly engaged in management tasks. When the enterprise, however, signed a building contract they hired dozens of construction workers from the formerly Soviet Central Asian republics or from Sakha villages, depending on the profile of the task: wooden buildings are mainly built by Sakha, whereas it is believed that Kyrgyz and Uzbek are better at constructing concrete buildings.

In 2010, there were 115 medium-sized and 4952 small enterprises registered in the Republic of Sakha. Of the latter 4029 were classified as microenterprises. In 2009, small enterprises contributed only 6.6 per cent of the regional GDP, but gave work to 40 per cent of the local workforce. In the Republic of Sakha, 80 per cent of small and medium sized enterprises are concentrated in four cities—Yakutsk, Neryungri, the base of the coal industry, Mirnyi, the centre of the diamond economy, and Lensk, another centre for diamond mining—indicating a direct link between a segment of the population with a high income and the service sector. From the regional enterprises, only 400 are agricultural. The statistics shows that in Sakha 44.4 per cent of small enterprises specialise in wholesale merchandise, 22.7 per cent in construction, 6.8 per cent in real estate, and 5.7 per cent in transport. The role of small-scale enterprises in different spheres can be high: 94.3 per cent of services (*bytovye uslugi*) and 100 per cent of the small bus transport market belongs to small enterprises, as does the majority of tourism, clothing repair, cargo transport, car and truck repair, and other services. (Stepanova and Nogovitsyn 2011: 56–57).[3]

The low profit levels of such enterprises can be explained by the underdeveloped infrastructure and high energy costs in the region. Another peculiarity of small-scale enterprises which scholars highlight, is their need for highly skilled specialist labour which is much lower than their demand for unqualified manual labour. Notwithstanding several federal and regional programs, the state has been unsuccessful in developing "innovative" and modern high-tech small enterprises. As some local economists explain, this is due to big mining and production companies, which have better financing and market prospects, and which already dominate these spheres and offer better salaries than private enterprises (ibid.: 2011: 59).

[3] According to the official statistics, the number has significantly changed over ten years. In August 2020, 35,752 microenterprises were registered in the Republic of Sakha, of which 10,029 were listed as legal companies, and the rest as what Russian law defines as 'individual enterprise' (IP) or one man companies. Additionally, 841 small enterprises were registered of which 692 were companies. See https://rmsp.nalog.ru/statistics.html?statDate=&level=0&fo=8&ssrf=14.

Entrepreneurs as a Networking Society

In January 2017, I met a person who is part of the board of the Association of Small and Medium Entrepreneurs of the Republic of Sakha. He told me about the Board's recent official meeting with the President of the Republic. In short, the meeting was 'extremely unproductive'. Entrepreneurs were interested in governmental support to obtain credit and more transparent tax policies, but the President demonstrated a clear disinterest in changing anything. 'And then they wonder why a big part of business operates in the shadow economy and does everything to be ignored by the government', he concluded bitterly. We both agreed that the relationship between state structures and the private economy is a two-way bond of mutual obligations, and it is only natural that many entrepreneurs 'muddle through' when they do not see any state support.

Below are several cases from the pool of entrepreneurs and other informants I have met in the last three years. Each of them has their own reasons to be an entrepreneur, but there are also common features and attitudes that unite them.

I interviewed Alena (all names are pseudonyms) in December 2013 and met her later in 2015. She was in her late thirties and worked as a bookkeeper in a municipal government structure in Yakutsk. Her husband did the same for a private enterprise and this was their major source of income. Yet, they conceived of themselves principally as managers of a fishing enterprise. The enterprise is located on the coast of the Arctic Ocean, in a town called Tiksi which served as a military post and trade port in the Soviet era. Alena's mother established the fishing enterprise in the mid-1990s and since then it has existed as a family enterprise. All employees of the small enterprise are family members, and the purpose of the enterprise has always been to guarantee income to kinsfolk. The northern fish is considered a delicacy in the Republic of Sakha and has traditionally been in high demand in more southern cities. There are several species that live only in rivers and lakes in the high North and people usually eat them in a deep-frozen form, cut into slices, called *stroganina*. But recent years have seen the market flooded. Thus, while ten years ago the northern fish was difficult to acquire in Yakutsk, today, markets and street corners are literally full of street vendors who sell such fish. As a result, prices have fallen and fishing enterprises are struggling to make ends meet. Alena took care of the marketing and all the paperwork (licenses, taxes, state subsidies) of their family enterprise, whereas her husband was in charge of the transportation of the catch to Yakutsk. This required certain skills and good connections. The fish was usually bought by state and private enterprises, and the price varied.

When Alena sold the catch to a state enterprise she was entitled to state subsidies, but she told me that private buyers paid more. Therefore, to have the optimal outcome, costs for transport had to be calculated carefully. Both—Alena and her husband—admitted that they did not earn much for managing the enterprise, as their priority was to pay salaries to their relatives in Tiksi and to obtain supplies for the fishing season. When we asked Alena why she was engaged in the enterprise at all, she gave me a long stare and said: 'How can I give up an enterprise that my mother established? And what will happen to all the people?' While managing the fishing enterprise, Alena had also tried other activities. For many years she ran a French-style café, offering pastries and different coffees. She quit when she had problems finding a skilled workforce to maintain the quality of the products. This café was also the route for many of her rural relatives to start a new life in the city. This was problematic because young relatives often left unexpectedly when they managed to find a more respectable and better-paid job. Alena's activities can thus be interpreted as a form of moral economy, where a sense of being obliged to kin people and her mother's legacy outweigh economic rationale.

I interviewed Marfa in December 2013. She was in her mid-fifties and had a laundry in an old Soviet concrete block district of Yakutsk. Previously a day-care teacher, she decided to open a kiosk, selling everything from beer to cheap jewellery and cleaning liquid, when the tumultuous nineties brought about delays in wages. Marfa struggled hard to support her family and raise her son. Over time, her business did well and she opened more kiosks. In the mid-2000s she was offered some laundry machines and so she switched her business. Located in the basement of a tenement building, the laundry had no sign whatsoever. As I learnt, Marfa found customers only via word of mouth or through personal contacts. In recent years, she acquired some valuable customers, such as restaurants needing to wash their tablecloths regularly. The enterprise hired only a few female pensioners and, as a rule, illegally, i.e. without registering them as taxable employees. Marfa viewed such an arrangement as a kind of charity, assistance for pensioners in need of money. Her son helped with managing the laundry and had opened a workshop in a garage on the property where they both lived. The garage operated illegally, or as 'black trade' (*tschernyi biznes*), to use the emic phrase. He repaired cars and motorcycles without paying any taxes. Marfa was also a co-owner of a shop in one of the northern villages. Another owner was her sister, and Marfa's task was to buy and organise the transportation of different goods to the village for sale. Both decided not to sell alcohol, despite the fact that this was the easiest way to make a profit in the remote northern settlements. 'We do not support it [alcohol consumption]! We support ZOZh!'.[4]

Ivan was in his late fifties and lived in the remote northern village of Saidy. I met him in the summer of 2014 during a research trip to the region. Saidy is located in the Verkhoiansk district and can only be reached by a 4–8 hour motorboat trip from the district centre during summertime. In winter, when the river is frozen, trucks use it as a winter road. The village is home to roughly 700 cattle herders, hunters, and fishermen, with a strong sense of community. Even by the standards of a remote Arctic village, Saidy stands out as an extraordinarily strong informal economy. Most of the hunting, transport services, cattle economy, and fishing, are carried out on an informal basis, leaving no trace in official records.

At the time of my fieldwork, the village had no official shops selling food and other consumables, and all three existing shops were technically illegal. A few years ago the community banned the sale of alcohol in the village, and as the head of the village administration told us: 'We decided to allow the shops to exist only if they do not sell any alcohol'. Ivan got into shopkeeping accidentally. His son-in-law, who was an entrepreneur in Yakutsk, learned how shops are run illegally in Saidy, without any license, registration, or other paperwork. He suggested that Ivan open a shop where he would deliver supplies directly from Yakutsk. Ivan's shop, like the rest of them, looked like the usual multifunctional stores everywhere, with shelves, price labels, and a freezer in the corner. Ivan's shop was well stocked with regular foodstuff, cigarettes, and clothes, and he even sold red caviar and expensive cheese. All the goods were shipped from Yakutsk and the profits were split between Ivan and his son-in-law. While Ivan was officially an unemployed pensioner, in reality, he was a full-time shopkeeper.

Ivan is just one example of the illegal shopkeeping that I encountered in the Verkhoiansk region. Interestingly, illegal shops were common in

[4] *Zdorovyi obraz zhizni* or a Healthy Way of Life is a campaign in Russia which promotes abstinence from alcohol and tobacco combined with active ways of life, such as sports.

villages where the community had banned the sale of alcohol. In villages, where the sale of alcohol was allowed, most shops were legal. Ivan's case demonstrates that in remote villages in the Russian Arctic, illicit practices are tolerated when they do not contradict the community's interests and moral pursuits. On the other hand, illegal shops also exemplify how the state bluntly ignores the well-being of peripheral communities and shows no interest in how people from such villages are supplied with food and other consumables. Most people in the district are aware of the illegal status of the shops, but the officials turn a blind eye to them, as otherwise, they would have to deal with the provision of the goods themselves.

Sveta was interviewed in December 2013. She was in her late thirties and worked as a sports teacher at a school in Yakutsk. She was also a fully licensed trader and entrepreneur. The license is needed to buy goods for her sister's shop in a village in the northern district of the republic. Thanks to the license, Sveta could officially hire trucks to send the ordered goods to the village. The document also entitled her to buy goods from wholesale traders. Alongside being the supplier for her sister, Sveta had her own small additional income in the village where most people are related by kinship ties:

> Sometimes people ask me to buy something. Like a fridge or furniture. Sometimes very specific clothes or … when they order these things, I usually send them to the village by the truck. I do it almost for free. Let's say, I take 1000 rubles (approx. 15 euro) to buy a fridge. It is not that much, rather a symbolic amount because a fridge costs a lot. And organising the transport takes time.

In general, Sveta engaged in her own small private business because she could send ordered goods using the trucks she filled with consumables for her sister's shop.

There was a practical agreement between the sisters, and the village community. Both women knew most of the village population, and were linked to them via various kinship ties or common biography: they went to school with them, played as children, were long-time neighbours, their parents had been friends, and so forth. Sveta could send bigger items, like fridges, to the village when there was free space in the truck. While Sveta's sister does not earn anything from this enterprise, and could use the place in the truck for things for the shop, she had to be cooperative in order to stay on good terms with the neighbours. As Sveta told me: 'It is sometimes difficult to distinguish whether you help your old childhood friend or make business'. To my question, why people preferred to order certain goods from her rather than buying them at the shop, Sveta replied that it was faster and easier to ask someone for that kind of favour: 'I can always run and buy that cupboard or jacket when going to the school or after the lessons. Otherwise it would be on the order list and it could take a few truck loads before it arrives'. She was adamant to continue to work as a school teacher: 'It is good to be in a state job. You have all the bonuses (*lgoty*) that one does not have as a private entrepreneur, such as free trips inside Russia for a vacation, insurance, sick leave compensation and more benefits'.

I met Sergei for the first time in summer 2014. He was a Russian man in his mid-fifties who moved to Yakutia several decades ago. Although retired, he worked in the security service of the summer residence of the President. Because the President uses this only a few times a year, Sergei lived mainly in the village of Bulguniakhtaakh, 110 km from Yakutsk, and engaged in various informal practices. As the village is located on the bank of the Lena river, Sergei organised, for a modest pay, fishing trips to numerous islands on the river. He took people to an island on his motorboat and picked them up after several days. He cleaned up the place after the tourists and kept them supplied with firewood. His house was right at the place where barges stop to pick up passengers who wish to travel to the other side of the river. Many people left their cars for a small sum in front of Sergei's house at the river, and he kept an eye on them. In the wintertime, people left their motorboats in the same place. Sergei told me that he began the business when friends asked him to watch over their vehicles. Now he operates an informal parking lot. In addition, he made some money by selling fuel to passing fishermen, fished himself, and sold his catch in the village. When I asked him how the people from the tax office regarded his activities, he smiled and responded: 'They also fish using my services!'.

These stories suggest that there is a blurred border between money making and favours. Social relations are important for the SMEs of the Republic of Sakha, to the extent where they affect profit. As research shows, maintaining the stability and durability of networks is often preferred to maximum gains from an economic transaction (Lomnitz and Pérez-Lizaur 1989; Schweitzer 1989). As we have seen, for many entrepreneurs in Yakutia, the idea of a nation and its well-being plays a significant role in their economic decisions. For many shopkeepers the decision not to sell alcohol was closely related to the idea that alcohol harms the Sakha people (see closer discussion by Peers and Kolodeznikov 2015). We also saw that entrepreneurs use the risk diversification strategy, being active in different spheres if possible. On the other hand, people are ready to engage in different things as long as these bring in money. And last but not least, many of the above-mentioned cases demonstrate that the state has little interest in the semi-legal activities of the entrepreneurs. The officials avoid taking responsibility for organising and controlling the SMEs, and also give up on the tax revenue that goes together with such control.

Trust, Morality, and Responsibility

Many scholars and commentators admit that Russia is not the friendliest country for entrepreneurship (Agarwal 2007; Berkowitz and DeJong, 2001; Gustafson 1999; Hedlund 1999). Russia is described as a society of strong interpersonal trust-based networks (Gibson 2001) in an economic environment where 'honesty does not pay' (Radaev 2005: 114). On the other hand, small and medium-sized businesses are adept at being engaged in an informal and illegal economy in order to 'escape state regulations on production and transactions' or to absorb risks (Beneria 1989: 184). The academic research shows that entrepreneurs tend to be sceptical about the state, but their scepticism is caused by the system of social norms and not only by the attitudes or politics of state institutions (cf. Nee and Young 1991: 297). In the Republic of Sakha, the relationship between entrepreneurship and the state is, however, unexpectedly calm. In some interviews, I asked the entrepreneurs whether the state tax officials ever show any interest in small enterprises. Marfa noted that the tax officials largely ignore small and medium-sized enterprises. 'They receive a certain percentage from any fine they extract from the enterprises. The situation is that the fines are much bigger when you deal with big companies. In our case it makes a few thousand rubles and is not worth it'. This was also confirmed by the former Minister of Economy of the Republic, Aleksandr Kugaevskii. He explained that by law the tax officials receive a certain financial bonus according to how much they contribute to the republic's budget.

As local economists from the North-Eastern Federal University of Yakutsk who have studied the republic's economy over decades explained to me, such policy should be seen in the context of the economic priorities of the republic. The main attention of the state institutions is on big companies as major cash cows. In the Republic of Sakha, it is mainly big companies that have access to the 'big money' or state orders and subsidies. This is an economic circle, pretty much operating as the classical *blat* system works. State construction or infrastructure contracts are distributed among a small number of big companies who have little interest in collaborating with smaller enterprises. These big companies also have revenues that are interesting for the tax officials. Smaller enterprises, by contrast, are of little interest because they do not produce much, and are just economic interlocutors. At the same time, due to the low purchase power of the population, incomes of small enterprises are insignificant from the point of view of the state (Egorov 2006: 188, 192). The entrepreneurs certainly have to cope with corruption, bribe their way through the bureaucracy, or 'support' a local police officer financially. They, however, have minimal access to state resources, and wish for the minimum attention from the state institutions in return. In my interviews with the owners of small and medium-sized enterprises, I sensed that they felt a certain disappointment about being 'left out' of the circulation of state resources. Local economists avoid direct estimations about the size of the 'shadow economy' in private business (Tarasov et al. 2013) and some of them have told me privately that no one really knows.

My general impression was that the entrepreneurs under study reduced their contact with state structures to a minimum, and in some cases even operated in unmarked buildings without any public advertisement. As the research shows, there may be a correlation between mistrust of the state and freedom of economic operation for SMEs in a post-Soviet setting (Polese and Stepurko 2016) Therefore, the practice applied by entrepreneurs is not 'to secure personal benefits for the members of one's circle at the expense of the state' (Hsu 2005: 321; Peng and Shekshnia 2001) but to avoid utilisation of state resources, and by doing so, to also avoid contact with state institutions. Such a 'guerrilla strategy' helps to maintain entrepreneurial 'flexibility' (Peng and Shekshnia 2001) which is realised by the diversification of risk by expanding into other spheres, and establishing economically non-related new enterprises. For example, combining a laundry with car repairs, or selling food products, yet staying small, helps to keep the business within the family without hiring outside managers. This also keeps the business simple, and open to an unskilled workforce, which helps incorporate trusted workers, primarily from among the clan or kin network. The case studies referred to above (and others not mentioned here) show that the incorporation of new members, or using existing kin-based contacts, is not dissimilar to Chinese *guanxi*.

Sakha kin relationships are sometimes very confusing. In general, Sakha families nurture strong reciprocal ties between different generations, linking relatives living in various villages and towns. The biological kinship network is usually extended by fictive kin, such as the best friend of a daughter, or a schoolmate of a father (cf. Ventsel 2005). For Sakha people, the district of origin is very important. People from one district have a moral obligation to support each other even when they meet each other outside of the district. As a rule, people born in the cities, still identify strongly with a district where their parents or grandparents come from. Traditionally, the first greeting in Sakha among strangers is 'Where do you come from?' In certain situations, this creates a space for manoeuvring because one can have roots in different districts. Therefore, everybody can become 'a de facto kinsman by virtue of a mutual act of choice' (Pitt-Rivers 2016: 444).

The informal practices of local entrepreneurs are impossible to ignore. Supposedly, all people of the republic have used transport services 'without tickets', repaired their property by finding workers through word of mouth, or bought fish or meat from the villages. As Sergei's example shows, even officials have a relaxed attitude to such informal practices. Aleksandr Kugaevskii told me in 2015 that the Ministry of Economy is very well aware of the 'shadow economy' but 'cannot provide employment to these people. Things are slowly changing, though'.

My feeling is that things are not changing due to the current economic crisis in Russia where the informal sector seems to only be increasing. As with Chinese *guanxi*, this happens through the constantly expanding networks, on the basis of finding 'common ground' (Puffer et al. 2010: 455). In contrast to the 'comparatively narrow definition of friendship' in *blat* networks, the *guanxi* practice absorbs new members and there are instrumentalised friendship, kinship, and fictive kin relationships (Hsu 2005: 321–322).

In a classic understanding of the moral economy, described by Thompson (1971) and Scott (1976), moral principles stand opposed to profit-maximising strategies. As we have shown, Sakha entrepreneurial networks stretch geographically over several thousand kilometres and cover certain 'grey areas' (Chang 2011: 328), like running illegal shops or hiring staff without informing the tax officials. These networks, however, contain a certain moral aspect that tends to contradict pure economic calculus. In the case of Alena, Marfa, Sveta, and Ivan, entrepreneurship involved a strong ideology of kinship and a certain reciprocity. In order to keep the business predictable and stable, they preferred to earn less profit. The obligation of responsibility for kin people's well-being is especially obvious in Alena's fishing enterprise, but is also evident in other cases. Most interesting is the decision not to sell alcohol by the shop owners. In general, the entrepreneurs were committed to what they believed were the community's and even national interests. All entrepreneurs could have made huge profits by selling alcohol either legally or illegally, as some others do in northern villages. Refusing to sell alcohol is a powerful statement. Alcohol consumption in indigenous communities is a topic which is hotly debated in the Republic of Sakha, and is connected to the values and discourses pertaining to the 'health of a nation' and 'maintaining the traditional way of life' (Krivoshapkin-Aiynga 1998; Lapparova 2002; Tyrylgin 2000). This merits more research, which goes beyond the scope of this chapter, but here 'supporting ZOZh' means adopting a larger framework of moral categories and shaping business strategies accordingly. In some cases, such as that of Ivan from Saidy, it means following the community's decision to forbid the sale of alcohol in the village, but in all cases opposing the sale of alcohol signals a commitment to shared values and a readiness to reject 'dirty money' which harms the community. Later on, this symbolic capital can be converted into economic capital, when a community supports the semi-legal business of the entrepreneur. Similar to Brantstädter's (2003) case of the moral economy in Chinese villages, following ZOZh is a ritualistic contribution to the sustainability of the community and shows the responsibility of the entrepreneur.

In the case of the Sakha, as in *guanxi* (Strickland 2010: 111), 'small gestures' are also slowly converted into monetary transactions. Marfa, Alena, and Sveta were engaged in a quasi-charity-like business model, offering employment and services to pensioners and relatives. Sergei was constantly expanding his various activities by charging a small amount of money for fish, or security services. At the same time, this moral aspect guarantees them loyal customers and a general silence in the community, even among the state officials as we have seen in Sergei's case. While his son-in-law was actually an outsider in the closely knit Arctic community, Ivan being 'embedded' in the village allows him to run an illegal business and earn money there.

Another expression of trust is the recruitment of new employees using the network. This is based on a code similar to *guanxi*, where a friend is a good source to find a trusted employee (Puffer et al. 2010: 454–456). This contradicts *blat*, where friends of friends are treated with suspicion and such links tend to be used in a more contractual manner, i.e. favours and their prices are negotiated in advance (see Humphrey 2012; Ledeneva 2008). The flexible border between friendship and kinship helps to expand the pool of employees, but also imposes certain obligations on the entrepreneur. Alena has little choice but to operate the enterprise because this is linked to the heritage of her mother, and provides regular income to her northern relatives. As is the case with the majority of ethnic Sakha, she has limited control over who was going to visit her in Yakutsk and ask for a place to sleep in her small apartment. As a head of the enterprise, she has to house and feed relatives.

Conclusion

The research on SMEs often stresses the importance of the social network and ethnic loyalty. The research on this topic demonstrates that the decision to operate 'between illegality and informality' is a conscious move from the side of the entrepreneurs (cf. Morris and Polese 2014). The case of SMEs in the Republic of Sakha differs from the case of Croatia (see Chapter by Banović et al. of this volume) because ethnic-based networks do not stretch beyond the region. Sakha entrepreneurs operate in their own homeland but also in a multi-ethnic setting. The ethnic component is in many respects related to the idea of morality, to the understanding that trade and entrepreneurship create a sense of well-being for certain social groups or the entire nation. Entrepreneurs prioritise the social and moral aspect of their endeavour. They provide jobs for disadvantaged groups, such as pensioners, and often abdicate the profit they could earn selling alcohol. As a rule, Sakha extended families are big, and kin networks expand over the vast territory of the republic. It is no secret that these networks function as economic networks (Ventsel 2005, 2011), where trust, obligation and rights are essential. The orientation towards family makes such economic activities less profitable but also fosters more cooperation and takes into account the interests of other kinsfolk. The entrepreneurs rely increasingly on informal, fictive, and extended kinship networks that have a strong element of moral economy. By providing jobs for kinspeople or pensioners, they also see these steps as a moral obligation to support close and distant relatives, or neglected groups in the society. Further, small and medium businesses rely significantly on constantly expanding flexible networks in attracting both customers and workers. One important strategy among such entrepreneurs is limiting the growth of the enterprise and dispersing their finances into different spheres whenever possible. This is a key part of maintaining direct control over the enterprise and preventing middle-level managers.

In order to understand the nature of the small-scale business, one should also have some understanding about the regional economy in the Republic of Sakha. This is an environment where, notwithstanding the rhetoric, the regional government shows little interest in supporting SMEs. The focus of local economic policy is on big companies who make up the lion's share of the local budget. Therefore, small enterprises, despite the fact they provide jobs for a significant part of the local population, are left to their own devices.

The response of a small and medium business is to steer clear of the state officials as much as possible. The entrepreneurs have developed strategies that circumvent state regulations but also forfeit access to the state's finances. The concept of *blat*, that is so popular in the academic discourse pertaining to the Russian economy of favours, has little relevance in examining SMEs in the Republic of Sakha. *Blat* means an informal economic relationship between entrepreneurs and state structures so that the former have access to state money and other resources. The SME in Sakha does not aspire to have their cut from the cake, meaning that they also have little obligation to pay taxes or conform to other laws. It could be even said that some entrepreneurs see the state as the enemy.

Acknowledgements This research was supported by the institutional research funding IUT34-32 of the Estonian Ministry of Education and Research.

Bibliography

Agarwal, M. (2007). "International Economic Management: Concerns of China, India, Russia" *China Report*, 43(2): 163–173.

Barsukova, S., & V. Radaev. (2012). "Neformal'naia ekonomika v Rossii: kratkii obzor [Informal Economy in Russia: Short overview]" *Ekonomicheskaia sotsiologiiq*, 13(4): 99–11.
Beneria, L. (1989). "Subcontracting and Employment Dynamics in Mexico City" In A. Portes, M. Castells, and L. A. Benton (eds.) *The Informal Economy. Studies in Advanced and Less Developed Countries*. Baltimore and London: The John Hopkins University Press, 173–187.
Benton, L. (1994). "Beyond Legal Pluralism: Towards a New Approach to Law in the Informal Sector" *Social and Legal Studies* 3: 223–242.
Berkowitz, D., & D. N. DeJong. (2001). "Entrepreneurship and Post-Socialist Growth" *William Davidson Working Paper* No. 406.
Brandtstädter, S. (2003). "The Moral Economy of Kinship and Property in Southern China" In C. Hann and "Property-Relations"-Group (eds.) *The Postsocialist Agrarian Question. Property Relations and the Rural Condition.* Münster: LIT Verlag, 419–440.
Cartwright, A. L. (2001). *The Return of the Peasant. Land Reform in Post-Communist Romania.* Aldershot, Burlington USA, Singapore, Sydney: Ashgate, Dartmouth.
Chang, K.-C. (2011). "A Path to Understanding Guanxi in China's Transitional Economy: Variations on Network Behavior" *Sociological Theory*, 29(4): 315–339.
Donovan, M. G. (2008). "Informal Cities and the Contestation of Public Space: The Case of Bogotá's Street Vendors, 1988—2003" *Urban Studies*, 45: 29–51.
Egorov, E. G. (2006). *Sever Rossii: ekonomika, politika, nauka* [The Russian North: Economy, Politics, Science]. Yakutsk: Sakhapoligrafizdat.
Estrin, S., R. Aidis & T. Mickiewicz (2006). "Institutions and Entrepreneurship Development in Russia: A Comparative Perspective" *William Davidson Institute Working Paper* No. 867, February 2007.
Gambold Miller, L. L. (2003). "Interdependence in Rural Russia: The Post-Socialist Mixed Feudal Economy" *Max Planck Institute for Social Anthropology Working Papers* No. 51. Halle/Saale: Max Planck Institute for Social Anthropology.
Gibson, J. L. (2001). "Social Networks, Civil Society, and the Prospects for Consolidating Russia's Democratic Transition" *American Journal of Political Sciences*, 45(1): 51–69.
Gustafson, T. (1999). *Capitalism Russian-Style.* Cambridge, New York: Cambridge University Press.
Hann, C. M. (2000). "The Tragedy of the Privates? Post-Socialist Property Relations in Anthropological Perspective" *Max Planck Institute for Social Anthropology Working Paper*, 2(1–23).

Hann, C. M. (2003). "Introduction. Decollectivisation and Moral Economy" In C. Hann and "Property-Relations"-Group (eds.) *The Postsocialist Agrarian Question. Property Relations and the Rural Conditions*. Münster: LIT Verlag, 1–46.

Heady, P. (2003). "Kinship and Economy in the Russian Countryside. A Provisional Model" In P. Heady and H. Grandits (eds.) *Distinct Inheritances. Property, Family and Community in a Changing Europe*. Münster: LIT Verlag, 297–312.

Hedlund, S. (1999). *Russia's "Market" Economy: A Bad Case of Predatory Capitalism*. London: UCL Press.

Hsu, C. L. (2005). "Capitalism Without Contracts Versus Capitalists without Capitalism: Comparing the Influence of Chinese Guanxi and Russian Blat on Marketization" *Communist and Post-Communist Studies*, 38(3): 309–327.

Humphrey, C. (2000). "An Anthropological View of Barter in Russia" In P. Seabright (ed.) *The Vanishing Rouble: Barter Networks and Non-Monetary Transactions in Post-Soviet Societies*. Cambridge, New York: Cambridge University Press, 71–92.

Humphrey, Caroline. (2002). *The Unmaking Soviet Life. Everyday Economies After Socialism*. Ithaca and London: Cornell University Press.

Humphrey, Caroline. (2012). "Favors and 'Normal Heroes'" *Hau: Journal of Ethnographic Theory*, 2(2): 22–41.

Ibañez-Tirado, Diana, and Magnus Marsden. (2020). Trade 'Outside the Law': Uzbek and Afghan Transnational Merchants Between Yiwu and South-Central Asia. *Central Asian Survey*.

Köllner, T. (2013). *Practising Without Belonging? Entrepreneurship, Morality, and Religion in Contemporary Russia*. Berlin, Münster, Wien, Zürich, London: LIT Verlag.

Krivoshapkin-Aiynga, A. (1998). *Evraziiski soiuz* [Eurasian Union]. Yakutsk: Bichik.

Kuznetsov, A., & O. Kuznetsova, O. (2005). "Business Culture in Modern Russia: Deterrents and Influences" *Problems and Perspectives in Management*, 2(25–31).

Lapparova, I. F. (2002). "Sotsial'nye posletsviia promyshlennogo osvoeniia Yakutii v raionakh prozhivaniia korennykh malochislennykh narodov Severa [Social Consequences of the Industrial Mastering in Yakutia in the Districts of Compact Settlement of Small Indigenous People of the North]" In G. A. Ivanov, T. E. Andreeva, and S. A. I. (eds.) *Malochislennye etnosy Severa Respubliki Sakha (Yakutiia) glazami molodykh uchenykh*. Yakutsk: IPMNS SO RAN, 15–20.

Ledeneva, A. V. (1998). *Russia's Economy of Favours. Blat, Networking and Informal Exchange*. Cambridge: Cambridge University Press.

Ledeneva, Alena. (2000). "Shadow Barter: Economic Necessity or Economic Crime?" In P. Seabright (ed.) *The Vanishing Rouble: Barter Networks and Non-Monetary Transactions in Post-Soviet Societies*. Cambridge, New York: Cambridge University Press, 298–317.

Ledeneva, Alena. (2006). *How Russia Really Works. The Informal Practices That Shaped Post-Soviet Politics and Business*. Ithaca and London: Cornell University Press.

Ledeneva, Alena (2008). "'Blat; and 'Guanxi': Informal Practices in Russia and China" *Comparative Studies in Society and History*, 50(1): 118–144.

Lomnitz, L. A., & M. Pérez-Lizaur. (1989). "The Origins of the Mexican Bourgeoisie: Networks as Social Capital" In T. Schweitzer (ed.) *Netzwerkanalyse. Ethnologische Perspektiven*. Berlin: Dietrich Reimer Verlag, 35–46.

Meagher, K. (2006). "Social Capital, Social Liabilities, and Political Capital: Social Network and Informal Manufacturing in Nigeria" *African Affairs*, 105/421: 553–582.

Morris, J. (2016). *Everyday Post-Socialism. Working-Class Communities in the Russian Margins*. London: Palgrave Macmillan.

Morris, J., & A. Polese, eds. (2014). *The Informal Post-Socialist Economy. Embedded Practices and Livelihoods*. London and New York: Routledge.

Nee, V., & F. W. Young. (1991). "Peasant Entrepreneurs in China's 'Second Economy': An Institutional Analysis" *Economic Development and Cultural Change*, 39: 293–310.

Oka, N. (2015). "Informal Payments and Connections in Post-Soviet Kazakhstan" *Central Asian Survey*, 34(3): 330–340.

Peers, E., & S. Kolodeznikov. (2015). "How to Enjoy a Teetotal All-Night Party: Abstinence and Identity at the Sakha People's Yhyakh" *Folklore*, 61: 117–134.

Peng, M. W., & S. V. Shekshnia. (2001). "How Entrepreneurs Create Wealth in Transition Economies" *The Academy of Management Executive*, 15(1): 95–110.

Pitt-Rivers, J. (2016). "The Paradox of Friendship" *Hau: Journal of Ethnographic Theory*, 6(3): 443–452.

Polese, A. (2014). Drinking with Vova: An Individual Entrepreneur Between Illegality and Informality. In J. Morris and A. Polese (eds.) *The Informal Post-Socialist Economy. Embedded Practices and Livelihoods*. London and New York: Routledge, 85–101

Polese, A., & T. Stepurko. 2016. In Connections We Trust. *Transitions* (14.04.2016): 1–4.

Puffer, S. M., D. J. McCarthy & M. Boisot. (2010). "Entrepreneurship in Russia and China: The Impact of Formal Institutional Voids" *Entrepreneurship Theory and Practice*, 34(3): 441–467.

Radaev, V. (1998). *Formirovanie novykh rossiiskikh rynkov: transaktsionnye izderski, formy kontrolia i delovaia etika* [Formation of New Russian Markets:

Transaction Costs, Forms of Control, and Business Ethics]. Moscow: Tsentr politicheskikh tekhnologi.
———. (2005). "Establishing Trust in a Distrustful Society: The Case of Russian Business" In H. Hohmann and F. Welter (eds.) *Trust and Entrepreneurship: A West-East Perspective*. Cheltenham, UK: Edward Elgar, 114–135.
Sakaeva, M. M. (2012). "Parlament kak "okno vozmozhnostei": issledovanie povedeniia predprinimatelei s deputatskim mandatom v khode realizatsii rynochnyx interesov" [Parliament as a 'Window of Opportunities': A Research on the Behaviour of Entrepreneurs with a Seat in the Process of the Fulfilment of Market Interests]. *Ekonomicheskaia sotsiologiia*, 13(3): 96–122.
Schweitzer, T. (1989). "Netzwerkanalyse als moderne Strukturanalyse" In T. Schweitzer (ed.) *Netzwerkanalyse. Ethnologische Perspektiven*. Berlin: Dietrich Reimer Verlag, 1–34.
Scott, J. C. (1976). *The Moral Economy of the Peasant. Rebellion and Subsistence in Southeast Asia*. New Haven and London: Yale University Press.
Shlapentokh, V. (2007). *Contemporary Russia as a Feudal Society. A New Perspective on the Post-Soviet Era*. New York, Houndmills: Palgrave Macmillan.
Ssorin-Chaikov, N. (2000). "Bear Skins and Macaroni: The Social Life of Things at the Margins of a Siberian State Collective" In P. Seabright (ed.) *The Vanishing Rouble: Barter Networks and Non-Monetary Transactions in Post-Soviet Societies*. Cambridge: Cambridge University Press, 345–361.
Stepanova, N. A., & R. R. Nogovitsyn. (2011). *Maloe predprinimatel'stvo severnogo regiona. v usloviakh perekhoda na innovatsionnyi put' razvitiia* [Small-Size Entrepreneurship of a Northern Region in the Conditions of Transition to an Innovative Path of Development]. Yakustk: Sfera.
Strickland, M. (2010). "Aid and Affect in the Friendships of Young Chinese Men" *Journal of Royal Anthropological Institute*, 16 (102–118).
Tarasov, M. E., E. G. Egorov, & G. P. Kulakovskii. (2013). *Vliianie tenevoi ekonomiki na ekonomicheskuiu bezopasnost' regiona* [The Impact of the Shadow Economy on the Economic Security of a Region]. Yakutsk: Severo-Vostochnyi federal'nyi universitet.
Thompson, E. P. (1971). "The Moral Economy of the English Crowd in the Eighteenth Century" *Past and Present* (50): 76–136.
Turaeva, Rano 2014. "Mobile Entrepreneurs in Post-Soviet Central Asia" *Communist and Post-Communist Studies*, xxx (1–10).
Tyrylgin, M. A. (2000). *Istoki fenomenal'noi zhiznesposonosti naroda sakha* [The Sources of Phenomenal Resilience of the Sakha Nation]. Yakutsk: Bichik.
Ventsel, A. (2003). "People Won"t Dance if They have Nothing to Eat: Do Economic Transformation and Centrally Planned Cultural Revival Fit Together?" *Polar Geography*, 27(2): 121–135.

Ventsel, A. (2005). *Reindeer, Rodina and Reciprocity: Kinship and Property Relations in a Siberian Village*. Vol. 7. *Halle Studies in the Anthropology of Eurasia*. Berlin: LIT Verlag.

Ventsel, A. (2010). "Creating Traditional Economy: State Reindeer Economy and Preservation of Ethnic Groups" *Review on Siberian Studies at Max Planck Institute for Social Anthropology, Ethnic Publishing House* (8): 78–103.

Ventsel, A. (2011). "Siberian Movements: How Money and Goods Travel in and out of Northwestern Sakha" *Folkore* (49): 113–130.

Ventsel, A. (2017). "Reluctant Entrepreneurs of the Russian Far North" In M. Laruelle (ed.) *New Mobilities and Social Changes in Russias Arctic Regions*. London and New York: Routledge, 45–56.

Azerbaijani Meykhana: Cultural Policy and Local Actors' Agenda

Aneta Strzemżalska

Introduction

Music plays an increasingly important role in national building in contemporary Azerbaijan (Naroditskaya 2002; Oldfield 2008; Huseynova 2016; Dessiatnitchenko 2017). Official authorities conduct an active cultural policy aimed at promoting the republic by popularizing musical phenomena valued by the local population (Strzemżalska 2020). Some of the ideas put forward by representatives of the state meet with the interest of citizens and affect their involvement in the branding of selected musical genres. On the other hand, some phenomena meet with the lack of approval of the state, which aims to change its social meaning, and even to disappear from the cultural space of the country.

Meykhana (meyxana) is an example of a musical genre in which the government and society find their common goals. Over the past few decades, meykhana gained great popularity and shifted from the cultural peripherals of Baku suburban areas toward the most prestigious locations in the capital of Azerbaijan. Due to the social boost given to the meykhana

A. Strzemżalska (✉)
Department of Anthropology, European University, St. Petersburg, Russia

A. Polese (ed.), *Informality, Labour Mobility and Precariousness*, International Political Economy Series,
https://doi.org/10.1007/978-3-030-82499-0_8

phenomenon, it has affected its previous uniqueness. Consequently, at least two categories can be distinguished: classical and musical meykhana.

Classical meykhana represents recitative competitions without musical accompaniment. The rhythm is beat out by tapping fingers against a table, finger-snapping, or by using a percussion instrument. The essence of the performance lies in the spontaneous composing of poems by a few or sometimes more than a few performers—meykhanachi (meyxanaçi). Their aim is to not only give an opinion about political and social events and everyday problems but to give an opinion in a joking manner. The linguistic rules necessary for the literary language are fading to the background during the competitions due to their hindering of the freedom of expression. Meykhanachis who are able to deliver the deepest expressions *ad lib* enjoy renown. The renown among the local population is of different kind to renown from the official authorities.

Classical meykhana is rare today, it is primarily spotted among the male population of the Absheron Peninsula (Baku and its neighborhoods), where it originated. In the remaining regions of Azerbaijan, musical meykhana has been popularized over the past three decades. Musical meykhana is a verbal contest accompanied by a synthesizer and sometimes other instruments. The new forms of the performances involve women and meykhana is performed not only in teahouses and weddings, as was the case in the classical form, but also on the radio (meykhana is read on public holidays), films and, most importantly, television. Alongside the increasing prestige of the phenomenon, recitative improvisations are considered as one of the genres of folk music by the musical community and the state. That is, they received the status of a national brand, a visiting card of Azerbaijan.

Considering the abovedescribed diversity of the cultural phenomenon, in this chapter, the question about under what political, social, and economic circumstances oppositional, underground art become the country's calling card is asked. Paying attention to the relationship between what is offered by the government and how meykhana is practiced and perceived by ordinary people, this paper examines how the interplay between formal and informal approaches can be used in study on national building process. Assuming that by not controlling or regulating certain social practices, the state sets the conditions under which informal practices can flourish (Haid 2017: 290), in this paper, I will show that informality has a concrete impact on top-level politics, including cultural policy. Meykhana for many years associated with creativity opposing the state, at the peak of its popularity, found itself in the spectrum of interest of official authorities. On the one hand, there were effective attempts to subordinate meykhana performers to the state's interests, that is, to promote meykhana as one of the main genres of folk art. On the other hand, meykhanachis, in order to improve their social status, took a number of steps to institutionalize and standardize their creativity, which coincided with the official authorities' desire for meykhana to become the country's calling card.

This chapter is organized as follows: at the beginning, the theoretical basis of the study is presented. That is, by mentioning the links between nationalism and cultural policy, the main directions of the latter are explained. In this regard, it is proceeded with the belief that the main method of state intervention in the musical culture of Azerbaijan is folklorization, which, through institutionalization and standardization, leads to the formation of a national culture using folk art. However, the government is not the only factor in determining the peculiarities of nationally specific genres. Local players, sometimes for their own purposes, and sometimes in the response to specific government actions, take specific measures to change the social status of meykhana. The variety of these practices is described in the last three subsections of this chapter, after a historical overview of the political dimension of meykhana.

The paper is based on the results of my fieldwork which I conducted in Baku and its environs from April 2012 to November 2016 (ten months in total). Initially, as a field method I used informal interviews in Russian and Azerbaijani languages with performers and listeners of meykhana, scientists, publicists, employees of cafes where performances take place, as well as with employees of government agencies responsible for cultural development programs in the country; I initially gave preference to interviews with key informants and collected 48 interviews. However, at some point, I realized that the dialogue with my informants was largely conditioned by the stereotypical ideas about meykhana widespread in the public space, over time I began to give preference to participatory observation. For a period of several weeks, I accompanied my informants during meykhana performances: I went to teahouse with them, I went to weddings and to the TV studio, I took part in rehearsals and social meetings with meykhanachis' friends and colleagues, during which the topic of meykhana was also discussed. Extracts from my field diary, written up each evening based on shorter notes made during the day, are quoted in the text below. Participatory observation, despite the fact that it has its limitations (meykhana performances to a large extent take place in a purely male environment and the performers and listeners of meykhana had to get used to my presence), I consider it the most effective, since it allows one to explore social phenomena such as they really are—without repeating ideologically processed discourses.

Informality, Nationalism, and Cultural Policy

Cultural diversity (ethnic, religious, and linguistic) as well as social layering of the countries that emerged as a result of the collapse of the USSR attract the attention of many researchers. Most of them, focusing on the study of official discourses, argue that nationalism in the post-Soviet space is that which refers to communities united by a common culture and differing from rival or hostile communities by differences in culture (Gellner 1983). However, primarily in countries with an authoritarian political system, where official discourse serves to create an imaginary reality, it happens that the most detailed analysis of all dominant narratives does not describe what is actually happening (Brubaker 1996; Polese and Horák 2015; Fox and Miller-Idriss 2008; Hobsbawm 1992; Isaacs and Polese 2015).

The smoothing out of reality occurs due to the fact that each nation is very multilayered, characterized by a large number of political players; each of them has their own interests and preferences that are causing tearing in the social fabric of the state and society (Jourde 2009). In this context, it is useful to draw attention to the political dimension of informality, which in this paper is understood as the capacity to complement formal rules and institutions when these do not take into account all the needs of a society (Polese 2015). Bearing in mind the complementarity of research on nationalism and everyday life, I move beyond dichotomizing formal and informal practices and begin the search for the patterns of coexistence and even collaboration between formal and informal practices. Moreover, a useful approach to the study of cultural space in all its diversity is focusing on the type of practices and actions that remind us of our national identity on a daily basis and strengthen the sense of national unity (Billig 1995; Edensor 2002; Skey 2011; Isaacs and Polese 2016; Fox 2016; Polese et al. 2018; Antonsich 2015; Pawłusz and Seliverstova 2016). These practices include the non-institutional activities of representatives of local players in the context of official cultural policy implementation (Schatz 2009). It is especially noticeable in communities with strong internal antagonisms, both social and political, when "different groups argue about how "we" should think about "us" and what should be "our" national goal (Reicher and Hopkins 2001).

The national idea determines the cultural policy of many countries, including Azerbaijan—the social fabric of society and its institutions are organized around it (Саламзаде and Абдуллаева 2019). The most important lead in terms of conceptualizing and implementing the republic's cultural policy is the Heydar Aliyev Foundation (Heydər əliyev Fondu). It was created in 2004 by the First Lady of Azerbaijan Mehriban Aliyeva (Mehriban əliyeva, 1964)[1] and today it predetermines the decisions of such institutions as the Ministry of Culture (Azərbaycan Respublikası Mədəniyyət Nazirliyi) and the Ministry of Youth and Sports (Azərbaycan Respublikası Gənclər və İdman Nazirliyi).

Projects aimed at popularizing ethnic music, that is, art associated with the musical traditions of a certain people, rank highest in the cultural policy of the Heydar Aliyev Foundation. It covers both traditional folk music, which is authentic—with original scales and instruments musical creativity, and folk music. The last one, also defined as "contemporary folk music," refers to popular music with pronounced elements of ethnic music. It is performed by both educated professional musicians and amateurs and, contrary to traditional music, is not exclusively acoustic. Its main difference from pop music lies, among other things, in the fact that it relies on folk traditions. It is above all created from the incentive to preserve these traditions (Ruehl 2018).

In this paper, tradition is intended to mean "invented tradition," that is a set of social practices of a ritual or symbolic nature, the connection of which with the historical past is mostly fictitious, but at the same time. they are perceived as unchanging and have always existed (Hobsbawm 2003). It allows for the management of identity construction, which often happens at the state level (ibid.: 9–11). Emphasizing the deep ethnic roots of modern nations and the belief that nationalists inherit myths and symbols from their ethnic ancestors led some scholars to use the term "ethnic tradition" in the meaning of "ethnic culture" (Armstrong 1982; Smith 2000). The term "culture" began to denote the phenomenon itself, and "tradition" started to be the mechanism of the phenomenon's functioning, a network (system) of connections between the present and the

[1] Since 2017, Mehriban Aliyeva has been the First Vice President of the country and the second most powerful person in the country.

past, with the help of which a certain selection of what is important for the creation of a national culture and thus for a national tradition is made (Чистов 1981: 106). At the same time, it is worth taking notice of the tradition's changing nature (Shils 1981) and the fact that tradition in conservative societies serves not only as a symbol of continuity, but also as a determinant of the limits of innovation, as the main measure of innovation's legitimacy, and also as a measure of social activity (Eisenstadt 1973: 51–52).

As a result of the implementation of the abovedescribed cultural policy in Azerbaijan, there is a coincidence of the concepts of "traditional," expressed in ethnocultural terms, and "national," which largely contributes to the representation of national identity (Spinetti 2005: 185–211). Consequently, any performers that wish to receive official recognition, are forced to rely on the traditions of folk art in their works. In this way, a politicized adaptation of folk art takes place, that is, folklorism (Stites 2004: 25). Folklorization, in turn, refers to government's way of controlling ethnic music, often through a combination of coercion and persuasion (Wong 2001: 249). It should be understood as a kind of nationalization, a situation where the rejection of the balance between the centralized policy of cultural construction and everyday musical practices leads to an almost total control of culture by the state and pro-government structures (Williams 1982: 158). The nationalization of musical art is accompanied by its institutionalization and standardization. Both processes, on the one hand, are making folk art professional, on the other hand, they are associated with fundamental changes in the artistic qualities of music and the context of performance, as well as a different perception of art by the audience.

Political Dimensions of Meykhana

Referring to both the memoirs of sources and the written records of meykhana, we can say that throughout the twentieth century, it was distinguished by a strong political engagement, although its ideological position in terms of support or opposition to the official authorities changed. The first written information about the genre, which dates back to the beginning of the twentieth century (10–30s), raises questions of the sociopolitical life of the second decade of the twentieth century, evaluates the historical events of that period and the actions of the characters (Imamaliyev 1993). For example, the text titled "Do You Know"

("Bacara bilmə sən") mentions the names of public figures of that time. It also stages the relations between Azerbaijan, Iran, Russia and Turkey (ibidem).

Similar description can be given to the verbal recitations published a little later, that is, from 1919 to 1930 (Səfərov 1925; Anatollu 1925; Anadollu 1927, 1930; Cabarzadə 1927). A common feature of these works lies in talking about local social realities: bribery, inflation, food and gas prices. They also implicitly emphasized the role of meykhana in the formation of Azerbaijani literacy and identity. For example, for this purpose, the authors, describing in detail the life of meykhanachi Mirza Bagir (Mirzə Bağır, 1810–1882), reported that at the age of eighteen, he was composing a sociopolitical bedie[2] about the division of Azerbaijan into southern and northern territories (Tağısoy and Zakariyya 2011).

Meykhana particularly enjoyed popularity as a national art during the Great Patriotic War, during the life and work of one the most important Azerbaijani meykhanachi Aliagha Vahid (əliağa Vahid, 1895–1965). In October 1938, Vakhid was appointed the literary director of the State Collective Farm Theater. Together with other professional meykhanachi, he performed miniature plays in which sociopolitical problems in the country were presented in a joking manner. Examples of this activity are depicted in the iconic Azerbaijani film about the life and work of Aliagha Vahid called "Gezelkhan" (Qezelxan, 1991, dir. Shahmar Alekperov). Here is an excerpt from a meykhana propaganda piece from this film:

> Hitler: You can't find a bad person like me in the whole world
> I have destroyed all the German people
> That's how terrible I am
> Like a dog I'm sorry for what I did
> People: Country couldn't care less about Hitler's conscience
> Fear settled in Hitler's soul
> Hitler: I used to beat myself in the chest, brag
> I was beaten like a donkey eating flour
> I'm not who I was, I'm an animal
> Like a dog I regret what I did

Consequently, meykhana of the first half of the twentieth century was part of the official cultural policy, had a propagandistic character and

[2] Bedie (bədihə)—another name for meykhana. See below for further discussion on this topic.

social background, and thereby served to form and strengthen the Azerbaijani nation. Due to the fact that the content of the works of that time presupposed social criticism, the general assessment of meykhana as a phenomenon was positive. The same can be said about the image of performers who were perceived as national heroes, fighters for the modernist ideas of state development.

The abovedescribed direction of meykhana development does not last long. After the war, the promotion of national art as a tool for the formation of local identity came into conflict with the communist merger politics, which was supposed to lead to a "complete unity of nations." Subsequently, the opportunities for the promotion of Azerbaijani art were not only reduced: on the one hand, it started to be considered a threat to the victory of communism, on the other hand, the Soviet government used local forms of art to realize its ideological goals. For example, in several films of that time, such as "I Remember You, Teacher" (Bizim Cəbiş müəllim, 1969, dir. Həsən Seyidbəyli) or "Indomitable Chicken" (Dəli Kür ilm, 1969, dir. Hüseyn Seyidzadə), meykhana is depicted as primitive street art or as "backward" ideas of a religious nature.

Despite the fact that at the official level meykhana was known as a semi-forbidden genre and evidence of cultural backwardness, the memories of the meykhanachi hero were still alive in the memory of the population. In this regard, the image of a performer of verbal recitations of the Soviet era is ambiguous: he is a vagabond, a dissident, and at the same time a champion of justice.

"Who were they (meykhanachi – AS) in Soviet times? These were underground people who, in general, loafers and used drugs. They did not fit the image of a Soviet person in any way. Roughly speaking, they were parasites. Therefore, they were heroes for the young. These were reckless heroes, and this, as it were, connected them with Sufism, where they play with their lives like with fire. They are just geniuses who speak to the rhythm about what others are afraid to voice,"—stated one source (M., m., 55 years old, musicologist).

The most popular meykhanachi hero of the Soviet period and the first years of the existence of independent Azerbaijan was Nizami Remzi (Nizami Rəmzi, 1947–1997). During his performances, he often touched upon moral issues, especially the unfair attitude of the official authority toward ordinary people, as well as everyday problems, which were not spoken about openly during that time. For these reasons, one of my sources equated Remzi's work with social and moral ditties, "free improvisation": "There was no glorification of the party, and this alarmed the Soviet leaders, but it was interesting for us to listen to" (N., M., 56 years old, teacher).

Remzi's work was very popular, not only because he raised relevant issues, but also because he did it in an attractive way. In the 1980s, the musician began performing meykhana, accompanied by original music. Today such art is called "musical meykhana" and is very widespread, but it was a real novelty back then. The show, during which the musician openly criticized the official authorities, attracted a mass audience. He gave concerts in such prestigious institutions as the Azerbaijan State Philharmonic Hall and the Palace of the Republic (now the Heydar Aliyev Palace). According to my sources, government officials "asked him to halt," but "it was impossible to stop Remzi, because he believed that the essence of meykhana was to touch upon politically and socially important issues and he could not compose it in any other way" (F., m., 58 years old, journalist).

Due to the unbridled love for freedom of speech, the performer did not enjoy his fame for long. Remzi, along with his friend, the talented meykhanachi Kebir Azeri (Kəbir Azəri, 1960–1997), died in a car accident. The reasons for the accident remain a mystery to this day, but many of my sources consider this incident not accidental: "It was after our Baba[3] ascended the throne. Nizami spoke at the New Year's program and insulted him. And the people said that he was punished because he criticized the authorities" (K., m., 52 years old, director).

Whether Remzi's death was an accident or a premeditated murder (we'll probably never know it), meykhanachi died like a political hero—the true defender of freedom of speech. Thanks to this, by some Azerbaijanis meykhana is considered not only independent, but even oppositional creativity. After Remzi's death, another group of performers took the central place among meykhan's performers, which, although independent for many years, under political pressure quickly began to cooperate with the official authorities.

[3] This refers to the former president of Azerbaijan, Heydar Aliyev, who ruled the country in 1993–2003.

State Control Attempts of Meykhana. Mashtaga Case

The second and strongest group of meykhanachi (both socially and politically, as well as in terms of the objective number of its members), which until the mid-90s of the last century was characterized by musical practices opposed to the official authorities, but whose professional success is different to the fate of Remzi, are the performers from the Absheron village of Mashtaga (Maştağa). During the Soviet era, this urban-type settlement located 35 kilometers from Baku was distinguished by economic self-sufficiency. Its inhabitants lived in comfort as a result of income from trade in the local bazaar, the largest in the eastern part of Azerbaijan. They also grew and sold flowers, vegetables, and fruit in Russian bazaars (M., m., 62, journalist). Economic independence created opportunities for the residents of Mashtaga to compose meykhana without censorship. Since the performers "spoke the truth," that is, among other things, criticized the Azerbaijani authorities, the local population assessed them as "fair."

Meykhanachi, as "defenders of justice," stood high in the social structure of Baku villages. People turned to them for advice and even asked them to play the role of alternative judges in contrast to state structures, the source recalls (T., m., 70 years old, cultural studies scholar).[4]

A similar situation was encountered during my fieldwork. While waiting for a meeting with the Khirdalan akhoond,[5] who was also valued by local performers as a connoisseur of meykhana and poetry, I had the opportunity to observe how he consulted a family whose member was imprisoned for tax evasion. The clergyman negotiated legal advice with local aksakals,[6] who included meykhanachi. The accusations against the arrested person were considered "slander," and official legislation was

[4] A similar social structure exists in Georgia. See in more detail: Voell, S. (2015) Moral Breakdown among the Georgian Svans: A Car Accident Mediated between Traditional and State Law in State and Legal Practice, in *The Caucasus: Anthropological Perspectives on Law and Politics*, edited by S. Voell, and I. Kaliszewska. Farnham: Ashgate, 95–112.

[5] Khirdalan (Xırdalan) is a city located in the western part of the Absheron Peninsula, 5 kilometers northwest of Baku. About 10 thousand inhabitants.

Akhoond is a Muslim scholar, his status is at the level of the imam of a mosque, although it happens that he is the head of a city or region's Muslims.

[6] Aksakal (ağsaqqallı, literally white-bearded)—in the Caucasus and among the Turkic peoples in Central Asia, a venerable elderly man, an elder.

perceived by them as "unnecessary, artificial obstacles to running private business" (F., m., 47, entrepreneur). Since the "victims" were convinced that the representatives of the official instances were not able to "understand the intricacies of the case," they looked with hope at the legal and moral support that the akhund, meykhanachi, and other aksakals were providing for them.

The above case demonstrates that the opinion of meykhanachi influences the public sentiments of ordinary residents of the Absheron neighborhood units. Due to the strong social position of the artists, from the very beginning of the existence of independent Azerbaijan, the official government pursued a policy aimed at winning the favor of the genre performers, which is most important for the Baku villages.

After several years of enmity: the inhabitants of Absheron villages did not want to obey the political and legal rules of official Baku, and the latter, blackmailing them, closed the gas supplies. The turning point of the conflict was the meeting of the leader of the Mashtaga meykhanachi, Aghasalim Childagh (Ağasəlim Çıldaq, 1930 k–2008) with the head of the republic, Heydar Aliyev, in the mid-90s of the last century. Over the course of the meeting, the performer "read a piece in honor of the ruler [of the country]" (U., m., 77 years old, journalist). The President of Azerbaijan, in turn, "expressed gratitude to Aghasalim and through him to the entire genre of verbal recitative," recalls a witness of the event (ibid.).

The consequences of this meeting can be considered crucial for the further fate of meykhana. In return for guarantees of social and economic security, meykhana reached the state level. According to my sources: "they have their own teahouses in the districts, and they didn't intervene there anymore," "the authorities provided them with gas at a very good price, and sometimes even free of charge" (A., m., 54, researcher and M., m., 62, journalist).

"The first Republic competition of meykhana performers (Respublika Meyxanaçılar Müsabigəsi) was held in 1992 in the building of the Baku circus. It was attended by several thousand people, and the performance of verbal recitatives was accompanied by playing exclusively on folk percussion instruments such as nagara (nağara). It was a real national holiday, but, after this competition, meykhana went aside again. It was again played only in the neighborhoods, the source recalls. – The situation changed sometime in 1995 or 96. Then they started to play meykhana on the radio"—adds the interlocutor (F., m., 48-year-old writer).

My sources associate the popularization of meykhana in the public space of the country precisely with a group of Mashtaga performers, primarily with Aghasalim Childagh. "Childagh became the favorite of Heydar Alievich.[7] Our president, like all Soviet rulers, was very fond of traditional culture, especially mugham, but also meykhana. Therefore, his (Childagh's) voice could be heard on the radio on national holidays. He was also invited to various official events, when foreign delegations came and he needed to entertain important guests, show them our culture, our folklore. Childagh was the best composer, he was a follower of Vahid himself!"—one Baku cultural studies scholar explains (T., m., 72-year-old, cultural studies scholar).

Aliagha Vahid (mentioned above) is considered the most recognized performer of classical meykhana. He was born in the Absheron village of Masazir (Masazır), located near Mashtaga. He was educated in a madrasah, where he became interested in poetry. He actively participated in literary circles, translated the works of the most famous Middle Eastern poets (the poems of Nizami, Fizuli, Khagani) from Persian into Azerbaijani and wrote ghazals[8] himself. Subsequently, he became the most recognized modern poet of ghazal (qəzəlxan), that is, the master of the ghazal.

By emphasizing the analogies between meykhana and ghazal—they are observed at the level of the poetic structure of both forms, meykhanachi try to increase prestige in their work and thereby grow their social status. Mashtaganians especially tried to do this since Childagh was the last actively performing student of Vahid and his immediate heir.

[7] In 2006, Aghasalim Childagh was awarded the Order of Glory, approved in 1993 by President Heydar Aliyev for his merits in the development of national culture.

[8] Madrasah (mədrəsə)—an educational institution that traditionally exists at the mosque, performing the function of a secondary general education school (its graduates had the right to enter the university). Classes in the madrasah were conducted in Persian, less often in the Turkic (Azerbaijani) languages.

Ghazals (qəzəl)—it is the main genre of lyric Azerbaijani (and Persian, sometimes for generalization called Middle Eastern) poetry. Ghazal is composed in different sizes of aruz, usually consists of 5–12 couplets—bayts. The two lines of the first bayt rhyme with each other, in the subsequent bayts, every second line rhymes with the first and second lines of the first bayt according to the scheme: aa ba ca da. See more details: Tağısoy N., Zakariyya Z. 2011. *Meyxananin poetikası*. Baku: Çaşıoğlu.

"De, Gəlsin!" TV Show

The dissemination of information in the public discourse of Azerbaijan about the similarities between Middle Eastern high poetry and meykhana has become another effective way of promoting the latter at the national level. It is under this slogan that the most popular television program airs in the republic—"De, Gəlsin!" (in direct translation "Come on, tell me!"). In its course, performers compete around the phrase given by the jury. The jury includes famous poets and local celebrities, and they evaluate the performances on a five-point scale. However, the victory is awarded not by the jury, but by the viewers. The meykhanachi who receives more phone calls and messages wins an expensive car, fame, and big money awaits him.

The first release of the TV program took place in 2001 and was directed by Emin Efendi (Emin əfəndi, 1978–), producer and manager of hip-hip groups famous for supporting Azerbaijani nationalism such as Dayirman and H.O.S.T. The following episodes, until the last in 2013, were held in a similar spirit—the producers of the program and the participants emphasized those characteristics that are also important for the official cultural policy. Among them are: the originality and uniqueness of the genre (it is argued that it only exists in Azerbaijan and there is no analogy to it in other countries), its Absheron roots (the birthplace of meykhana—Mashtaga), the belief that it is part of traditional Azerbaijani culture, that it is old (has always existed), and also has a relationship to Middle Eastern classical poetry (the connecting figure is Aliagha Vahid) and various forms of belief (primary connection to Sufism).

Participation of Mashtaga performers in the first episodes of "De, Gəlsin!" was the key to success. One of the creators of the program said that it was not easy to convince them:

E.: I practically lived with them. I did not leave the house. We ate together. I played with their children. I was separately convincing each meykhanichi that it was worth taking part in the competition, but they did not want to.

E.: They are all finicky. Artists. One smokes. Another one does not want to. They do not care about anything. It was very difficult for them to meet our requirements.

Me: Why didn't they want to?

Me: Requirements?

E.: You understand that you can't say everything on television. You must behave a little differently. And they didn't want to. They are peculiar people. They love to spend time in their area. People rarely come to the city. Especially the residents of Mashtaga. They don't want to interact with others. They accept me because I am from Nardaran (from nearby village – AS)

Me: Why did they agree?

E.: I had been pointing out the advantages to them for so long. I promised money and fame. (Yu., m., 42 years old, TV presenter).

The program "De, Gəlsin!" attracted the attention of a mass audience. The director of the TV channel that broadcasted it told me that "the show beats popularity records, hundreds of people vote for their favorite performers" (S., f., 32 years old, director). The winners became local stars and were invited to weddings, private parties, and other lucrative events. "Meykhana ignites because it has real money," a representative of the local show business admitted in an interview with me (K., f., 55-year-old, manager).

The success of "De, Gəlsin!" marked the fact that meykhana moved from a narrow circle of performers and listeners to a wide, national audience. At the same time, the number of social players who wanted (and still want) to change the character, artistic form, and even the function of competitions has also significantly increased. Such projects will be discussed in the next subsection.

Let's Make Better Meykhana: Bottom-Up Attempts

Meykhana performers can be divided into several groups, both geographically and ideologically (Strzemżalska 2020). Describing this classification, it is important to remember that each of the groups of performers is characterized by the fact that it has its own teacher—ustad and sometimes there are several of them. Meykhanachi that are younger in age and social status learn from them and obey them.

The abovedescribed television show, which materialized the rules of a new cultural policy for Azerbaijan, changed the old social hierarchy. Meykhanachi, who became popular due to television programs, saw the place of meykhana in a modern country differently. The most striking figure among them was Namiq Məna, laureate of the "De, Gəlsin!" in 2006.

Namig Mena (Namiq Məna, real surname Həsən, 1972–) was born and raised in the city of Sumqayit (Sumqayıt), a large industrial center of Azerbaijan, to which his ancestors came from Zangezur (today's Armenia). Since Namig showed acting skills and was fond of poetry from early childhood (he even wrote poetry himself), he participated in meetings of Absheron literature lovers whenever possible. There he met the Mashtaga meykhanachi. "Namig fit in our crowd. He is very capable, he quickly picked up the rhythm, the melody and we accepted him, although he is in general a stranger"—in my field diary I wrote down the statement of one of the Mashtaga performers (K., m., 46, meykhanachi).

Mena, having learned meykhana from the Mashtagins, later became their main enemy. The bone of contention was Mena's conviction that meykhana needs to be "improved," that is, its social status needs to be improved. Mena and his associates intend to do this in several ways, among others, changing the name of the genre[9]:

"Mey" is wine. "Khana" is a house, a room. So, meykhana is a "drinking establishment," a "booze parlor." This is an unworthy name for a national genre. All in all, it used to be called "bedie." This is evidenced by many sources—said Namig Mena in an interview with me, referring to the work of scientists and writers such as Aziza Jafarzade (Cəfərzadə 2002) and Manaf Suleymanov (Süleymanov 1987) among others (N., m., 48, meykhanachi).

From the beginning of the second decade of this century, when the first postulate to change the name of the competition was put forward, Mena began to speak of himself as "poet" (literally bədihəçi-şair). He, like other performers, considered himself a follower of the aforementioned gazelkhan Aliagha Vahid, and, at the same time, the direct successor of Nizami Remzi. That is, he mainly composed a musical meykhana. Considering the genre features of meykhana, such a combination seems mutually exclusive, but it positioned Mena highly among the meykhanachi:

"Mashtaganis are conservatives. For a long time they did not allow the musical accompaniment of meykhana, they said that they followed only Vahid, for whom meykhana is verbal recitations, clever, funny expressions spoken in verse to a certain rhythm"—stated one interlocutor (R., m., 66, researcher). "It is very difficult to imitate aruz in musical meykhana. Only

[9] More on this topic: Strzemżalska A. (2020) Slam in the Name of Country: Nationalism in Contemporary Azerbaijani Meykhana, Volume 79, Issue 2, Summer 2020, pp. 323–344.

Remzi did it well. He was a real master, but the Mashtaganis criticized him for his musical meykhana. There were constant, fierce disputes over this between him and Aghasalim Childagh. Even though they invited each other home (they were friends – AS), they fought for first place among meykhanachi"—recalled the second source (F., m., 48, poet).

In addition to the manner of performance, there is another similarity between Mena and Remzi's work: Remzi was the first to invite women to participate in competitions, both in the character of the performer and the audience. Mena follows this principle, arguing that "bedie" is a part of "national culture" and it should be available to all residents of the country, not only men (M., m., 48 years old, meykhanachi). It was the accessibility of meykhana for the mass audience that dictated profound changes in its poetic form and content.

Mena's concerts take place in the most prestigious places in the country, where cultural events can only be held with permission from the official authorities. Such places include the narrow and steep streets of Icheri Sheher (İçəri şəhər), that is, the old city of Baku. In the early summer evenings, there are many people walking; women, men, and children of different ages enjoy themselves to the sound of meykhana, which Mena sings right on the street. His works, accompanied by a synthesizer and sometimes string instruments, are more like songs than recitatives. People join the vocalist and perform the chorus with him. "Until recently, meykhana was associated with criminal activity, with a vulgar language. Perhaps it still exists, but we are glad that we can listen to meykhana live in the city center," says Sevda who, walking around Baku along with her cousins and children, accidentally attended a concert (S., f., 48 years old, bank employee).

Not all listeners of Mena's performances agree with Sevda. Many lovers of meykhana emphasize that the "beautiful," "correct" language and "neutral plots" used by Mena in his works are not "real" meykhana:

> "It's like a conversation, like a conversation with each other. Well … We, for example, scold each other in the meykhana. So, well … we do not scold, like this, without meaning. Meaning should always be present. But the main thing in meykhana is to argue. For example, in a normal life when you don't say much, but people still are offended by you. And in meykhana it is not cursing, it is just already accepted that way, you know. It turns out badly without swearing. There should be a little bit of spice. It is necessary to joke. If you answer sharply, they will answer you even sharper. This

sharpness is the whole appeal of meykhana. If there is no sharpness there, it is no longer interesting. This is the whole meaning of meykhana"—one popular meykhanachi interpreted (S., f., 48 years old, singer).

Mena's smoothed-out form and content, its correct literary language and inclusion of love themes differs from traditional meykhana, but in recent years, particularly it becomes noticeable at national and international events during which Azerbaijani culture is presented. In a conversation with me, Mena mentions that he performed for the presidential couple at the Novruz holiday. As evidence of this fact, he shows a video taken by a phone and photographs of him with the main officials of the republic (M., m., 48 years old, meykhanachi). In addition, one can find many of Mena's studio recordings on the Internet during which he promotes the national culture of his country. He was most often invited on state television after 2013, when the program "De, Gəlsin!" had stopped broadcasting. From time to time, Mena also gives concerts "in military units on the front line" (T., m., 42., journalist), and directly cooperates with the Heydar Aliyev Foundation. For example, in 2011, he hosted the final concert of the II International Mugham Competition sponsored by the Foundation, and in 2009, he spent a month in America promoting Azerbaijani culture and poetry.

Despite the political engagement of "bedie" and Mena's close ties with high-ranking officials, the artist and his associates argue that "their verbal recitatives" are a kind of popular, not folk culture. For this reason, they are opposed to the Mashtaga meykhanachi, who, in turn, tried to give the official status of folklore art to verbal recitations. So far, the status is ambiguous. For this purpose, a letter was sent to the National Academy of Sciences of Azerbaijan (Azərbaycan Milli Elmlər Akademiyası). In response to it, the council of the Institute of Folklore of the National Academy of Sciences of Azerbaijan issued a statement that it is impossible to speak of meykhana as a folklore genre with full confidence. The researcher explained his decision as follows:

"Verbal recitations are based on aruz. Aruz occurs only in high poetry and therefore one cannot equate meykhanaMeykhana with folklore art. Yes. And it (meykhanaMeykhana – AS) is not an anonymous genre. It has authors, right? <...> 'Yes...' And one more thing, medium length meykhanaMeykhana is a work of fifteen to twenty stanzas in length. Each stanza has four to six lines. It is almost impossible to repeat such lengthy

> statements. They are not remembered as quickly as other folklore poems, for example, bayats (bayatıs), lullabies (layla) or proverbs (atalarsözü). Therefore, the reproduction of meykhanaMeykhana is almost impossible. Usually a few lines, or a stanza that made a special impression, are remembered. But then they are not read in the context in which they were originally performed" (A., m., 54 years old, researcher).

Carefully observing the activities of the Mashtaganis, Mena and his colleagues tried to "improve" the recitatives in a different way. Convinced that quality reading of "bedie" requires professional training, they strove to open a new specialty at the theater department of the Azerbaijan State University of Culture and Arts (Azərbaycan Dövlət Mədəniyyiyt və İncəsənət Universiteti). So far, their efforts have not found encouragement from the authorities.

Conclusion

The reasons for the dispute between supporters of "meykhana" and "bedie" may seem superficial, but they are important for determining the power relations between representatives of official structures and ordinary people who also form political alliances. In this regard, the communities formed around meykhana are especially distinctive, since the social status of the phenomenon has changed dramatically over the past century. The results of this study demonstrate that changes in the social status of performances (meykhanachi were underground hero poets who developed the "tradition" of reading improvisational poetry; they were the defenders of justice associated with the semantic locus of the "thief in law," brave political opponents, but also pop stars and local celebrities) are directly related to where the state places them in the cultural space—in the center or on the periphery. From the second decade of the twentieth century until the end of World War II, meykhana occupied an important place in official cultural policy, and at that time performances played a significant role in popularizing and strengthening Azerbaijani nationalism. After the improvisational (and therefore uncontrollable) nature of performances ceased to fit into the framework of the cultural policy of Soviet Azerbaijan, meykhana went underground and became an opposition phenomenon. Then, in response to the needs of the new political realities in the 90s, the social status of the meykhana was changed one more time and since that performances have occupied the central point of the musical space of Azerbaijan.

The place of meykhanachi in the social ladder was largely predetermined (and still is) by the state. The state has the ability to create images of performers that meet the requirements of cultural policy and their presence is very convenient for the official authorities—depending on the situation, they launch those meanings of meykhana that seem to be more effective for the implementation of state interests. However, it should be noted that local players are also actively involved in the process of meykhana nationalization. They not only fill up the official discourse aimed at meykhana folklorization, but also through the establishment of professional meykhana schools tend to its institutionalization and standardization.

In recent years, the state has managed to balance between hostile meykhana communities. However, we must remember that this is a transitional situation and no government can keep full control over cultural phenomena. Already today we observe that the socially and politically rich background of meykhana is a convenient starting point for the implementation of local projects for the development of the genre. In this regard, the mechanism of combining mass culture and "traditional" (folk) components—elements of local culture—seems especially interesting. Given the widespread synthesis of traditional and popular music in Azerbaijan, it can be noted that any musical genre that is influenced by certain official cultural policy and social circumstances can be considered "traditional." Consequently, traditional music is understood and practiced more flexibly than is commonly judged, while popular music takes on a new dimension as a means of creating history (a retrospective definition of tradition), not only as a form of social activity aimed at substantiating the future.

References

Anadollu, ə. 1927. Qupletlər "Şapalaq," 3-ncü hissə. Baku: Publisher unknown.

Anadollu, ə. 1930. Kupletlər, Çuvalduz, II təbi. Baku: Azərnəşr.

Anatollu, ə. 1925. Qupletlər, Bütün dünya işçiləri birləşin! Baku: Ali siyasi- məarif idarəsi nəşriyyatı.

Antonsich, M. 2015. The 'Everyday' of Banal Nationalism—Ordinary People's Views on Italy and Italian. *Political Geography* 54, 32–42.

Armstrong, J. A. 1982. *Nations Before Nationalism*. Chapel Hill: University of North Carolina Press.

Billig, M. 1995. *Banal Nationalism*. London: Sage.

Brubaker, R. 1996. *Nationalism Reframed: Nationhood and the National Question in the New Europe*. Cambridge: Cambridge University Press.

Cabarzadə, B. 1927. Qupletlər. Baku: Publisher Unknown.

Саламзаде, Э. А., Абдуллаева, Р. Г. 2019. Мультикультурализм – действенная модель культурной политики Азербайджана. Наследие веков 3, 62–68.

Cəfərzadə, ə. 2002. Bədahətən nə Deməkdir? Mədəniyyət, 5–18.

Dessiatnitchenko, P. 2017. Musical and Ontological Possibilities of Mugham Creativity in Pre-Soviet, Soviet, and Post-Soviet Azerbaijan. Ph.D. Dissertation. Toronto: University of Toronto.

Edensor, T. 2002. National identity, popular culture and everyday life. Oxford: Berg Publishers.

Eisenstadt, S. N. 1973. *Tradition, Change, and Modernity*. New York, Sydney, Toronto: John Wiley.

Fox. 2016. The Edges of the Nation: A Research Agenda for Uncovering the Takenforgranted Foundations of Everyday Nationhood. *Nations and Nationalism* 23(1), 26–47.

Fox, J. E., Miller-Idriss C. 2008. Everyday Nationhood. *Ethnicities* 8(4), 536–556.

Gellner, E. 1983. Nations and Nationalism. Ithaca: Cornell University Press.

Haid, C. G. (2017). The Janus Face of Urban Governance: State, Informality and Ambiguity in Berlin. *Current Sociology* 65(2), 289–301.

Hobsbawm, E. 2003. *Introduction: Inventing Traditions, in the Invention of Tradition*, edited by E. Hobsbawm, T. Ranger, Cambridge: Cambridge University Press, 1–14.

Hobsbawm, E. J. 1992. *Nations and Nationalism since 1780: Programme, Myth, Reality*. Cambridge: Cambridge University Press.

Huseynova, A. 2016. *Music of Azerbaijan: From Mugham to Opera*. Bloomington: Indiana University Press.

Imaməliyev, A. R. 1993. *Abşeron meyxanaları*. Baku: Nicat nəşriyatı.

Isaacs, R., Polese A. 2015. Between 'Imagined' and 'Real' Nation-Building: Identities and Nationhood in Post-Soviet Central Asia. *Nationalities Papers* 43(3), 371–382.

Isaacs, R., Polese, A. 2016. *Nation Building in the Former USSR: New Tools and Approaches*. London: Routledge.

Jourde, C. 2009. The Ethnographic Sensibility: Overlooked Authoritarian Dynamics and Islamic Ambivalences in West Africa. In *Political Ethnography: What Immersion Contributes to the Study of Power*, edited by E. Schatz. Chicago: University of Chicago Press, 201–216.

Naroditskaya, I. 2002. *Song from the Land of Fire: Continuity and Change in Azerbaijanian Mugham*. New York: Routledge.

Oldfield, A. 2008. *Azerbaijani Women Poet-Minstrels: Women Ashiqs from the Eighteenth Century to the Present*. Lewiston, NY: Edwin Mellen Press.

Pawłusz, E., Seliverstova, O. 2016. Everyday Nation-Building in the Post-Soviet space: Methodological Reflections. *Studies of Transition States and Societies* 8(1), 69–86.

Polese, A. 2015. Informality Crusades: Why Informal Practices Are Stigmatized, Fought and Allowed in Different Contexts According to an Apparently Ununderstandable Logic. *Caucasus Social Science Review* 2(1), 1–26.

Polese, A., Morris, J., Seliverstova, O., & Pawłusz, E. 2018. Introduction: On Informal and Spontaneous National Identities. In *Identity and Nation Building in Everyday Post-Socialist Life*, edited by A. Polese, J. Morris, E. Pawłusz & O. Seliverstova. London: Routledge, Routledge Contemporary Russia and Eastern Europe Series, vol. 75, 1–14.

Polese, A., Horák S. 2015. A Tale of Two Presidents: Personality Cult and Symbolic Nation-Building in Turkmenistan. *Nationalities Papers* 43(3), 457–478.

Reicher, S., Hopkins, N. 2001. *Self and Nation*. London: Sage.

Ruehl, K. 2018. What's the Difference Between Folk and Acoustic Music? A Brief History of How Acoustic Pop Music Became Known as "Folky", in LiveAbout.com. Available at: https://www.liveabout.com/folk-music-vs-acoustic-pop-music-1322446. Accessed 29 September 2020.

Schatz, E. 2009. Ethnographic Immersion and the Study of Politics. In *Political Ethnography: What Immersion Contributes to the Study of Power*, edited by E. Schatz, Chicago: University of Chicago Press, 1–18.

Səfərov, S. 1925. *Satiragit Tjatrosunda deyilmiş Vahidin Meyxənələri*. Baku: Publisher unknown.

Shils, E. 1981. *Tradition*. Chicago: University of Chicago Press.

Skey. 2011. *National belonging and everyday life*. Basingstoke: Palgrave.

Smith, A. D. 2000. *The Nation Is History: Historiographical Debates about Ethnicity and Nationalism*. Oxford: University Press of New England.

Spinetti, F. 2005. Open Borders. Tradition and Tajik Popular Music: Questions of Aesthetics, Identity and Political Economy. *Ethnomusicology Forum* 14(2), Music and Identity in Central Asia, November, 185–211.

Stites, R. 2004. The Ways of Russian Popular Music to 1953. In *Soviet Music and Society Under Lenin and Stalin: The Baton and Sickle*, edited by N., Edmunds. London: Routledge Curzon, 19–32.

Strzemżalska, A. 2020. Slam in the Name of Country: Nationalism in Contemporary Azerbaijani Meykhana. *Slavic Review* 79(2), Summer, 323–344.

Süleymanov, M. 1987. *Eşitdiklərim, oxuduqlarım, gördüklərim*. Bakı: Azərnəşr.

Tagısoy, N., Zakariyya, Z. 2011. *Meyxanənin poetikası*. Baku: Çaşıoğlu.

Williams, R. C. 1982. The Nationalization of Early Soviet Culture. *Russian History* 9 (2/3), 157–172.

Wong, D. 2001. *Sounding the Center: History and Aesthetics in Thai Buddhist Performance*. Chicago: University of Chicago Press.

Чистов, К. В. 1981. Традиции, «традиционные общества» и проблемы варьирования. Совет. этнография 2, 106–112.

Political vs. Everyday Forms of Governance in Uzbekistan: The Illegal, Immoral, and Illegitimate

Abel Polese, Rustamjon Urinboyev, Mans Svensson, Laura Adams, and Tanel Kerikmäe

Introduction: The Art of Not Claiming Your Rights

In May 2009, the taxi in which two of the authors of this paper were travelling from Tashkent to the Ferghana Valley, in eastern Uzbekistan, was stopped by the traffic police for a document check. While this might sound relatively commonplace in our world, the situation is noteworthy for its genesis and epilogue. First, in contrast to what common sense

---

This chapter was originally published as Political vs. Everyday Forms of Governance in Uzbekistan: the Illegal, Immoral and Illegitimate Politics and Legitimacy in Post-Soviet Eurasia.

A. Polese (✉)
School of Law and Governance, Dublin City University, Dublin, Ireland
e-mail: abel.polese@dcu.ie

Department of Law, Tallinn University of Technology, Tallinn, Estonia

A. Polese (ed.), *Informality, Labour Mobility and Precariousness*, International Political Economy Series,
https://doi.org/10.1007/978-3-030-82499-0_9

would suggest, the car was allegedly stopped because one of the authors had his seatbelt fastened. Seatbelts were made compulsory only in 2010 so, in the words of the driver "wearing a belt was a sign that a (rich) foreigner was sitting in the car" and would thus attract the attention of some hungry traffic policemen. This was even more likely because the car was riding through the *Kamchik* pass that, being the only entry point into Fergana Valley, was highly surveilled. Second, the policeman asked to open the car's trunk to check its content but, once made aware that the foreigner wanted to be present at the check, he got inhibited and gave up the check. He asked the driver to follow him instead for a private discussion on the car's documents and, upon return, the driver angrily reported to have had to pay about the equivalent of 7 Euro for this "lack of respect" by the foreigner. By demanding to be present during the luggage check, the foreigner de facto challenged the undiscussed authority of the Uzbek policemen, who then decided to retaliate with a fine on the driver who was driving that very foreigner.

The situation would fit well Alice in Wonderland: if not wearing a seatbelt will keep you clear of troubles, then exerting your rights to be present when your luggage is checked will land you into trouble. This situation is not uncommon in transition states and societies like Uzbekistan, and in the post-Soviet world in general, and this informs the two areas of research this article engages with. The first, rooted in legal and economic studies of corruption, has examined the relationship between corruption and state failure in one or more spheres of management (Acemoglu and Verdier 2000; Rose-Ackerman 1999; Laffont 2006; Della Porta and Vanucci 1999; Heidenheimer and Johnston 2011; Kerikmäe and Särav 2016). The second, starts from a critique of the standpoint above, that is regarded as linked to a renewed "moralism" in political economy (Kotkin and Sajó 2002: 26) and have inspired a thick analysis of "corrupted"

R. Urinboyev · M. Svensson
Department of Sociology of Law, Lund University, Lund, Sweden

L. Adams
Freedom House, Washington, DC, USA

T. Kerikmäe
Tallinn Law School, Tallinn University of Technology, Tallinn, Estonia

practices and situations leading to contextualization of a given transaction within a ritual form and emotional valence of the transaction (e.g., propriety vs. humiliation) (Werner 2000; Gupta 2005; Haller and Shore 2005; Rivkin-Fish 2005; Polese 2008; Urinboyev and Svensson 2013). Indeed, the uncritical use of the World Bank's definition "the abuse of a public function for a private advantage" has pushed scholars, and definitions into a slippery territory. Elaborations of this definition have gone as far as to deny the Maussian definition of a gift and de facto invade the competencies of reciprocity and social interaction studies. In response, an emerging streaming of research has attempted a conceptualization of informality that can help defining a boundary between corrupted and non-corrupted practices. The starting idea being that an informal practice is something concealed from the state (Routh 2011) and corruption is only a subcase (Polese and Stepurko 2017). Indeed, the wide range of concealed practices can include extra-legal (de Soto 2000) or socially accepted practices (Granovetter 1984) that are tacitly accepted or, at least not punished by the authorities.

This article is a further study in this direction. Starting from the above debates, it is intended to contribute to understanding of the governance processes in transition states and societies through focusing on informal everyday practices and sites where state officials and citizens come into contact on a daily basis. We are inspired by recent findings in what has been called "the everyday dimension" (Edensor 2002; Fox 2016; Skey 2015) and encompasses daily and "banal" (Billig 1995) practices of resistance and renegotiation of state and top-actors political measures (Scott 1985, 2012). By looking at the way official state narratives are challenged by silent, unorganized, often unawares, gestures of resistance at the bottom of a society, we look at the political side of these activities. Footing on a framework suggested by Scott's definition of infrapolitics (2012), we propose to incorporate informal practices in a definition of informality that is more inclusive, and better explains the anatomy of a modern state, whose functioning rests on a combination of formal and informal practices (Morris, 2019).

By doing this, we suggest that governance in transition states and societies may be regarded as a space where formal institutions and citizens (or informal institutions) compete for power and resources and thereby produce informal, alternative "legal orders" and mechanisms that regulate public life in a given area. We will suggest that such a space of informal negotiation is vital in contexts where collective mobilization and public articulation of social claims is not a preferred, or even available, strategy for citizens. This is ultimately connected with our understanding of everyday forms of governance that is a meaningful, and useful, site for understanding sociopolitical developments in the process of the construction of the governance system. But that system cannot be limited to what is formally the political sphere but touches a large spectrum of behaviors and situations that are not always acknowledged (Navaro-Yashin 2002). The everyday dimension is of particular importance here when trying to understand the governance trajectories as it allows to look critically, and from a broader perspective, at situations where individual and state perception of events, but also individual and state morality, diverge.

Methodology

This paper is based on extensive ethnographic field research conducted between 2009 and 2017 in Tashkent city and Fergana region of Uzbekistan for a total of fifteen months. The fieldsites were chosen based on contacts and social networks available to the authors. During these field trips ethnographic material was collected mainly through participant observation and informal interviews both in the city of Tashkent and in the Ferghana region. Our fieldsite *Oqtepa*,[1] where we conducted observations is one of the mahallas in *Shabboda* village in rural Ferghana and has a population of more than 2000 people. We regularly visited mahalla's "gossip hotspots" as guzar (community meeting space) and *choyxona* (teahouse), weddings, childbirth, circumcision ceremonies, funerals, and monthly get-together events (*gap*) where most residents come together and exchange information on a daily basis. Most of the residents in this mahalla were *dehqonlar* (farmers) involved in cucumber and grape production. However, due to our research focus, we were particularly interested in two mahalla residents, *Sardor* and *Rahmon*, who were both state officials and the center of "everyday mahalla talk."

The starting hypothesis was that the way the state is lived at the everyday level is different from the way it is perceived in official documents. In particular, we were interested in exploring the role of state in everyday life and people's perception of good governance. We looked at, inter alia, the way ordinary people talked about, and lived, the role of

[1] The name of the *mahalla* has been changed to protect the anonymity of our informants.

state. This was done in order to grasp to what extent people were willing, or likely, to conform to laws when dealing with state officials, and how people think about the proper social behavior and law. Our material has informed the preparation of several other studies (Urinboyev 2011, 2013, 2014, 2015, 2016, 2017; Urinboyev and Svensson 2013, 2014, 2017; Urinboyev and Polese 2016) that have been used also to triangulate the presented data and the interpretation of the evidence presented in the next sections. For this specific paper, in addition to several ethnographic vignettes, we present some examples of most relevant observations and two interviews that focus on informal practices in two-state institutions: (1) the practice of traffic safety enforcement; and (2) the practice of maternity services. The first interview is centered around *Ahmet,* a traffic policeman, and the second focuses on *Umida*, a midwife at maternity hospital in Ferghana. These two key informants were selected on the basis of their daily involvement with the law, knowledge, willingness, and communicability. The interviews lasted from 45 minutes to 2 hours, depending on the informants' status, location, and time available. Safety precautions both for researchers and informants forced us to avoid all types of apparent documentation methods, such as recording and taking notes. Instead, immediately following an interview, we wrote down our recollections. To protect the anonymity of our informants, their names have been changed.

Between "Good" and "Bad" Governances: Introducing the Everyday

Governance debates have preoccupied scholars from a variety of disciplines and as high profile as Francis Fukayama's (2013: 350) who defines it, at the most basic level, as "government's ability to make and enforce rules, and to deliver services' and 'about the performance of agents in carrying out the wishes of principals, and not about the goals that principals set." This, in some respects, may be regarded as one of the basic conditions for democratic consolidation in a country. It is indeed by asking whether good governance can lead to democracy and executive constraint by rule of law (rather than the other way around) that Fukuyama turns on his head the first set of criteria most international organizations use when talking about comparative governance issues, questioning the causal relationships between these various elements. In turn, this inspired our question on whether good governance reforms in a country such as Uzbekistan can productively precede democracy.

Stemming from the above definition, good governance can be regarded as relative to state capacity, relevant to public administration structures as opposed to politics and policies. It entails, in other words, a focus on output measures with no assessment on the policy goals that the government is pursuing. Fukuyama's approach also follows the good advice to disaggregate the state, acknowledging that some state entities will be more efficacious than others, and that states vary across regions, especially across the urban and rural divide. However, although Fukyama's notion of governance provides useful insights when measuring the state capacity it is limited in its capacity to provide a clear understanding of the country, and of other similar cases worldwide. Attempts to use existing theoretical tools to conceptualize alternative forms of governance have passed through an examination of paradigms engaging with failed states and limited statehood (Risse 2012). Uzbekistan, nonetheless, can hardly be considered fitting this framework. Even if not fully in control of all levels of political governance, in contrast to a number of other cases, the Uzbek state has retained control, and governance capacity, in a number of cases with full control of institutions and domestic and national sovereignty. In other words, the Uzbek state is certainly a powerful actor when it comes to using coercion and preventing political instability, but it is weak in terms of enforcing 'rule of law' and service delivery. These are usually phenomena associated with democratic developments of a country. In spite of attempts to look for the roots of good governance in the level of democracy present in a country, governance and democracy are not necessarily correlated. Indeed, not only some authoritarian states are remarkably well-governed (see China in Fukuyama's example, 2013), but state performance at parliamentary level can actually find corrupted or non-transparent practices beneficial. Resting on a number of traditional ultraliberal and new institutionalist arguments (de Soto 2000; Leff 1964), this has been empirically demonstrated in post-Soviet spaces, where blackmailing states have been performing relatively well, if compared to other cases of the world (Darden 2008).

Engaging with alternative definitions of governance, and its perception (Polese et al. 2017), a different standpoint can be adopted. For one thing, the World Bank (2018) has defined governance as the mixture of traditions and institutions by which authority in a country is exercised. This includes the process by which governments are selected, monitored, and replaced; the capacity of the government to effectively formulate and implement sound policies and the respect of citizens and the state for the institutions that govern economic and social interactions among them (Kaufmann et al. 2004). However, in contrast with Fukuyama's definition, it leaves room for the use, and agency, of informal governance actors. This is not a minor approach linked to practices of development. A growing body of literature has been engaging with the concept thus demonstrating the relevance of the concept not only in practice, but in theory, and its usefulness to understand not only local but global mechanisms of governance (Christiansen et al. 2003; Peters 2007; Puetter 2006). The rise of new informal governance mechanisms can also be observed on a global level where the states increasingly rely on informal practices and arrangements to govern global policy issues (Bueger 2015). This is due to the fact that formal rules are often inadequate and insufficient when dealing with the complex and dynamic issues and challenges of global politics that require innovative solutions and coordination. Recent works have started to explore informal governance as a mode of international cooperation, arguing that governance relies not only on formal rules and institutions but to a significant degree also on informal practices and arrangements (Christiansen and Neuhold 2012; Westerwinter 2016; Stone 2009).

Starting from an understanding of governance that includes also non-state actors and takes into account the relationship between a state and its citizens, not only its coercive capacity, we can attempt to create a framework that goes beyond the conventional definitions of governance and thereby employ additional tools and concepts if we are to better understand the local dynamics and governance trajectories in Uzbekistan. A concept that thus could better fit our aims is 'good enough governance', forwarded by Merilee S. Grindle in the context of development assistance, and rediscovered in the context of failed U.S. programs to promote democracy abroad (Grindle 2011; Krasner 2013). In particular, Krasner argues that a well-functioning democracy is hard to maintain and exceptionally difficult to promote from outside, and a more effective way to promote long-term freedoms is 'to improve the prospects for security and economic growth in the short run—rather than pressing for direct democratic reforms' (Krasner 2013). Elections may make things worse in terms of prosperity and freedom, and Krasner argues that U.S. policy has been aiming at the wrong target in pushing countries like Iraq and Afghanistan toward consolidated democracy. Instead, Krasner argues that we should start thinking in terms of 'good enough governance', which consists of a state being able to provide for the physical integrity of its citizens (but not their more broadly defined human rights) as well as providing some basic public services, especially in the area of health; to provide public order including a norm of property rights, and some degree of security from transnational threats; to constrain corruption and provide some form of checks on the arbitrary power of the state through media or civil society; and since free and fair elections can produce instability and trigger violence, elections "would only be useful *if* they ratified and legitimated agreements that had been reached *ex ante* among political leaders with regard to control over security forces and the distribution of spoils" (Krasner 2013).

As a further help, one can use informality as a further theoretical tool to understand how alternative forms of governance works. Defined as an aggregate of practices, norms, and exchanges that take place beyond formal/state regulation (Routh 2011; Ledeneva 2006; Polese et al. 2015), informality is a phenomenon that has left the economic sphere of life of a state to be considered part of governance (Polese and Morris 2015). Informal practices can, indeed, develop in a way to complement or supplement the state (Morris 2011, 2012, 2013, 2016; Polese, Kovacs, and Jancsics 2018; Polese et al. 2016) in one or more of its functions as shown in the scheme below.

Across regions, the concept of informal governance has been evoked by a variety of scholars and in a variety of meanings. For this article, we use Ledeneva's understanding (Ledeneva 2013) of informal governance that can be regarded as intertwined with policy-making mechanisms within a state. In line with what Scott calls infrapolitics (2012) we see informal actions, performed at the everyday level by ordinary citizens, as ultimately affecting the way policy measures become acceptable, and accepted, by a wider variety of actors. Empirical studies have suggested that everyday negotiations of policy measures have an effect beyond the initially though micro one (see Scott 1985) and that the agency of middle actors or ordinary people can substantially reshape the way political measures are implemented, mitigate state-led coercive means and eventually lead the state to either limit its sphere of influence or renegotiate its initial intentions (Kerikmäe, Nyman-Metcalf and Papageorgiou 2013; Ó Beacháin 2012; Ó Beacháin and Kevlihan 2013; Polese 2010; Polese et al. 2017; Richardson 2008; Troitino 2013).

Already Migdal has demonstrated the way state and society are interrelated (2001) reproducing a mechanism that Giddens had attempted to define with his structuration theory (1984). Footing on these works, we propose here that people have a wide amount of tools, and agency, to renegotiate, reinterpret, or even distort official political measures in a manner that allows governance to be contributed to by grassroot and bottom actors. When bottom actors end up moving in a same direction and endorse similar values they can, formally or informally, engage with the production of the political (Navaro-Yashin 2002; Polese and Kevlihan 2013). In other words, an informality framework can be used to identify, or at least acknowledge, spheres of life or areas of the state–society relationship where laws and rules are renegotiated by non-formalized actors. Informality thus becomes a space, in the geographical meaning of the term, where the encounter between citizens and state institutions generates an opportunity for informal governance to emerge (Polese, Reviakhashvili, and Murru 2018). The relatively large area allowing resistance and renegotiation, by individuals or communities allows reorganization of governance actors, practices, and their results, proposing alternative models and dynamics that are central to the current article.

Places of Resistance to State-Led Governance

Ledeneva's understands "sistema" as the aggregate of informal practices, unwritten rules and personal relationships that can be used to manage governance of virtually all aspects of a state (Russia, in her case, 2013). In her view, the relationship between formal and informal actors, and transactions, can lead to an equilibrium with rules relatively stable and easy to understand to an insider. Eventually, this generates an alternative system of governance that, illogic or unintelligible to the casual observer, makes perfectly sense to those who are embedded into it. The next two sections are an attempt of applying the sistema logic to our fieldwork and show the existence of parallel, or alternative (Wanner 2005), moral orders and systems of references that, taken together, are useful to understand how governance in Uzbekistan works. During fieldwork, a team including some of the authors of this paper happened to remark a few inconsistencies that are then reported, and analyzed, before engaging with two of the interviews collected, that are reported in the next section.

Fieldwork Notes

The first "incident" happened on the first day of the fieldwork when, after breakfast the team members happened to walk around *Alay a* nearby bazaar, where they were approached by a group of money changers who asked if they needed to buy Uzbek sum. This is not unusual in the post-Soviet region, or even in the post-socialist, and is usually used in two cases. One is to rip off foreigners and the other, more common, is a way to compete with the official exchange rate. In Uzbekistan, but also in Turkmenistan until a few years ago, the official exchange rate was utterly theoretical, meaning that it would be impossible to exchange currency at the official rate, set by the national central bank. The significant gap between how much local money should cost and how much it costs in reality, in addition to making local prices unbelievably high, creates an opportunity for whoever is able to establish a demand-supply chain and change at an exchange rate higher than the one offered in local banks. What is surprising here, to someone with experience in transitional countries, is that money from the black market was offered in spite of policemen widely present on the territory of the bazaar who seemed not interested by this kind of illegal transactions.

This was not the only time the team encountered this kind of apparent contradiction. When going to official meetings with the administration of one of the universities in Tashkent, regular taxi services were used. Someone would call a number and the taxi would come shortly afterward. Surprising was the situation where no taxi were available at a given moment and the team went onto the street in the hope to get a random taxi just driving around. Not only they were able to secure a car in the space of a minute, in fact several cars with no taxi sign on the roof stopped to offer their services. Virtually, anyone with a car is a potential taxi. They just need to stop and negotiate a price. This ultra-liberalization of the taxi sector seemed not to interest the traffic police or any other police, who are supposed to take care of street codes or at least get interested in shadow transactions. Not only has working as a taxi driver become a major source of self-employment in Tashkent, and beyond, what is interesting is that these drivers interact with traffic policemen, who are fully aware of their informal job. Taxi drivers were seen often shaking hands with policemen, who could sometimes stop them for breaking traffic rules but rarely, if ever, rather that for working illegally as drivers. Drivers generate income by working informally for people who need to be carried around.

Policemen generate income by fining, and receiving cash, when these drivers break the street code.

The evident contradiction here is the conflict between an official state narrative, that designates the boundaries of legality and hires officers to enforce state rules. However, these officers seem not interested in keeping social and political order they are hired for. Not only do they go against the state, in that they turn a blind eye when coming across transactions that, at least in theory, damage the state, and indirectly, its citizens. They actually ascribe to parallel or alternative moral orders (Stepurko et al. 2013; Wanner 2005) that are apparently unintelligible to anyone external. Eventually, the divergence between legality and social acceptability goes as far as to define a different moral order. A member of the team met with *Sardor*, a high-level state official working as the deputy chief of Ferghana region police, and *Rahmon,* a district-level traffic policeman (a low-level official). In spite of the first having a more prestigious position than the second, in the everyday mahalla life, *Sardor* was often described as "communist." This is usually used in a negative shade to indicate law-abiding state officials who do not share their political influence and resources with their kin and mahalla. As a high-level police official, *Sardor* had an enormous power and he could easily divert the resources of the state to mahalla, but he always rejected the requests of mahalla and asked them to solve their problems through formal channels. Because of his attempts to keep his public office separate from private sphere, *Sardor* was regarded as "communist" in the words of many mahalla residents we encountered. On the other hand, low-level official *Rahmon* was a "man of respect" and enjoyed very high social status and reputation in mahalla. Unlike *Sardor*, *Rahmon* provided patronage to mahalla residents, for instance by helping mahalla residents to avoid or manoeuvre around the state law. *Rahmon* was especially praised for his ability to act as a bridge between high-level state officials and ordinary residents in terms of negotiating the amount of informal payment for job or university admission issues, and bending state laws to meet the interests of mahalla residents. Therefore, when invited to weddings, *Rahmon* was always offered a "best table" and served more quickly than others. Thus, according to mahalla's norms, *Sardor* was neither a good person nor a good state official due to his law-abiding behavior and unwillingness to help mahalla people, while *Rahmon* was the "*pride of mahalla*," due to his sensitivity to the needs and concerns of mahalla.

Most, not to say all, normative interpretations of the functioning of a state take as a starting position the fact that the state is more important than its citizens. It is the state to grant social cohesion, provide welfare, security to its citizens and therefore individual welfare is not as important as the collective one stemming from the state official positions. A state is, however, peopled (Jones 2011) and history has provided a large array of cases where projects by the state for its citizens have blatantly failed (Scott 1998). It is therefore possible to assume that a state may be "wrong" in cases where one of its policies, designed for a bigger welfare of its citizens, is not bringing substantial benefits in practice. This can be due to at least two reasons. One is that the policy is based on an inaccurate or inadequate assessment of a given situation and is, thus, addressing issues that will not necessarily mean improvement of a given situation. The other one is that a policy, even if right in theory, fails in its practical implementation because middlemen officers fail to carry out the right measures. They could fail to understand the intentions and the priority areas of that policy (Morrison 2008) or simply "boycott it," which is the case presented above and further developed in the narrative below.

Micro Forms of Governance and the Everyday

Ahmet is a policeman who agreed to be interviewed and report his work routine, which is far from being simple. Policemen have to socially interact with hundreds of people daily and have no real working hours. To get a job with the police, informants reported that an upfront payment of 6–7000 USD to top officials of the traffic police was necessary. Ahmet also explained that, although his official salary is around 900,000 som,[2] (a bit more than 100 USD), he does barely sees more than 10% of the amount. To touch more than 100,000 som is virtually impossible since money is hardly ever paid out. There is, however, a well-functioning "out-sourcing" system based on perks. His administration provides policemen with "traffic tickets" that have to be "sold" to drivers to earn their salary. In other words, this means that they have full liberty on whom to fine and for how much and that, most or all the money collected will count toward his salary. The most common infraction is to drive without wearing a seat belt since this is visible even without checking the car's documents.

[2] *Som* is the national currency of Uzbekistan. 1 USD is equal to 8070 *som* (April 2018 exchange rate).

However, technical checks are also popular and, once the car is found not meeting safety standards, the driver will be fined by 12,500 som. This, in practice, means that 70–80 tickets per month earn you as much as you are due by the government who will turn a blind eye on how many tickets you issue and how much money you make in a given month. In a separate paper, it was suggested that this can be considered somehow useful to society. Policemen get a salary and they fine drivers who do not abide by the rules so social order is maintained (Polese 2008). However, a complication is introduced by the quotas demanded by traffic police line managers. As Ahmet reports, they get pressure to issue at least 20 tickets per day if they do not want their job to be on the line. At this pace, we are talking of 600 tickets per month, much higher than the amount needed to make a salary so it is easy to imagine that line managers expect a kickback on the tickets issued every day. Besides, they are expected to issue 20 tickets regardless of whether drivers break traffic rules or not, which means in the long run to impose fines on diligent drivers. In addition, traffic policemen are also expected to pay directly a "gruz" (burden) between 50,000 and 100,000 som to their line managers. At the end of the day, policemen are widely scorned in the country and considered highly corrupt but, in Ahmet's words:

> Ordinary people do not know about these problems and therefore hate us. It is politics. We cannot talk about these problems openly. I know many people look upon traffic police as the most corrupt profession in Uzbekistan. Since we do not receive any salary from the state for our work, the money we earn through selling tickets is completely legal. I am also an ordinary man, like everybody else; I have a family, kids to feed! Instead of giving salary, our bosses force us to earn our salary through selling tickets to drivers. So tell me, how should I feed my kids when the state does not pay me any salary? Were I on a standard payroll I would not bother selling tickets to law-abiding drivers.

The register used to define transactions is interesting to our reflections. When describing his informal practices, the traffic policeman tends to use the expression "selling tickets to drivers" rather than saying "imposing a fine on drivers." This shows that the work ethics of traffic policeman is guided by the unwritten rules. In addition, his perception of morality did not seem to take into account the fact that imposing, and pocketing, fines were somehow illegal, at least normatively. Rather than "objective" morality, we could discuss here perceptions of morality (Gill 1998; Polese 2014; Wanner 2005; Werner 2000) in two specific contexts, one is when certain actions, considered illegal by state morality, help a citizen to survive (Blundo and Sardan 2013; Jávor and Jancsics 2016; Rasanayagam 2011; Urinboyev and Svensson 2013; Round et al. 2008) and the other, stemming directly from this, is the case when individual and state morality does not overlap. Our interview illustrates that informal practices may have different meanings and logic within different levels of society. For instance, the elite level corruption, rent-seeking, and "clan struggles" described by scholars such as Ilkhamov (2004) and Collins (2006) are not the same as the everyday "getting things done" practices of low-level traffic policemen that recall the difference between "what I need" and "what I want." Ledeneva warned that this boundary is fluid and that better salaries, usually the first solution imagined by international organizations, do not necessarily help in this case. However, an income that is considered too low can land you into a situation of ineffectiveness where a state cannot impose its moral standards since it does not function, in fact, as a state (Polese 2014).

Governance of the Health Sector

A second informant, that we will call Umida, reported a similar attitude from a different environment. In a similar fashion, she tended to justify what people perceived as "corrupt" practices by the fact that ordinary citizens failed to understand the complexity of the system they were part of. In her view, most of the shortfalls of the system were ascribable to the state or "the system" (possible what Ledeneva called *sistema*). She recalled the Soviet period as a moment when the state provided everything for hospitals and physicians received a good salary. However, after independence, state funding significantly decreased financing for hospitals, equipment became scarce, hospitals over-crowded, electricity and gas cuts became common and, eventually, it all collapsed. Although a number of researchers might not agree with this idyllic view on lack of informality during Soviet times (Grossman 1977, 1989; Critchlow 1988; Treml and Alexeev 1994), it is possible that the situation has worsened after the collapse of the USSR. Umida mentioned that, given that all maternity hospitals are state-owned in Uzbekistan, giving birth should be free of charge, but it is rarely the case and most people pay for maternity services.

Money is passed through handshakes, slipped into a pocket, or left on the table but it is rarely called other than "suyunchi" (literally "joy" in English), where the father or relatives of the new-born baby give cash (or sometimes expensive gifts) to the midwife and nurses who deliver the baby. Usually, it is in the range of 50,000–100,000 som for easier cases and can get up to 500,000 som for birth complications. Her perception is that:

I know my actions are illegal according to law, but real-life circumstances force me to accept suyunchi from patients. Law and real life are completely different things. You will understand what I mean after I explain my work conditions. First, it is very difficult to get a job at a maternity hospital. For instance, if you want to work as a nurse at our hospital, you must pay a bribe of at least 500 USD to top health officials. Second, our salaries are extremely low. A midwife's monthly salary is 800,000 som, around 100 USD, and a nurse's salary is 550,000 som (65 USD). Isn't it frustrating when you pay a 500 USD bribe in order to get a job with a 100 USD salary? Our salary is very low, but I have to feed my kids. I studied for seven years to become a midwife, but I do not receive a high enough salary to live on from the state. Due to my good education, I believe I should earn more money than people who sell potatoes at the bazaar.

She goes on explaining that she finds prices extremely high even at the local bazaar, where meat costs in the region of 35,000 som per kilo and a sack of flour is about 80,000 som. She also regularly needs to buy clothes for her kids so her official salary (800,000 som) is hardly sufficient to survive a few days. Her argument is not too distant from what, in similar settings, has been used to justify the fact that public workers are not public workers all the time but only for the corresponding part of their salary and with the moral manipulation suggested by a colleague that a voluntary bribe is more a gift (see Polese 2008, 2014):

> Since the state does not reward me properly, I have a full right to supplement my salary through suyunchi. I do not force anyone to give suyunchi, but people themselves voluntarily reward me. This is the only way to feed my kids and I do not see any other alternatives. I will not be able to feed my kids if I follow the law. Therefore, it is quite understandable that we expect people to reward us for our efforts.

Trying to make sense of the complicate web of relationship we can start from street-level bureaucrats who, de facto, bring harm to the state with their attitude and behavior. They hijack money from ordinary citizens directly into their pockets bringing harm to society and fail to declare revenues taking away from state revenues. They are not alone in this. At a higher level of the scale, not only their managers do not seem to worry about their behavior, they actually encourage it and benefit from it, fostering harm to the society, to the state and weakening the perception of the state among citizens. From its side, the state does not really struggle to regain symbolic power, authority, or any respect from its citizens. But who is the state? A state and its institutions are a theoretical entity that works only thanks to the work, and agency, of people working in it and united by the idea, and ideology, that they contribute to something bigger than their own welfare. If we take a definition of governance involving traditions and institutions by which authority is exercised we can easily consider both formal and informal attitudes and practices as being part of it. In fact, the way state officers at various levels interact and interpret formal rules can be considered governance, or better a form of governance. True that it has little in common with the way governance is conceived in a number of countries considered effective. But true also that this is the way it is and that, as anarchy is what people make of it (Wendt 1999) also the state is what people make of it.

We agree that a state that pays salaries sufficient to subsistence and where civil servants feel little pressure to "outsource" payments, by burdening other citizens with traffic tickets or other tools sounds more solid than the one described above. However, the question is where should such a transformation start. In Fig. 1 we have shown that a state can actually purchase a given practice. Not long ago, some research on informal payments in health institutions suggested that these payments make the health sector de facto private (Polese 2006a, 2006b). In fact, this is not too distant from the situation where in countries like Australia going with a private insurance allows you to be received when you want by the doctor of your choice (Polese 2014) whereas public insurance gives you little rights. The boundary between legal and illegal is defined by normative criteria that change across time and space. Likewise, welfare and good living for citizens are defined according to normative standards that sometimes are simply not applicable to a given situation. What should be one's attitude once this is faced? Deny the whole system of

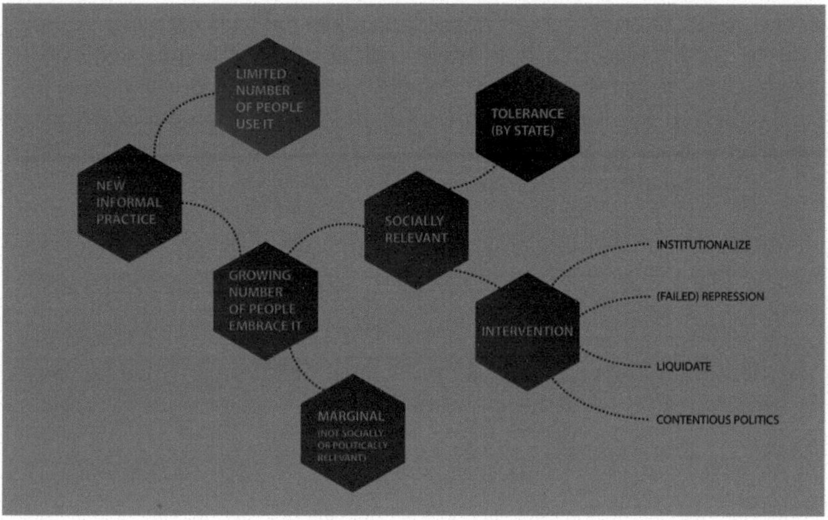

Fig. 1 Evolution of informal practices (Polese 2016)

inter-relations or trying to make sense of it and propose it as an alternative? We are aware that this does not answer but rather introduces new ones. However, our goal here was to show that governance is constructed by bottom actors as much as by top ones (cf. Morris, forthcoming, on deregulation by street-level bureaucrats) and that good governance is not necessarily an objective concept.

Conclusion

To what extent are the stories of midwife and traffic policeman comparable? The professional sphere, amounts of informal payments, ways of bending law, and other details may vary, but both stories reveal the existence of informal, alternative (to the state) governance system that emerge in the course of everyday life, as state officials and citizens engage in transactions of one sort or another, navigating the particular constraints, risks, loopholes or possibilities when getting the things done, securing the basic needs or in the encounters that a taxi driver negotiates with traffic policemen on the amount of bribes. Our study in this sense shows that the everyday (micro-level) informal practices and arrangement complement the formal governance system, making up for the incapacity of the state that fails to secure the basic needs of its citizens. Although the state in Uzbekistan might look omnipotent in terms of its coercive capacity, it is considerably weak in terms of enforcing rule of law and service delivery. As a result, the people of Uzbekistan have created alternative, informal ways to run the state infrastructure. Since the state does not pay them, the state officials such as traffic policemen and midwives do not feel any moral obligation to follow the laws of the state, and therefore opt for informal means of survival and service provision. However, these informal coping strategies are considered by international bodies such as the Transparency International and World Bank as instances of "bad governance," whereas they are more like an ad hoc solution that has been devised by ordinary people in order to keep the infrastructure running. These practices could thus be regarded as an "everyday forms of governance," as this is the system in place that keeps, for example, the health care system and traffic enforcement running.

We assume that good governance leads to long-term benefits to the common good: bad drivers are punished and become safer drivers; fines go to benefit the community, not a single family; insurance companies and municipal authorities cooperate to produce a broader sense of generalized trust in society. But what are our assumptions about the corruption scenario? We have argued that this kind of "petty corruption" is far from disorderly and immoral, but are these informal practices also contributing to the common good, as culturally appropriate means of exchange that reinforce sociability and interdependence, strengthening the social fabric? Our answer is a qualified yes, but qualified because it seems equally true that people are making the best of a bad situation where the state has largely withdrawn from all but its capacity to maintain political stability and interethnic peace functions.

Is everyday governance in Uzbekistan the 'good enough governance' that is the best many societies can hope for? Does it provide for individual physical security, public order, limited economic growth, security from transnational threats, basic health and sanitation services (all yes so far), checks—or not—on high-level corruption, and minimal mechanisms of accountability such as media and civil society (Krasner 2013). Krasner argues that democracy promotion activities often create instability, but that good enough governance creates the kind of stability that eventually gives birth to a moderately prosperous middle class, which in turn will push for democratic reforms on its own. Uzbekistan's strong coercive state has produced security for most citizens and overall stability, but it cannot ensure rule of law and its weak welfare state has withdrawn from the social sphere and in its place informal governance systems have emerged. The coexistence of these two strong/weak state systems may be the main explanation for regime's ability to maintain political stability. It remains to be seen if kleptocracy and the top-down closing of spaces for civil society and diverse expression will be an Achilles heel that prevents Krasner's scenario of an expanding middle class from coming to fruition in Uzbekistan.

Acknowledgements This chapter was originally published as *Political vs Everyday Forms of Governance in Uzbekistan: the Illegal, Immoral and Illegitimate Politics and Legitimacy in Post-Soviet Eurasia.* / Urinboyev, Rustam; Polese, Abel; Svensson, Måns; Adams, Laura L; Kerikmae, Tanel. In: *Studies of Transition States and Societies*, Vol. 10, No. 1, 28.06.2018, p. 50–64. Reproduced with permission of STSS.

References

Acemoglu, D., and T. Verdier. 2000. The Choice Between Market Failures and Corruption. American Economic Review, 90, 194–211.

Billig, Michael. 1995. Banal Nationalism. London: Sage.

Blundo, G., and J.-P.O. De Sardan. 2013. Everyday Corruption and the State: Citizens and Public Officials in Africa. London: Zed Books.

Bueger, C. 2015. The Rise of Informal Governance: Complementing or Challenging the UN System? MUNPlanet.

Christiansen, Thomas, A. Føllesdal, and S. Piattoni. 2003. Informal Governance in the European Union: An Introduction. In T. Christiansen and S. Piattoni (eds): Informal Governance in the European Union. Cheltenhem, UK: Edward Elgar.

Christiansen, T., and C. Neuhold. 2012. International Handbook on Informal Governance. Cheltenham: Edward Elgar Publishing.

Collins, K. 2006. Clan Politics and Regime Transition in Central Asia. Cambridge: Cambridge University Press.

Critchlow, J. 1988. Corruption, Nationalism, and the Native Elites in Central Asia. The Journal of Communist Studies, 4(2), 142–161.

Darden, K. 2008. The Integrity of Corrupt States: Graft as an Informal State Institution. Politics and Society, 36(1), 35–59.

Della Porta, D,. and A. Vannucci. 1999. Corrupt Exchanges: Actors, Resources, and Mechanisms of Political Corruption. New York: de Gruyter.
De Soto, H. 2000. The Mystery of Capital: Why Capitalism Triumphs in the West and Fails Everywhere Else. Basic Civitas Books.
Edensor, T. 2002. National Identity, Popular Culture and Everyday Life. Bloomsbury Publishing.
Fox, J.E. 2017. The Edges of the Nation: A Research Agenda for Uncovering the Taken-for-Granted Foundations of Everyday Nationhood. Nations and Nationalism, 23(1), 26–47.
Fukuyama, F. 2013. What Is Governance? Governance, 26(3), 347–368.
Giddens, A. 1984. The Constitution of Society. Cambridge: Polity Press.
Gill, M.S. 1998. India: Running the World's Biggest Elections. Journal of Democracy, 9(1), 164–168.
Granovetter, M. 1984. Small Is Bountiful: Labor Markets and Establishment Size. American Sociological Review, 49(3), 323–334.
Grindle, M.S. 2011. Good Enough Governance Revisited. Development Policy Review, 29, s199–s221.
Grossman, G. 1977. The Second Economy of the USSR. Problems of Communism, 26, 25–40.
Grossman, G. 1989. Informal Personal Incomes and Outlays of the Soviet Urban Population. The Informal Economy: Studies in Advanced and Less Developed Countries, 150–172.
Gupta, A. 2005. Narratives of Corruption Anthropological and Fictional Accounts of the Indian State. Ethnography, 6(1), 5–34.
Haller, D., and C. Shore. (eds.). 2005. Corruption: Anthropological Perspectives. London; Ann Arbor, MI: Pluto Press.
Heidenheimer, A.J., and M. Johnston. 2011. Political Corruption: Concepts and Contexts. New Brunswick and London: Transaction Publishers.
Humphrey, C. 2012. Favors and "Normal Heroes": The Case of Postsocialist Higher Education. HAU: Journal of Ethnographic Theory, 2(2), 22–41.
Ilkhamov, A. 2004. Archaeology of Uzbek Identity. Central Asian Survey Central Asian Survey J1—Central Asian Survey, 23(3/4), 289–326.
Jávor, István, and David Jancsics. 2016. The Role of Power in Organizational Corruption: An Empirical Study. Administration & Society, 48(5), 527–558.
Jones, R. 2011. People—States—Territories: The Political Geographies of British State Transformation. Malden, MA: John Wiley & Sons.
Kaufmann, D., A. Kraay, and M. Mastruzzi. 2004. Governance Matters III: Governance Indicators for 1996, 1998, 2000, and 2002. Washington, DC: The World Bank, World Bank Policy Research Working Paper No. 3106.
Kerikmäe, T., and S. Särav. 2016. Legal Impediments in the EU to New Technologies in the Example of E-Residency. Baltic Journal of Law & Politics, 8(2), 71–90.

Kerikmäe, T., K. Nyman-Metcalf, and I. Papageorgiou. 2013. Is This Really What I Voted for? Legitimacy of European Integration. Baltic Journal of Law & Politics, 6(1), 45–62.

Kotkin, Stephen, and András Sajo. (eds.). 2002. Understanding and Misunderstanding Corruption. Political Corruption in Transition: A Skeptic's Handbook. Central European University Press, 25–31.

Krasner, S.D. 2013. Seeking 'Good-Enough-Governance'—Not Democracy. Reuters Blogs—The Great Debate.

Laffont, J.J. 2006. Corruption and Development. Understanding Poverty. Oxford Scholarship Online, 161–168.

Ledeneva, A.V. 2006. How Russia Really Works: The Informal Practices That Shaped Post-Soviet Politics and Business. Ithaca, NY: Cornell University Press.

Ledeneva, A.V. 2013. Can Russia Modernise? Sistema, Power Networks and Informal Governance. Cambridge: Cambridge University Press.

Leff, N.H. 1964. Economic Development Through Bureaucratic Corruption. American Behavioral Scientist, 8(3), 8–14.

Migdal, J.S. 2001. State in Society: Studying How States and Societies Transform and Constitute One Another. Cambridge: Cambridge University Press.

Morrison, C. 2007. A Russian Factory Enters the Market Economy. London: Routledge.

Morris, J. 2011. Socially Embedded Workers at the Nexus of Diverse Work in Russia: An Ethnography of Blue-Collar Informalization. International Journal of Sociology and Social Policy, 31(11/12), 619–631.

Morris, J. 2012. Unruly Entrepreneurs: Russian Worker Responses to Insecure Formal Employment. Global Labour Journal, 3(2), 217–236.

Morris, J. 2013. Beyond Coping? Alternatives to Consumption within a Social Network of Russian Workers. Ethnography, 14(1), 85–103.

Morris, J. 2016. Everyday Post-Socialism. London: Palgrave Macmillan

Morris, J. 2019. The Informal Economy and Post-Socialism: Relational Perspectives on the State, Labour and Social Embeddedness. Demokratizatsiya The Journal of Post-Soviet Democratization, 27(1), 9-30.

Navaro-Yashin, Y. 2002. Faces of the State: Secularism and Public Life in Turkey. Princeton University Press.

Ó Beacháin, D. 2012. The Dynamics of Electoral Politics in Abkhazia. Communist and Post-Communist Studies, 45(1–2), 165–174.

Ó Beacháin, D., and R. Kevlihan. 2013. Threading a Needle: Kazakhstan Between Civic and Ethno-Nationalist State-Building. Nations and Nationalism, 19(2), 337–356.

Peters, B. Guy. 2007. Forms of Informal Governance: Searching for Efficiency and Democracy. In T. Christiansen and T. Larsson (eds): The Role of Committees in the Policy-Process of the European Union. Cheltenham: Edward Elgar.

Polese, A. 2006a. De jure oder de facto? Die Gesundheits-und Bildungssektoren in der Ukraine. Ukraine Analysen, 16, 2–4.
Polese, A. 2006b. Border Crossing as a Daily Strategy of Post Soviet Survival: The Odessa—Chisinau Elektrichka. Anthropology of East Europe Review, 24(1), 28–37.
Polese, A. 2008. If I Receive It, It Is a Gift; if I Demand It, Then It Is a Bribe': On the Local Meaning of Economic Transactions in Post-Soviet Ukraine. Anthropology in Action, 15(3), 47–60.
Polese, A. 2010. The Formal and the Informal: Exploring 'Ukrainian' Education in Ukraine, Scenes from Odessa. Comparative Education, 46(1), 45–62.
Polese, A. 2014. Informal Payments in Ukrainian Hospitals: On the Boundary Between Informal Payments, Gifts and Bribes. Anthropological Forum, 24(4), 381–395.
Polese, A. 2016. Corruption State Trust and Informality in Ukraine. In R. Kesek, M. Borora and Z. Jozwik (eds): Addressing Security Risks at the Ukrainian Border Through Best Practices and Good Governance. Amsterdam: IOS Press (NATO Science for Peace and Security Series).
Polese, A., B. Kovacs, and D. Jancsics. 2018. Informality 'In spite of' or 'Beyond' the State: Evidence from Hungary and Romania. European Societies, 20(2), 207–235.
Polese, A., B. Kovacs, J. Morris, and L. Reviakhshvili (eds.). 2017. Post-socialist Informalities: Power, Agency and the Construction of Extra-Legalities from Bosnia to China. London: Routledge.
Polese, A., and J. Morris. 2015. My Name Is Legion. The Resilience and Endurance of Informality Beyond, or In spite of, the State. In J. Morris and A. Polese (eds): Informal Economies in Post-Socialist Spaces: Practices, Institutions, Networks. London: Palgrave.
Polese, A., J. Morris, and B. Kovacs. 2015. Introduction: The Failure and Future of the Welfare State in Post-Socialism. Journal of Eurasian Studies, 6(1), 1–5.
Polese, A., L. Rekhviashvili and J. Morris. 2016. Informal Governance in Urban Spaces: Power, Negotiation and Resistance Among Georgian Street Vendors, Geography Research Forum, 36(2), 15–32.
Polese, A., L. Rekhviashvili, and S. Murru. 2018. Negotiating Public Space Informally: Language Politics and Language Use Practices in Odessa High Schools. Space and Culture (forthcoming).
Polese, A., and R. Kevlihan. 2015. Locating Insurgency, Informality and Social Movements on a Spectrum: Is There a Theory Linking Them All? 9Th Pan-European Conference on International Relations of EISA, Naxos 23–26 September 2015.
Polese, A., and T. Stepurko. 2017. Love the Briber, Hate the Bribe, Transitions online, 12 January

Puetter, Uwe. 2006. The Eurogroup. How a Secretive Circle of Finance Ministers Shape European Economic Governance. Manchester: Manchester University Press.

Rasanayagam, J. 2011. Informal Economy, Informal State: The Case of Uzbekistan. International Journal of Sociology and Social Policy, 31(11/12), 681–696.

Richardson, T. 2008. Kaleidoscopic Odessa: History and Place in Contemporary Ukraine (Vol. 35). Toronto: University of Toronto Press.

Risse, Thomas. 2012. Governance Configurations in Areas of Limited Statehood: Actors, Modes, Institutions, and Resources. SFB-Governance Working Paper Series No. 32, April 2012.

Rivkin-Fish, Michele. 2005. Bribes, Gifts and Unofficial Payments: Rethinking Corruption in Post-Soviet Russian Health Care. In Dieter Haller and Shore (ed): Corruption: Anthropological Perspectives. London: Pluto Press.

Rose-Ackerman, S. 1999. Corruption and Government: Causes, Consequences, and Reform. Cambridge, New York: Cambridge University Press.

Round, J., C.C. Williams, and P. Rodgers. 2008. Corruption in the Post-Soviet Workplace: The Experiences of Recent Graduates in Contemporary Ukraine. Work, Employment and Society, 22(1), 149–166.

Routh, Supriya. 2011. Building Informal Workers Agenda: Imagining Informal Employment in Conceptual Resolution of 'Informality'. Global Labour Journal, 2(3).

Scott, J.C. 1985. Weapons of the Weak: Everyday Forms of Peasant Resistance. New Haven, CT: Yale University Press.

Scott, J.C. 2012. Infrapolitics and Mobilizations: A Response by James C. Scott. Revue française d'études américaines, (1), 112–117.

Skey, M. 2015. 'What Nationality He Is Doesn't Matter a Damn!' International Football, Mediated Identities and Conditional Cosmopolitanism. National Identities, 17(3), 271–287.

Stepurko, T., M. Pavlova, I. Gryga, and W. Groot. 2013. Informal pAyments for Health Care Services–Corruption or Gratitude? A Study on Public Attitudes, Perceptions and Opinions in Six Central and Eastern European countries. Communist and Post-Communist Studies, 46(4), 419–431.

Stone, R.W. 2009. Informal Governance: International Organizations and the Limits of US Power. Unpublished Manuscript, University of Rochester.

Treml, V.G., and M. Alexeev. 1994. The Growth of the Second Economy in the Soviet Union and Its Impact on the System. The Postcommunist Economic Transformation, 221–247.

Troitino, D.R. 2013. European Integration: Building Europe. New York: Nova Science Publishers.

Urinboyev, R. 2011. Law, Social Norms and Welfare as Means of Public Administration: Case Study of Mahalla Institutions in Uzbekistan. The NISPAcee Journal of Public Administration and Policy, 4(1), 33–57.

Urinboyev, R. 2013. Living Law and Political Stability in Post-Soviet Central Asia. A Case Study of the Ferghana Valley in Uzbekistan. Ph.D. Dissertation, Lund Studies in Sociology of Law. Lund University, Lund.

Urinboyev, R. 2014. Is There an Islamic Public Administration Legacy in Post-Soviet Central Asia? An Ethnographic Study of Everyday Mahalla Life in Rural Ferghana, Uzbekistan. Administrative Culture, 15(2), 35–57.

Urinboyev, R. 2015. Local Government Capacity in Post-Soviet Central Asia. Public Policy and Administration, 14(3), 177–199.

Urinboyev, R. 2016. Migration and Transnational Informality in Post-Soviet Societies: Ethnographic Study of 'Po rukam' Experiences of Uzbek Migrant Workers in Moscow. In A.-L. Heusala and K. Aitamurto (eds): Migrant Workers in Russia: Global Challenges of the Shadow Economy in Societal Transformation. London: Routledge, 70–93.

Urinboyev, R. 2017. Establishing an "Uzbek Mahalla" via Smartphones and Social Media: Everyday Transnational Lives of Uzbek Labor Migrants in Russia. In M. Laruelle (ed): Constructing the Uzbek State: Narratives of Post-Soviet Years. Boulder, CO: Lexington Books, 119–148.

Urinboyev, R., and A. Polese. 2016. Informality Currencies: A Tale of Misha, His Brigada and Informal Practices Among Uzbek Labour Migrants in Russia. Journal of Contemporary Central and Eastern Europe, 24(3), 191–206.

Urinboyev, R. and M. Svensson. 2013. Living Law, Legal Pluralism, and Corruption in Post-Soviet Uzbekistan. Journal of Legal Pluralism and Unofficial Law, 45(3), 372–390.

Urinboyev, R., and M. Svensson. 2014. Rethinking Corruption in Post-Soviet Uzbekistan: Ethnography of 'Living Law'. In K. Papendorf, S. Machura, and A. Hellum (eds): Eugen Ehrlich's Sociology of Law. Zürich/Berlin: LIT Verlag, 207–238.

Urinboyev, R., and M. Svensson. 2017. Corruption, Social Norms and Everyday Life in Uzbekistan. In: I. Kubbe and A. Engelbert (eds): Corruption and Norms: Why Informal Rules Matter? Basingstoke: Palgrave, 187–210.

Wanner, C. 2005. Money, Morality and New forms of Exchange in Postsocialist Ukraine. Ethnos, 70(4), 515–537.

Wendt, A. 1999. Social Theory of international Politics. Cambridge: Cambridge University Press.

Werner, C. 2000. Gifts, Bribes, and Development in Post-Soviet Kazakstan. Human Organization, 59(1), 11–22.

Westerwinter, O. 2016. The Politics of Informal Governance. Presented at the SVPW Jahreskongress, Basel: University of St. Gallen.

World Bank. 2018. The Worldwide Governance Indicators [online]. Available from: http://info.worldbank.org/governance/wgi/index.aspx#home [Accessed 4 May 2018].

Competing

Mixed Perceptions of State Responsibility Among Informal Sector Participants in MENA

Anil Duman

Introduction

Informality is growingly accepted to be an encompassing concept touching on all aspects of societies and how they are governed, which goes well beyond the contours of economic transactions (Hart 2006; Polese et al. 2017). Nevertheless, in majority of the economics literature, the regulatory role of state is essential in drawing the line between informal and formal economic activities (Lewis 1955; de Soto 1989; Perry et al. 2007). In three primary theoretical approaches to informal sector,[1] different explanations for the existence, and persistence of such economic activities are highlighted. While the modernist school argues that informal sector is a remnant of the traditional production systems

[1] Informal economic activities and informal sector are used interchangeably throughout the chapter.

A. Duman (✉)
Central European University, Vienna, Austria
e-mail: dumana@ceu.edu

A. Polese (ed.), *Informality, Labour Mobility and Precariousness*, International Political Economy Series,
https://doi.org/10.1007/978-3-030-82499-0_10

and would disappear as the economy progress (Lewis 1955), the neoliberal approach emphasizes the imperfections in governance and claims that high taxes, corruption, and excessive state interventions make economic agents to voluntarily exit the formal sector to minimize their costs of operation (de Soto 1989). Structuralist perspective, on the other hand, suggests that informal–formal divide is tenuous since informal enterprises are functionally integrated into the global value chains via subcontracting and outsourcing. People work in these largely unregulated and poorly paid jobs because they are excluded from the formal sector (Castells and Portes 1989; Davis 1994).

All three schools have varied expectations about the causes and consequences of informal sector, and consequently, empirical research following each approach adopted distinct measurements. In numerous studies, both for workers and entrepreneurs, registration into the existing social security and tax system are considered to be the primary distinguishing features of informal economy (Acar and Tansel 2016). There are also, several researchers who take low productivity and small enterprise size as the main characteristics of informal economic activities (Perry et al. 2007; Kanbur 2017). Nonetheless, empirical literature reveals mixed evidence for all three theoretical models. On the one hand, there is no secular decline in the size of informal sector with economic development and technological upgrading. On the other hand, corruption and quality of governance indicators are not fully explanatory for the variance across countries. And despite the importance of social protection for the most vulnerable groups, demographic, occupational, and regional features remain to have profound effects on selection into informal sector (Chen 2012; Kanbur 2017; Duman and Duman 2020). As a result, heterogeneity of the informal economic activities is gradually recognized in the literature and socioeconomic implications are begun to be discussed by identifying the segments within the informal sector.

Regardless of the difference in conceptualization of informal sector and subsequent measurements, it has been a fundamental part of the economy and livelihoods of large segments of the population across the world as well as in the Middle East and Northern African (MENA) region. As opposed to the theoretical expectations of informal economic activities shrinking with development and modernization, there is no clear evidence of reduction in many countries. In most emerging and developing economies, informal employment remains to be around 44%, although some of these economies have experienced high rates of growth and institutional transformations over the last three decades (World Bank 2019). Similarly, informal sector is quite extensive and persistent in the MENA region, which suggests that it can be a salient factor in determining people's opinions about certain policies and preferences. While the causes and consequences of informal sector have been broadly analyzed in the literature, its political implications and particularly the variation of policy preferences within informal sector participants have received less attention. Not only informal sector participants can have different policy preferences in comparison to their formal sector workers, but they also might greatly vary depending on how well or worse-off they are in these positions.

My objective in this chapter is twofold; first to compare the policy preferences and opinions of formal and informal sector participants in a number of MENA countries.[2] The selection of countries and time frame is based on the availability of internationally comparable data, which is derived from the Arab Barometer.[3] As mentioned above, MENA region has one of the highest share of informal economic activities in the world, and the size of the sector has been quite stable over decades, which suggest that its effects are not ephemeral and can be influential on citizens' perceptions. Besides, the research on informal sector in MENA countries is extremely limited, especially with regards to its political consequences. Given the limited availability of data at the international level, the definition of informal economic activities in this chapter is entirely based on the status in the labor market, namely, if the employees have contracts or not. There are several researchers who also adopt the same definition of informal employment and categorize workers according to their registration into social security and legal protections offered by contractual relations (Acar and Tansel 2016; Hussmanns 2005).

Second, I extend the existing discussions on the relationship between informality and public opinion by considering the diversity among the informal employees. I propose that informal sector participants who voluntarily opted for such activities would have greater confidence in governments even though public sector failures such as corruption and heavy bureaucratic burden might lead to higher share of informal

[2] These countries are Algeria, Egypt, Iraq, Jordan, Kuwait, Lebanon, Libya, Morocco, Palestine, Sudan, Tunisia, and Yemen.

[3] A detailed description of the survey is presented in Sect. 3.

economic activities. People who are at the upper tier of the informal sector, in other words, who exited from the formal sector, might perceive the formal–informal practices as complementary and sustain their relatively rewarding informal activities. Hence, their confidence and trust in governments are expected to be positive despite their informal attachment to the economy. On the other hand, people who are at the lower tier of the informal sector with low paid and insecure jobs might have more negative opinions about the state, especially with regards to its inability to meet the redistributive expectations. As a result, excluded informal sector workers are anticipated to have lower confidence and trust in their governments compared to their exited peers. Formal and informal sector economic activities would be viewed as substitutes and state's inability to provide adequate safety nets for the most disadvantaged groups would be highly problematic for this group.

From the individual level analysis, I found that there is no significant impact of formal–informal divide on political opinions and economic evaluations in MENA region once country dummies are included. People who have formal or informal jobs do not differ in their views on corruption and trust. Similarly, there is no variation with regards to individual evaluation of government performance on job creation and redistribution when only formal–informal divide is regarded. However, my results indicate that diversity within the informal sector is crucial as income mediates the relationship between political opinions and employment. The likelihood of asserting that corruption is high and possibility of not trusting the government goes down drastically for people in the upper tier. These findings support the idea that individuals who exited the formal economic activities trust their governments whereas bottom tiers of the informal sector are much more critical. When job creation and redistribution are considered, the effect of income is more complicated and in the informal sector, only after a certain threshold negative evaluation go down. This can also be taken as evidence for exclusion-exit approach since high earning informal employees are much less likely to assess the governments for not creating jobs and lack of redistribution even though they do not have legal and social protection. These findings can be beneficial for our understanding of informal sector and points to the significance of heterogeneity among the informal sector participants. They can also be helpful to policymakers as they might need to develop targeted policies to motivate the formalization of economic activities by exited and excluded individuals.

The chapter is organized as follows. In the second part, I briefly summarize the literature on the causes and consequences of informal economic activities, especially in relation to how informal employment and political opinions are linked. I present data, operationalization, and empirical findings from MENA region in part three. This section also includes a discussion of the results and data limitations. In the final part of the chapter, brief concluding remarks and policy implications are offered.

Overview of Models of Informal Sector

Informal sector is an essential part of work and production in almost all countries, and informal practices are very common both for economic transactions and governance. Yet, political consequences of informal work are relatively an understudied topic in social sciences, and big chunk of the existing work doesn't offer systematic examination of how informal sector participation shape preferences and attitudes. The relationship between an individual in the informal sector and state is expected to be highly distinct from the formal sector counterparts as the work and income of the former are outside the regulatory and protective framework. This looser attachment to state and its existing institutions can be either voluntary or involuntary, and the distinction establishes the basis of the well-known theoretical approaches to informal sector. On the one hand, there are people excluded from modern economy and publicly provided benefits, who have to maintain their livelihoods engaging in informal work. On the other hand, certain firms and individuals decide to exit from the formal institutions according to the evaluations of private costs and benefits (Hirschman 1970; Perry et al. 2007). While it is easier to theoretically distinguish these groups, it is harder to identify the exact reasons for people exiting the formal arrangements. Similarly, the factors that generate exclusion of particular individuals or groups from the more lucrative economic opportunities and social protection are challenging to pin down empirically.

According to modernization school, there are rigid divisions between formal and informal activities. With productivity gains and technological upgrading, informal sector is expected to get smaller, and duality of the economic structure would recede with industrialization and urbanization (Lewis 1955; Portes and Hoffman 2003). However, empirical support for dualism is rather weak and informal economy remains to be persistent in various countries even when they experience growth and higher income levels. In Latin America, Africa, and Arab countries, informal sector makes up almost 50% of the economy, and over time there is no secular decline in its size and even some increase across several countries. For example, it has been found that informal employment expanded in Egypt over the last twenty years both because of very limited mobility between the sectors and increased risks of unemployment among the formally employed (Elsayed and Wahba 2019).

At the macro level, there are a number of reasons behind the tenacity of informal sector such as economic and political turmoil, long-lasting inequalities, and mixture of low-quality institutions and weak enforcement. In line with the neoliberal approach to informal sector, which focuses on governance imperfections, empirical literature demonstrates that the association between the size informal economic activities and regulatory burden is positive. Additionally, tax evasion is higher in countries that have above-average rates and weak enforcement, which can be taken as another marker of informal economic activities (Vuletin 2008; Ordonez 2014). For example, corruption is argued to be one of the biggest obstacles against formalization in non-Gulf countries of MENA region, and survey evidence shows that level of trust to governments are rather low (World Bank 2019). In Morocco, people find taxation extremely burdensome and conceal their earnings to avoid tax payments. The same study also discusses that corruption is another major determinant of why individuals go for informal economic activities (Bourhaba and Hamimida 2016). Hence, corruption and other deficiencies of governance mechanisms not only harm people's loyalty to the system but they also lower the ability of particular groups in the society to have access to public services.

While inefficiency of bureaucracy and low-quality institutions are prominent for informal sector participation, there are also subjective determinants of how people evaluate the performance of states. Noncompliance and disregard of the regulations are contingent on perceptions of effectiveness and fairness of state actions (Hirschman 1970; Saavedra and Tommasi 2007). If individuals believe that tax and transfer system in a country is ineffective or unfair, they are less likely to contribute and might try to free ride or find alternative solutions to their collective action problems. In other words, citizens might be interested in regulating a given situation in spite of the state when they deem the existing governance mechanisms inappropriate (Polese et al. 2017). At the same time, inefficiency of formal institutions and unequal outcomes lead to various individuals to be excluded from the system. Inequality of opportunity as well as lack of access to basic services mean that certain parts of the society are left voiceless and cannot channel their objections to the system through formal means (Perry et al. 2007). In such cases, citizens try to make up for the absence of formal governance structures, which is defined as informality beyond the state (Polese et al. 2017). As can be understood from these conceptualizations, the factors causing informality to grow, either in the form of exit or exclusion, also alter the relations between people and the state. In view of this, it can be argued that political views and evaluations of informal sector participants would be highly distinct.

Besides the social contract between citizens and states, and perceptions about the functioning of governments, there are material factors affecting the political capacities of formal and informal sector participants. Informal employees are discussed to be less mobilized as their resources are limited and they face greater economic risks Unionization and other types of collective organizations are barely present among the informal sector workers. Thus, bargaining power against employers is extremely low and heavily depends on individual endowments, which elevate the existing inequalities. Like other regions, informal employment is widespread among younger and uneducated people in MENA countries, which mean that their individual bargaining options are also highly deficient (Angel-Urdinola and Tanabe 2012). Furthermore, informal sector participants fall outside of the regulatory coverage placing them in even a more disadvantaged status. As a result, their strategies to cope with labor market uncertainties are far more restricted than the formal sector workers. For example, during the latest pandemic, the poorest of the informal employees are hit hardest economically, and at the same time, they are not able to get public assistance or healthcare (Holtmeier and Alami 2020). Hence, informal sector workers, even when they have similar income levels to their formal counterparts, need to individually insure against labor market risks since they fall outside the scope of social security. I argue that the effect of income on policy demands and political opinions are varied across sectors, and for high earners, informal sector participation does not necessarily lead to more negative perceptions despite governance failures.

In relation to the materially disadvantaged position of informal sector participants, it is argued that mobilization and class identity formation among this group are quite weak. On the one hand, informal workers are employed in small firms and self-employed are usually solo proprietors. On the other hand, heterogeneity of informal economic activities and reasons for being in the informal sector cause atomization and lack of shared interests (Baker and Velasco-Guachalla 2018; Kurtz 2004). Moreover, it is claimed that a big part of the informal sector consists of petty entrepreneurs, who might oppose interventionists policies of the state and align with capitalist values (de Soto 1989). For example, self-employed in Tunisia asserted that they are content with the revenues they generate from their informal economic activities and do not have any incentives to join the formal sector or preference for regulation (Trabelsi 2013). Also, it is shown that better-perceived government performance in the form of corruption control and freer elections in Arab nations increase the support for democracy but clientelistic practices have no such effect (Benstead et al. 2019). These ideational factors and diversity of interests make it difficult for people to unite around common goals and muster enough power to shape policymaking. However, implications of scarce mobilization and capitalist values are distinct for the poorer and wealthier segments of the informal sector. The former is likely to suffer from inability to collectively act and does not have the private resources to shield against labor market risks. In contrast, latter group might prefer a hands-off approach due to their entrepreneurial inclinations and support the government policies as they are able to benefit from the system economically.

There are few studies that examine the relationship between informal employment and political attitudes for other regions in the world. It has been shown that labor market dualization is not a significant predictor of social policy preferences and individual level opinions in Latin America (Baker and Velasco-Guachalla 2018; Berens 2015a). However, it has been also demonstrated that the size of the informal sector in a country lowers the support of formal sector workers for public welfare goods due to lack of contribution by the informal employees (Berens 2015b). Additionally, income has differential effect on the redistributive preferences of formal and informal sector participants, and informal employees are shown to be less driven by material factors (Berens 2015a). Even though this line of research is immensely helpful in understanding the effects of informal work on individual-level choices, there can be still major differences between formal and informal sector participants depending on the social policy under investigation. Besides, these studies are exclusively focused on Latin America and inclusion of another region, MENA, is useful to check the validity of these arguments and highlight any possible inter-regional differences.

The diversity of the informal sector is progressively acknowledged in the literature; however, most of the studies examine the economic consequences of segmentation. I aim to expand this literature by considering the perceptions of individuals on a number of political and economic issues both by examining the differences between formal and informal employees, and variance among the informal sector participants. In Arab countries, labor market segmentation is also very acute, particularly due to rather large share of public employment and strict regulations for formal private employment. The system exacerbates the inequalities and put informal sector workers in extremely precarious positions and weakens the links between business owners and the society (Hertog 2017). Examining the relationship between informal employment and political opinions at the individual level for MENA region can add to our understanding of how citizens' perceptions are shaped, and if dualism carries any weight for political views. Second, I attempt to elucidate the relationship between various segments of informal work and individual views on policy. Economic aspects of diversity are well researched and prominent earning differences as well as variation in personal characteristics between upper and lower tiers are uncovered in the literature (Chen 2012; Kanbur 2017; Duman and Duman 2020). Nevertheless, political attitudes of lower and upper-tier informal sector participants are not systematically examined, which is one of the main objectives of the chapter.

Informal Sector and Its Effect on Individual-Level Opinions

Data Sources

All the data for empirical analysis is obtained from the 5th wave of the Arab Barometer, which is a source of high-quality public opinion surveys in the MENA region. This is the most comprehensive and standardized international survey covering a range of questions about demographic features of the participants as well as their views on a number of issues ranging from political institutions to international relations. Survey was first launched in 2006 and the last wave was conducted in 12 countries between the years of 2018 and 2019.[4] For each country, the survey is nationally representative and covers the citizens aged 18 and above. Individual is the unit of analysis, who are selected randomly from households and each household is represented only by one individual (for technical details, see arabbarometer.org, n.d.). From the last wave, in total, I have 26,780 observations for Algeria, Egypt, Iraq, Jordan, Kuwait, Lebanon, Libya, Morocco, Palestine, Sudan, Tunisia, and Yemen. Kuwait's classification for occupational status, contract, and income is uniquely different than the rest of the sample, which led me to drop it from the estimations. Given that size of informal sector in Gulf countries is much lower than the other MENA countries due to immigrant labor and public employment, exclusion of Kuwait should not bias the results.

In order to account for public opinion about various issues, I use four different questions in the survey as dependent variables. The first question asks the extent of corruption within the national state agencies and institutions in the respondent's country, and the answers vary between not at all to large extent. For trust, the participants are requested to declare their level of trust (a great deal of trust to no trust at all) for a number of institutions including central government, parliament, religious leaders and local government. Arab Barometer deals with economic assessments of the citizens extensively and there are several queries on the past, current, and future of the economy of the country. However, these are very broad, and I chose to focus on two issues that might be more relevant for informal sector participants; government's performance on creating jobs and narrowing the gap between rich and poor. For each, the respondents can rank how the current government is doing from very good to very bad. Since all the dependent variables have ordered categorical values, I utilize ordered logistics regressions to inspect the effect of informal employment and its diversity. In order to control for heterogeneity at the country level, I run logistics regressions with country dummies. For robustness checks, I also recorded the dependent variables based on the least favorable answers and run logistics regressions with binary values.

The main independent variable is informal sector employment, which is not easy to operationalize because of data problems and consistency across

[4] The data is cross sectional, and, in this study, the most recent available year is used. To my knowledge there is no panel data at the international level that covers informal sector participants.

countries. In the survey, there is a question directly asking about whether the respondents have a contract for their current position if they stated that they are employed. I formed a dummy variable based on this question for informal employment. No contract is used to operationalize informal employment in other studies as well since it involves the most crucial aspects. Not having a contract as a wage worker means not being able to benefit from social security benefits and at best little protection by the legal framework. I also included self-employed in an alternative measure of informal sector for robustness checks. To pinpoint the diversity within the informal sector and distinguish between excluded and exited workers, I use income scales and grouped the bottom and top three categories as lower and upper tiers, respectively. Additionally, a list of control variables is included in the estimations, which are common to the literature and make my findings comparable to the previous studies. These are age, gender, highest level of education, residing in rural area, religiosity, being employed in the public sector (Baker and Velasco-Guachalla 2018; Berens 2015a, b). Since the survey does not have a question on political ideology, I constructed a variable, statists, based on responses about the necessity of citizens supporting the government's decisions, even if they disagree with them. Individuals who agree with this statement are expected less likely to challenge governments' decisions and performance. Lastly, I add a control variable on the number of employees in the household to capture the overall employment composition of the household, which can be blended and lead to varied opinions.

Overview of Informal Employment and Perceptions

As can be seen from Fig. 1, informal employment measured by not having a contract is very prevalent in the region where the average is more than 50%. Even though the ratio is well above the developed country levels, it is still comparable to various economies in Asia and Africa. There is also a lot of variation within the region as informal employment oscillates between nearly 7% in Morocco to almost 58% in Libya. In addition, the proportion of self-employment, which is sometimes taken as a proxy for informal sector, is around 43% on average. Once again, there are important differences among MENA countries where the share of self-employed is 56% in Morocco and approximately 13% in Libya. These figures are very much in line with previous studies on the region, which show that some of these economies represent the most informal economies in the

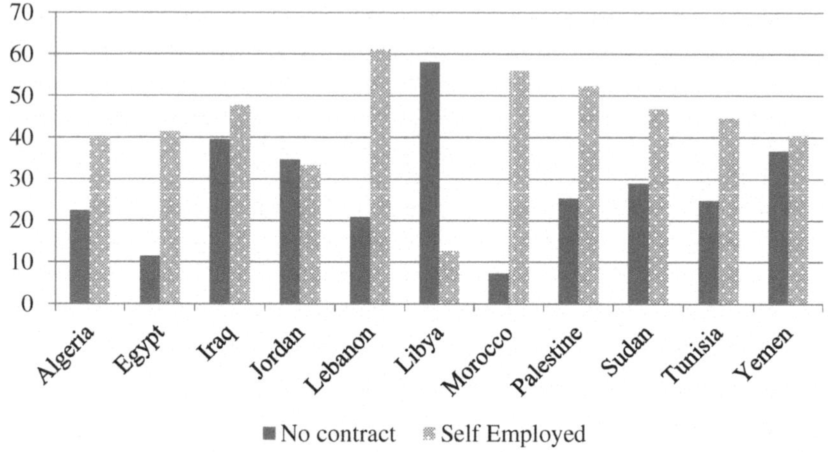

Fig. 1 Share of employees with no contract and self-employed (*Source* Author's calculations based on Arab Barometer)

world (Loayza and Wada 2011). Particularly, in countries with larger agricultural employment such as Morocco and Libya, total number of informal sector employees is much higher. In contrast, urbanization and bigger public sectors in Egypt and Lebanon reduce the proportion of informal economic activities in these economies (Angel-Urdinola and Tanabe 2012). However, on average informal sector is very sizable in MENA region and can be a relevant factor for the differences in policy preferences and opinions.

I present the degree of trust and corruption perceptions across informal workers, formal workers, and self-employed in Table 1. The share of people affirming that they have great trust in government is around 13% for the formal sector employees while it is merely 8% for the informal ones. On the other hand, the ratio of individuals who declared that they do not trust the government at all is almost 47% for the informal workers and 38% for the formal workers. While the self-employed have lower trust than the formal sector participants, they still rank above the informal workers. Hence, it can be said that there are divergences across employment types with regards to trust in governments across MENA countries. A similar picture arises when I consider perception of corruption. Overall, the share of respondents asserting that there is no extent of corruption in their respective nations is quite low, nearly 2% for both formal and informal employees and 2.4% for self-employed. However, the opinions get more varied if large extent of corruption is taken into account. Among the informal sector employees, 62% claim that corruption is high whereas the same ratio is 56% for self-employed and almost 59% for formal sector employees. These figures suggest that informal sector participation in MENA region can be a potentially explanatory factor behind the political views.

In Table 2, respondents' opinions on the current economic situation, and government's performance on creation of jobs as well as narrowing

Table 1 Trust and corruption perceptions by type of employees (%)

	Great trust	*Some trust*	*Little trust*	*No trust*
No contract	8.04	20.72	24.3	46.93
Formal	13.26	23.49	25.55	37.69
Self-employed	9.64	21.73	26.19	42.44
	Large extent	*Medium extent*	*Small extent*	*No extent*
No contract	62.42	26.58	9.07	1.92
Formal	58.89	27.03	12.03	2.04
Self-employed	56.12	30.14	11.32	2.42

Source Author's calculations based on Arab Barometer

Table 2 Economic performance evaluations by type of employees (%)

	Very good	*Good*	*Bad*	*Very bad*
Current economy				
No contract	1.88	22.63	41.15	34.35
Formal	3.52	23.96	38.37	34.14
Self-employed	2.32	21.69	37.43	38.56
Creating jobs				
No contract	2.94	16	39.52	41.55
Formal	3.11	18.22	37.51	41.17
Self-employed	2.03	12.81	37.36	47.8
Narrowing gap between rich and poor				
No contract	3.11	14.36	41.57	40.95
Formal	2.62	18.18	40.98	38.22
Self-employed	2.13	14.41	40.21	43.25

Source Author's calculations based on Arab Barometer

the gap between rich and poor are shown. As can be noticed, there is almost no difference with regards to the economic performance among employees in the formal and informal sector. In both cases, 34% of the survey participants indicated that the current economic situation in their country is very bad. For the self-employed, the same ratio is almost 39% in MENA region. It should be also noted that the respondents who evaluated current economic situation as very good are merely 2% for the informal employees and the self-employed, and 3.5% for the formal sector workers. Similar results are obtained for job creation and redistribution where the difference arises due to self-employment rather than formal-informal divide. For example, nearly 41% of the informal and formal workers indicate that the government is functioning poorly with regard to creating jobs. While there is a slight gap between these groups when redistribution is considered, still the share of people asserting that the government is performing very badly is 41% versus 38% for informal and formal sector participants, respectively. These figures combined with the above results point toward that informal sector participation might have more weight in explaining political opinions instead of economic ones in MENA region. However, it should be kept in mind that other factors that can shape people's perceptions on these issues are not taken into account in these aggregate pictures.

Figure 2 presents the share of respondents in MENA region across income tiers, who asserted that there is great extent of corruption in their country and government is doing very poorly in terms of narrowing the gap between rich and poor. At the bottom end, it can be observed that informal sector workers have more negative views about the level of corruption as opposed to the workers with similar incomes but are employed in the formal sector. While 75% of the low earning informal sector employees report that there is very high corruption, the same ratio is 70% for the formal sector counterparts with similar earnings. Interestingly, for the top income tier, it is the opposite where 53% of the informal and 63% of the formal sector workers declare that there is large extent of corruption. The second part of the figure demonstrates the views on government's performance on redistribution. Once again, there are stark differences not only across formal and informal employees but also according to their income levels. Within the bottom earners, formal sector workers have more positive evaluation of government's performance with 36% ranking it as very bad. This ratio goes up to nearly 44% for the informal sector workers receiving comparable incomes. In contrast for

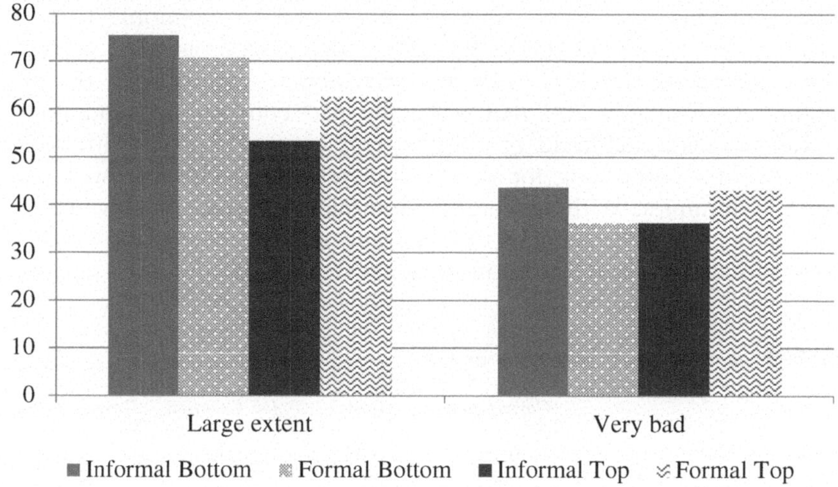

Fig. 2 Great extent of corruption and very bad performance on redistribution (%) (*Source* Author's calculations based on Arab Barometer)

top earners in informal employment, government's efforts to decrease the gap between rich and poor are more satisfactory, only 36% stating a very poor performance. For formal sector-top earners, the evaluation is much more negative, 43% of the survey participants. Both of these findings imply that there is significant diversity within the informal employees, and incomes play a distinct role in mediating the opinions about government performance on several economic policies across formal and informal sectors.

Empirical Analysis

The discussions in the previous section indicate that informal and formal sector employees have diverse opinions about the governments, especially with regards to trust and corruption. On average, people without contracts have lower trust and higher perception of corruption in MENA region. Moreover, the findings underline the importance of heterogeneity within the informal employees. While these tables give us a broad picture of how different types of employment might be affecting the views about governments at the individual level, they do not control for other individual-level factors that can shape political opinions. Table 3 presents the estimation results for all the dependent variables; corruption perceptions, level of trust in government, performance on job creation performance and narrowing the gap between rich and poor. As can be seen, coefficient of informal employment changes with specification after controlling for various individual-level characteristics and country dummies. Without country dummies, informal employment is explanatory for all dependent variables and increase corruption perceptions, reduce trust and lead to more negative evaluation of governments'

Table 3 Estimation results for four DV

	Corruption	*Trust*	*Job Creation*	*Redistribution*
Informal	−0.10	0.04	−0.06	0.01
	0.07	0.07	0.06	0.06
Age	0.01*	0.00	0.00	0.00
	0.00	0.00	0.00	0.00
Gender	0.06	0.11	0.02	−0.08
	0.07	0.07	0.06	0.07
Education	0.12***	0.06**	0.09***	0.13***
	0.03	0.03	0.03	0.03
Rural	−0.03	0.07	0.00	0.01
	0.08	0.08	0.07	0.08
Statist	−1.14***	−1.35***	−1.09***	−1.15***
	0.07	0.07	0.06	0.07
Religiosity	0.04	0.02	0.14**	−0.16**
	0.07	0.07	0.06	0.06
Public sector	−0.42***	−0.32***	−0.18***	−0.17**
	0.08	0.07	0.07	0.07
# of employees in the HH	0.00	−0.03	0.06**	−0.02
	0.03	0.03	0.03	0.03
Middle tier	−0.1	−0.16	−0.02	0.08
	0.15	0.13	0.11	0.12
Upper tier	−0.15	−0.24	−0.12	−0.08
	0.1	−0.16	−0.02	0.08
# of observations	4164	4134	4465	4332
Pseudo R^2	0.1	0.14	0.07	0.08
Prob > chi2	0.00	0.00	0.00	0.00

Notes All columns present the coefficients from ordered logit regressions with country dummies. Standard errors are in the second row. *, **, and *** denote statistical significance at the 0.1, 0.05 and 0.01 levels, respectively performance for job creation and redistribution.[5] For example, informal employees report that the extent of corruption is lower in comparison to formal employees by 0.1 points. But they also declare lower level of trust in the government. On the other hand, informal employees' evaluations of government with respect to job creation are more positive, around 0.06 points but their ranking is more adverse if redistribution is taken into account. Nevertheless, the coefficients are insignificant regardless of the dependent variable employed in the models, which suggest that there is no distinction between formal and informal workers' political views. In this sense, my findings are comparable to other studies that also infer no statistically significant effect of informal sector on political attitudes and policy preferences (Baker and Velasco-Guachalla 2018; Behrens 2015a).

Even though informal employment appears to be unrelated to political opinions when country dummies effects are included, still there can be prominent differences among the informal sector workers. I try to capture the diversity of informal employees by categorizing them according to the income groups. Before I move on to the discussion of how informal sector heterogeneity can shape perceptions, it is noteworthy to look at the control variables. All variables are highly robust across specifications and keep their signs and significance levels. Education, statism, and being employed in the public sector are explanatory across all four dependent variables. While people believe that there is more corruption and have lower level of trust as the education goes up, the link is the opposite between these and public sector employment. As mentioned, in Arab countries, labor market segmentation extends to public–private sector, which can be the cause of this solid relationship. Moreover, if individuals think that people should support government decisions even when they disagree, as expectedly, they are more trusting, and their perceptions of corruption are less adverse. For job creation and redistribution, higher education reduces the likelihood of positive evaluations. Public employment and statism have the opposite effect on the same dependent variables.

In Table 4, I examine the relationship between the diversity of informal employment and dependent variables by including an interaction variable

[5] The results are not presented for the reasons of brevity but can be asked from the author.

Table 4 Estimation results with interaction of income and informal employment

	Corruption	*Trust*	*Job creation*	*Redistribution*
Informal	0.01	−0.02	−0.24**	−0.01
	0.13	0.12	0.12	0.13
Informal lower	1.14***	−0.08	0.13	−0.27
	0.36	0.32	0.26	0.27
Informal middle	−0.07	0.11	0.26*	0.06
	0.15	0.14	0.14	0.14
Age	0.01**	0.00	0.00	0.00
	0.00	0.00	0.00	0.00
Gender	0.06	0.11	0.02	−0.08
	0.07	0.07	0.06	0.07
Education	0.12***	0.06**	0.09***	0.13***
	0.03	0.03	0.03	0.03
Rural	−0.03	0.07	0.00	0.01
	0.08	0.08	0.07	0.08
Statist	−1.14***	−1.35***	−1.09***	−1.15***
	0.07	0.07	0.06	0.07
Religiosity	−0.04	0.02	0.14**	−0.16**
	0.07	0.07	0.06	0.06
Public sector	−0.43***	−0.32***	−0.18***	−0.17**
	0.08	0.07	0.07	0.07
# of Employees in the HH	0.00	−0.03	0.06**	−0.02
	0.03	0.03	0.03	0.03
Middle tier	−0.93***	−0.29	−0.11	−0.16
	0.32	0.27	0.23	0.23
Upper tier	−1.01***	−0.32	−0.09	−0.29
	0.33	0.29	0.24	0.24
# of observations	4164	4134	4465	4332
Pseudo R^2	0.1	0.14	0.07	0.08
Prob > chi^2	0.00	0.00	0.00	0.00

Notes All columns present the coefficients from ordered logit regressions with country dummies. Standard errors are in the second row. *, **, and *** denote statistical significance at the 0.1, 0.05 and 0.01 levels, respectively

between income and sector. As can be seen, there are crucial differences when heterogeneity of informal employees is considered. While for corruption, trust and redistribution informal–formal divide continues to be insignificant, for job creation, it positively affects performance evaluations of the government. In comparison to formal sector workers, informal sector participants are 0.24 more likely to find governments unsuccessful in terms of generating employment. Moreover, the interaction term is also significant for corruption and informal sector workers who belong to the lower tier report high levels of corruption by 1.14 points as opposed to their upper-tier counterparts. When it comes to the effect of income, it can be also observed from Table 4, richer individuals have more positive views about corruption. All in all, the findings from regressions suggest that in MENA countries, there are vital differences among the informal sector participants and informal sector should not be treated as a binary category.

In the next steps, I further examine diversity within informal sector and if there are opinion differences among income segments. People, who are involuntarily participating in informal economic activities, or in other words, the excluded from the formal sector, are more likely to be at the bottom of earnings distribution. On the other hand, people who voluntarily left the formal sector or exited can be in relatively better-off positions, and hence are represented more in the middle and upper tiers. As I argued earlier, there can be noticeable differences between lower and upper-tier informal sector workers in terms of their political attitudes. Excluded individuals might have negative opinions about governance mechanisms and believe that the state is not fulfilling its social and economic responsibilities. In contrast, exited individuals might see informal sector as rewarding and have higher confidence and trust in government. Figure 3 demonstrates the marginal effect of income tiers on corruption perceptions and trust. It can be observed that the higher the income is the lower the likelihood of people thinking that there is great extent of corruption for both groups. The marginal effect of belonging to the lower tier in the informal sector on adverse views about corruption is 0.69 points as opposed to 0.55 point for a similar person in the upper tier. For trust, the decline from the lower to upper tiers is even more pronounced for informal sector employees, from 0.6 points to 0.38 points. This implies that informal sector categorization is not sufficient to capture the effect on political views and within sector differences should be borne in mind.

Figure 4 repeats the above exercise for government's performance on job creation and narrowing the gap between rich and poor. As opposed

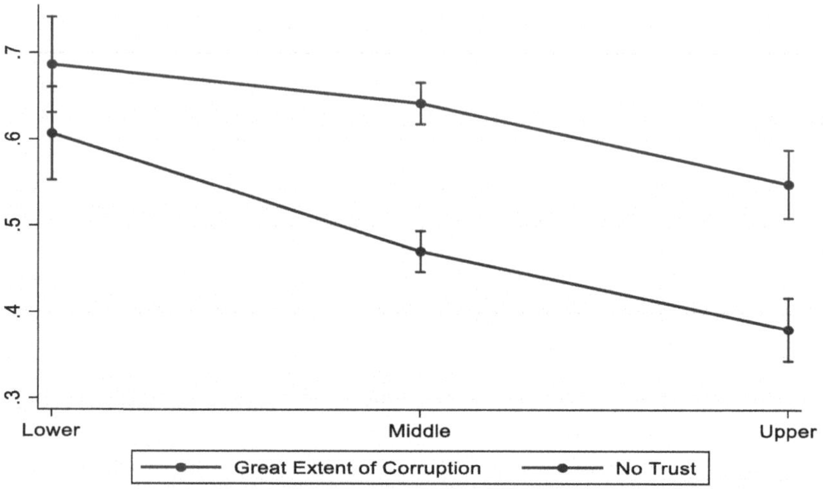

Fig. 3 Marginal effect of income tiers on corruption and trust in informal sector (*Note* Based on the marginal effect of income tiers on the outcomes of large extent of corruption and no trust responses obtained from ordered logistics estimations for the informal sector sample. Marginal effects are statistically significant at 0.1 level for all income tiers)

to the views on political issues such as corruption and trust, there is no immediately discernable association between tiers and economic evaluations for informal sector workers. An informal sector employee in the lower tier has a similar probability of assessing the government's performance on job creation negatively, 0.35 versus 0.39 points. On the other hand, middle-tier informal employees are much more likely to perceive job creation with 0.44 points. For redistribution, a similar picture arises where lower and upper-tier informal sector participants have comparable opinions while middle group assert more negative opinions about governments' performance on narrowing the gap between rich and poor. Hence, it can be said that with regards to economic policy, there is a threshold level of income beyond which informal employees begin to have positive assessments. This is partially in support of exclusion exit approach since the upper-tier informal sector participants do not differ very much from their formal counterparts even though regulation and social protection are largely missing.

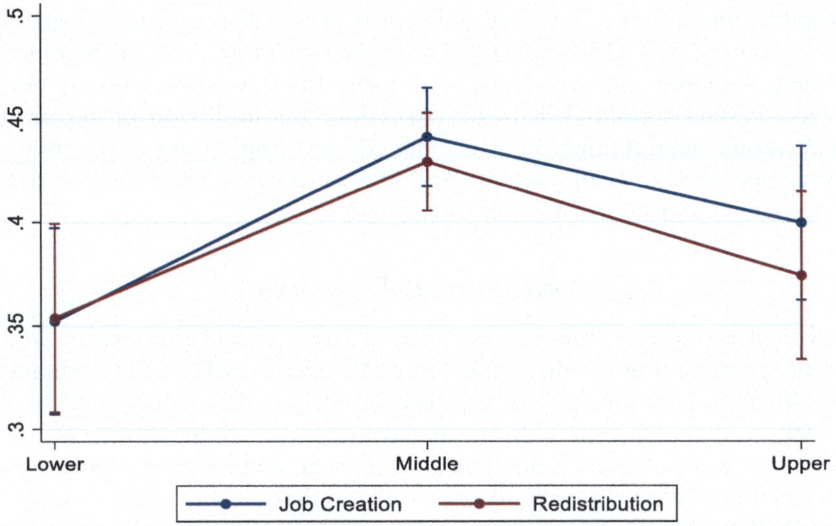

Fig. 4 Marginal effect of income tiers on job creation and redistribution in informal sector (*Note* Based on the marginal effect of income tiers on the outcomes of very poor performance of job creation and narrowing down the gap between rich and poor obtained from ordered logistics estimations for the informal sector sample. Marginal effects are statistically significant at 0.1 level for all income tiers)

Overall, my empirical analysis reveals that the broad formal–informal divide does not have a significant impact on political opinions and economic evaluations in MENA region at the individual level.[6] However, once the diversity of informal sector is considered, the findings indicate that there are clear distinctions among informal sector participants in terms of which income group they belong. Upper-tier participants in the informal sector are much less likely to perceive high extent of corruption and low levels of trust, which is in line with my hypothesis that exited workers do not necessarily judge government actions negatively and trust their governments as they benefit from informal activities.

[6] The lack of association can be due to the poor measurement of informal employment, yet the results are robust to the inclusion of self-employed in the operationalization. Also, I derive similar signs and coefficients when binary dependent variables are utilized.

Notably, the impact of diversity among the informal sector participants is more complicated when economic issues are taken into account. Income, at first, heightens the possibility of assessing the government poorly but after a certain threshold, it begins to reduce the likelihood of negative evaluations. Even though this is not directly verifying my argument about lower-tier workers being more critical of government, it still implies that heterogeneity of informal employment matters.

Concluding Remarks

Informal economic activities constitute a large part of production and exchange in various countries including the ones in MENA. The studies that focus on informality are relatively scarce for the region, and they mostly consider macro level factors determining the size of informal sector or micro-level characteristics behind informal employment. However, informality matters not only for economic livelihoods of people but also for their relationship with state and how they perceive political institutions. Moreover, reasons for joining informal sector and outcomes are largely heterogeneous, and this heterogeneity can be explanatory for political opinions and policy preferences. This chapter examined the relationship between informal employment and opinions on corruption, trust, job creation and redistribution at the individual level for MENA countries. In line with several other studies, I also find that formal–informal distinction alone has no major effect on citizen's views about various economic and political issues. While the absence of a relationship can be an outcome of narrow definition of informal sector, it also points towards the diversity of informal employment. Given the divergent reasons for undertaking informal economic activities, people might have very distinct perceptions about the role of state in regulating these pursuits and providing labor market opportunities in the formal sector.

To understand the impact of informal sector diversity on citizens' opinions about state's role in economic and other policy domains, I also examine how different income tiers interact with informal employment. As proposed in the chapter, people, who exited the formal sector and manage to receive high incomes from informal economic activities, are expected to be less critical of governments given their personal gains despite the institutional deficiencies. In contrast, for people who are excluded from the formal sector and are able to only maintain basic living standards, trust and confidence in government would be quite low. My findings support these propositions as there are stark differences between individuals from varied income tiers in the informal sector. There is a clear negative relationship between incomes and negative opinions of government performance for the informal sector participants in MENA region, and income has a complex effect on the assessment of government's performance on job creation and redistribution. Perceptions on corruption and degree of trust among the informal employees from upper tier are much lower than the ones from bottom tier, which indicate that exited and excluded individuals have distinct assessments of political institutions in their respective countries. Hence, diversity within the informal sector participants is explanatory for citizens' views on institutional quality and government's performance.

While dualization is not notably associated with opinions at the individual level, my chapter's findings about the importance of heterogeneity can be useful from an academic and policymaking perspective. On the one hand, researchers should not merely focus on informal–formal divide and rather take the diverse positions within the former to better understand how people view institutional quality and effectiveness of policies. On the other hand, distinct policies might be needed to give incentives to the excluded and exited individuals to formalize their economic activities. Expanding social security coverage, broadening job opportunities in the formal sector and labor market reforms that are inclusionary would help people at the bottom tiers of informal employment. For people, who voluntarily exited the formal sector in order to avoid hefty taxes or cumbersome regulations, policies need to be focused on these areas rather than employment generation. Certainly, these inferences are based on available data, which doesn't allow us to classify employees into detailed groups. Moreover, international surveys in MENA region do not enable us to look at the preferences over various types of social policy, and whether there are distinctions for targeted and universal programs. In the future, I aim to supplement cross-country research by in-depth examination of informal sector participants and their relation to state.

References

Acar, E.O. and A. Tansel (2016). "Defining and Measuring Informality: The Case of Turkish Labor Market". *Sosyoekonomi*, 24(28): 147–174.

Angel-Urdinola, D. and Tanabe, K. 2012. "Micro-Determinants of Informal Employment in the Middle East and North Africa Region". WB Social Protection and Labor Discussion Paper, No. 1201.

Arabbarometer.org. n.d. "Survey Data". Available at: https://www.arabbarometer.org/survey-data/. Accessed [10 July 2020].

Baker, A. and Velasco-Guachalla, V.X. 2018. "Is the Informal Sector Politically Different? (Null) Answers from Latin America". *World Development*, 108: 170–182.

Benstead, L.J., Atkeson, L.R. and Shahid, M.A. 2019. "Does Wasta Undermine Support for Democracy? Corruption, Clientelism, and Attitudes Toward Political Regimes". In I. Kubbe and A. Varraich (eds.) *Corruption and Informal Practices in the Middle East and North Africa*. London: Routledge, pp. 77–101.

Berens, S. 2015a. "Preferences on Redistribution in Fragmented Labor Markets in Latin America and the Caribbean". *Journal of Politics in Latin America*, 7(3): 117–156.

Berens, S. 2015b. "Between Exclusion and Calculating Solidarity? Preferences for Private Versus Public Welfare Provision and the Size of the Informal Sector". *Socio-Economic Review*, 13(4): 651–678.

Bourhaba O. and Hamimida M. 2016. "An Estimation of the Informal Economy in Morocco". *International Journal of Economics and Finance*, 8(9): 140–147.

Castells, M. and A. Portes. 1989. "World Underneath: The Origins, Dynamics and Effects of the Informal Economy". In A. Portes, M. Castells, and L.A. Benton (eds.) *The Informal Economy: Studies in Advanced and Less Developed*. Baltimore, MD: John Hopkins University Press, pp. 11–37.

Chen, M.A. 2012. "The Informal Economy: Definitions, Theories and Policies". WIEGO Working Paper, Vol. 1, No. 26.

Davis, D.E. 1994. *Urban Leviathan: Mexico City in the Twentieth Century*. Philadelphia, PA: Temple University Press.

de Soto, H. (1989). *The Other Path: The Invisible Revolution in the Third World*. New York, NY: Harper & Row.

Duman, A. and Duman, A. 2020. "The More the Gloomier: Informal Wage Penalty Over Time in Turkey". Unpublished mimeo.

Elsayed, A. and Wahba, J. 2019. "Political Change and Informality: Evidence from the Arab Spring". *The Economics of Transition*, 27(1): 31–66.

Hart, K. 2006. "Bureaucratic Form and the Informal Economy". In B. Guha-Khasnobis, R. Kanbur, and E. Ostrom (eds.) *Linking the Formal and Informal Economy: Concepts and Policies*. Oxford: Oxford University Press, pp. 21–36.

Hertog, S. 2017. "The Political Economy of Distribution in the Middle East: Is There Scope for a New Social Contract?". In G. Luciani (ed.) *Combining Economic and Political Development: The Experience of MENA*. Boston: Brill-Nijhoff, pp. 88–113.

Hirschman, A.O. 1970. *Exit, Voice, and Loyalty: Responses to Decline in Firms, Organizations and States*. Cambridge, MA: Harvard University Press.

Holtmeier, L. and Alami, M. 2020. "Informal Workers in Arab World Hit Hardest by Coronavirus, Unlikely to Get Help". Available at: https://english.alarabiya.net/en/features/2020/04/03/Informal-workers-in-Arab-world-hit-hardest-by-coronavirus-unlikely-to-get-help. Accessed [10 July 2020].

Hussmanns, R. 2005. "Measuring the Informal Economy: From Employment in the Informal Sector to Informal Employment". ILO WP, No. 53.

Kanbur, R. 2017. "Informality: Causes, Consequences and Policy Responses". *Review of Development Economics*, 21(4): 939–961.

Kurtz, M. 2004. "The Dilemmas of Democracy in the Open Economy". *World Politics*, 56(2): 262–302.

Lewis, W.A. 1955. *The Theory of Economic Growth*. London: George Allen & Unwin Ltd.

Loayza, N. and T. Wada. 2011. "Informal Labor in the Middle East and North Africa: Basic Measures and Determinants". Available at: http://documents1.worldbank.org/curated/en/657651468052140219/pdf/675900WP0P11320y0paper0text09021011.pdf. Accessed [10 July 2020].

Ordonez, J. 2014. "Tax Collection, the Informal Sector, and Productivity". *Review of Economic Dynamics*, 17(2): 262–286.

Perry, G.E., Maloney, W.F., Arias, O.S., Fajnzylber, P., Mason, A.D. and Saavedra-Chanduvi, J. 2007. *Informality: Exit and Exclusion*. Washington, DC: World Bank.

Polese, A., Kovacs, B. and Jancsics, D. 2017. "Informality 'In Spite Of' or 'Beyond' the State: Some Evidence from Hungary and Romania". *European Societies*, 20(2): 207–235.

Portes, A. and Hoffman, K. 2003. "Latin American Class Structures: Their Composition and Change During the Neoliberal Era". *Latin American Research Review*, 38(1): 41–82.

Saavedra, J. and Tommasi, M. 2007. "Informality, the State and the Social Contract in Latin America: A Preliminary Exploration". *International Labour Review*, 146(3–4): 279–309.

Trabelsi, K. 2013. "Current State of the Informal Economy in Tunisia as Seen by Its Stakeholders: Facts and Alternatives". Available at: https://www.solidaritycenter.org/publication/current-state-of-the-informal-economy-in-tunisia-as-seen-through-its-stakeholders-facts-and-alternatives-june-2014/. Accessed [10 July 2020].

Vuletin, G. 2008. "Measuring the Informal Economy in Latin America and the Caribbean". IMF Working Paper, No. 08/102.
World Bank. 2019. "Growing in the Shadow: Challenges of Informality". In *Global Economic Prospects*. Washington, DC: World Bank, pp. 129–195.

State Collusion or Erosion During a Sovereign Debt Crisis: Market Dynamics Spawn Informal Practices in Lebanon

Joseph P. Helou

Lebanon, often acclaimed as a mosaic of religious diversity and political plurality in the middle east, is home to infamous political elites whose sectarian practices, undergirded by a consociational power-sharing arrangement, dominate vast portions of the political economy, fragment state institutions, and incentivize communal allegiances at the expense of national affinities (Lijphart 1984; Salamey 2014). These elites' pervasive practices persist as progenitors and derivatives of sectarianism, which, according to Salloukh et al. (2015: 3), "produces and reproduces sectarian subjects and modes of political subjectification and mobilization through a dispersed ensemble of institutional, clientelist, and discursive practices." When Lebanese elites harness a cobweb of patron–client ties where they negotiate for shares and positions for their political favorites in the private and public sectors, intercede on behalf of their proteges with government ministries to ensure healthcare coverage and dispense

J. P. Helou (✉)
Department of Social and Education Sciences, Lebanese American University, Beirut, Lebanon
e-mail: joseph.helou01@lau.edu.lb

A. Polese (ed.), *Informality, Labour Mobility and Precariousness*, International Political Economy Series,
https://doi.org/10.1007/978-3-030-82499-0_11

state projects to their favored contractors (with their proposals voiced through the use of incendiary, partisan and sectarian language), they reorient individual loyalties away from state institutions and symbols toward their services (Khalaf 1968; Johnson 1986; Cammet 2014; Helou 2015; Salloukh et al. 2015). But the problem lies in the enormous political-economic costs of this blatantly unsustainable model which has political elites debunking institutional watchdogs, concentrating wealth in the hands of their clients and cronies, favoring the lucrative gains of the financial and real estate sectors at the expense of productive capacity, monopolizing markets and locking Lebanon in an unsustainable debt trap (Leenders 2012; Cammet 2014; Traboulsi 2014; Salloukh 2019). In so doing, Lebanese elites' fragmentation, manipulation and contestation of state institutions influence citizens' behavior by creating a reliance on informality, which, in turn, reshapes state institutions (Giddens 1979; Migdal 1994). Therefore, the novelty in this empirical case and theoretical formulation is that it takes stock of how sectarian elites' distortions in Lebanon's public finances, particularly imbalances in taxes, the national currency's foreign exchange rate and inflation, influence the incubation of informality.

This chapter presents an empirical case study of the Lebanese state's public finance mismanagement that ushered in the sovereign debt crisis of 2020 and tilted citizens' economic preferences toward informality. It explores how consistently negative balance of payments, soring national debt, dwindling foreign currency reserves and the political system of sectarianism that thrives on government spending to finance budget deficits made Lebanon an economic time bomb. But while this ticking time bomb's diffusion occurred with Lebanon's failure to settle its external debt, particularly Euro-bond payments in 2020, the crisis's severe domino effect plunged the country into an economic abyss, demoted its creditworthiness on international markets, restricted depositors' withdrawals from dollar accounts as the central bank solidified a system of unspoken capital controls to retain foreign currency notes and imposed narrow spending limits on credit cards issued by Lebanese commercial banks. Such financial dilapidation, coupled with severe monetary dislocations and various currency exchange rates floating on the market, culminated in a clear divergence of citizens' economic behavior away from state institutions toward the espousal of informality. Whereas citizens' embracement of informality includes licit practices, which occur outside state regulations but without constituting part of the shadow or criminal economies (see Pugh et al. 2004; Pugh 2004, 2005a, b), such as foreign currency exchange along various unofficial rates (occupying an essential part of the discussion herein), the discounting of banker's cheques, purchasing real estate and buying jewelry to convert part of depositors' frozen bank accounts to mobile assets or cash U.S. Dollars, their behavior negotiated a new social contract with the state, contested state institutions with regards to monetary policy and highlighted the significance of informality in securing stable media of value for business transactions. Therefore, we seek an appreciation of the market dynamics that have incentivized citizens' espousal of informality (Morris and Polese 2014; Polese et al. 2018), while noting that these interest-maximizing citizens pursue informality in an incoherent institutional space that displays variable incentives for formal behavior and that is not necessarily undergirded by any clear monetary logic (see Ledeneva 2013; Polese et al. 2014; Steenberg 2016). Because every political-economic environment produces different incentives for citizens' behavior, such as institutionalized democracy (Van Tatenhove et al. 2006; Ayres 2017), post-conflict settings (Danielsson 2016a, b; Distler 2018), and post-communist economies transitioning to free markets (Polese 2010, 2014; Polese et al. 2016, 2018; Polese and Rekhviashvili 2017), we seek an understanding of Lebanese citizens' economic responses to the state's mismanagement of finance.

We build on the model set out by Polese et al. (2018) who explain that citizens engage in informal practices "in spite of the state"—where state institutions regulate a situation but citizens deem it to be insufficient—and "Beyond the State" where failed state regulations oblige citizens' compensation of that deficit. Here citizens engage in informality "in spite of the state" by dynamically contesting the basic meaning and value of money (Parry and Bloch 1989). When Lebanese citizens react to their country's sovereign debt crisis by embracing the aforementioned informality that orients their economic activity outside formal state institutions and commercial banks toward the free market—often in defiance of state regulations, their economic choices help negotiate the contours of a new economic governance model that is mediated by foreign exchange offices, jewelers and real estate developers whose services help citizens earn a trusted medium of value. While informality, as this volume clearly acknowledges, risks leaving workers—and here a wide segment of the middle class—in a precarious situation, weakening the state's capacity for remedial economic measures and discouraging investment and credit provision in entire sectors, it emerges in response to state misgivings in public finances, commercial bank's inability to dispense dollar bills to clients and a huge trust deficit on behalf of clients, which oblige them to negotiate their finances with new actors on the market via informality. Therefore, the mutually transformational role of state institutions and citizens' espousal of informality—catching both in a vicious circle—can be minimized when an adequate set of economic reforms permit depositors' access to their frozen bank accounts, the state's restoration of its capacity to provision credit and market stability that reinforces the national currency as a trusted medium of exchange (with stable or predictable rates).

The current research employs a mixed approach in exploring the topic under scrutiny by combining participant observation with semi-structured interviews. Slightly before the October 17, 2019 Revolution, a growing gap between the official Lebanese pound's exchange rate and the rate offered by foreign exchange offices started taking shape, which reoriented citizens away from official institutions—commercial banks—toward the market rates. During this period and until this day, I have been privy to tens of personal conversations with individuals whose experiences inform an understanding of the kind of informality molded on the Lebanese market and confirm many of my findings. I also gained very valuable information concerning market volatilities, fluctuations, and currency exchange, in addition to the consequences of such phenomena on citizens' economic behavior, from informal interviews with currency traders, particularly individuals who own foreign exchange offices. I was able to speak to these individuals—typically in an informal setting in a café—and in regular visits to their offices for observation throughout the period stretching from October 2019 to November 2020. My extensive experience in building reliable trust networks in Lebanon facilitates the recruitment of primary sources whose experiences illuminate aspects of this research. Here I also draw on semi-structured interviews with the president of the Lebanese Businessmen's association in the world, a financial expert and a broker of some of the informal practices under scrutiny. Brokers, such as foreign exchange officers, jewelers or middle men, forge the informal practices pertaining to cheque discounting and the conversion of frozen bank deposits to cash and other assets that I will analyze in more detail in subsequent sections. Due to the repetitive imposition of COVID-19 associated lockdowns and the need for the maintenance of social distancing, these approximately 75-minute interviews were digitally recorded via video-conferencing platforms (Zoom) and then transcribed by this author.

This chapter divides the topic into three distinct sections. Firstly, "sectarian elites, public finance distortions and the sovereign debt crisis" reveal how the sectarian strategies of the Lebanese political elites distort public finances and culminate in the sovereign debt crisis of 2020. Secondly, parallel currency markets, traders and informal practices" show how Lebanese citizens' contestation of state authority and institutions occurs via informality on the market and as a coping mechanism against risk. Thirdly, 'the way out" concludes with a discussion of potential avenues out of Lebanon's economic predicaments, policy recommendations, and the value of market reforms.

Sectarian Elites, Public Finance Distortions, and the Sovereign Debt Crisis

While sectarian representation in public administration positions persists unabated in Lebanese politics as of the Ottoman empire's Mutassarifieh, Greater Lebanon under the French mandate and the post-1943 independence of the Lebanese state (Salibi 1988; Harris 1997; Traboulsi 2012), it becomes more obstructive to the country's political and economic life after the 1975–1990 civil war with the integration of ex-militia leaders in the political process. These postwar elites' skewed implementation of the Taif accord, which consolidates sectarian parity in the Council of Ministers, Parliament, and all high-ranking public administration positions (Taif Accord 1989: I.2A.5A,B,C; Picard 1996: 65; Dabbagh et al. 1997: Article 95), fragments Lebanese state institutions, fuels politically charged interpretations of constitutional prerogatives and invites Syria's intervention—specifically in the period 1990–2005—in every nook and cranny of Lebanese politics (El-Khazen 2003; El-Husseini 2012; Salloukh et al. 2015). The post-Taif political process, undergirded by convoluted elite practices, harnesses a system characterized by an extreme dispersal of power and associated quasi-permanent gridlock in the decision-making process, the predominance of political bargaining among the President, Prime Minister and Speaker of the house—manifest in the sharing of the spoils of government—and the intensification of clientelistic practices such as appointments across state bureaucracies (Leenders 2012). Its associated constitutional amendments shatter checks and balances by curbing the Maronite President's constitutional prerogatives, transferring executive authority to the council of ministers headed by a Sunni Prime Minister and expanding the role of the Shia speaker of the house (Picard 1996; Dabbagh et al. 1997: Article 57; Salam 2003; Leenders 2012: 128–129).

Lebanese political elites also bank on their political influence to gain critical positions within the country's media scene and private sectors. For instance, close scrutiny of the ownership patterns of private Lebanese television stations reveal indisputable connections between OTV and President Michel Aoun or the Free Patriotic Movement, Future TV, and Prime Minister Hariri, the News Broadcasting Network and Speaker Berri, AlManar and Hezbollah, MTV and the influential Murr family, Lebanese Broadcasting Corporation and the political heavyweight Pierre Daher, and AlJadeed and the financially capable businessman Tahseen Khayat (Melki 2007; Dabbous 2013; Helou 2020). These elite connections to television stations epitomize the country's sociopolitical composition by revealing the intimate nexus between wealth and political influence, advancing political mobilization of supporters along partisan lines and deepening sectarian cleavages in an already divided society. This pervasive political elite influence also subdues markets by consolidating ownership of various business sectors in the hands of a select few political favorites. In a paper prepared for the Lebanese Ministry of Economy and Trade, Gaspard (2003) finds that more than 50% of all markets in Lebanon are monopolies or oligopolies. Such monopolies as in the National Cement Industry renders the price of a bag of cement in Lebanon threefold its price on international markets (Wood et al. 2020). Lebanese political elites harness a system that not only safeguards monopolies, disincentivizes national production, and provides an economic edge for imports over locally produced items, but also ensures the persistence of these monopolies by refraining from anti-trust legislation, avoiding the establishment of a national authority to promote competition and diminishing the capacity of institutional watchdogs (Wood et al. 2020). Illustrating this point is the famed Lebanese economist Elie Yachoui who writes that 1000 out of 25,000 importers in Lebanon account for 90% of imports (AlSafir July 19, 2004). Thus the postwar Lebanese economy is controlled by financial, import, and real estate networks epitomized in an increasingly monopolistic class system and the growth of holding companies (Traboulsi 2014: 30).

But perhaps the Lebanese political elites and their proteges across state bureaucracies perpetrate the worst blunders in their adoption of the policies of artificially setting the Lebanese Pound's exchange rate at 1500 to the Dollar, the reliance on the sales of real estate and massive consumption of imports, which disincentivize production, offset U.S. Dollar inflows to Lebanon and repress positive GDP growth. With such a lopsided economic structure as a backdrop, the Dollar peg finances a lavish lifestyle for middle-class Lebanese citizens above what an otherwise real floating rate affords, trumps the productive sectors of the Lebanese economy by making imports more competitive with local products and drains foreign currency reserves as Lebanese citizens afford expensive vacations abroad and imported goods payable in U.S. Dollars. Illustrating this point is Dan Azzi who mentions "... how you make the economy grow at 7% annually to outpace the debt, which, if you read economics textbooks, it makes sense. Except in Lebanon, such a growth means feeding our consumption of imports, an increase in the balance of payments deficit, a drainage of dollar reserves as they make their way outside the country etc. This is the structure of the Lebanese economy... So, this gave us the preconceived notions of looking at the traditional measures such as debt-to-GDP ratio... Very few people were looking at the balance of payments approach, the fiscal deficit –is the government spending more money than it was making– and stuff like that."[1]

The following table reveals the size of various indicators in relation to Lebanon's Gross Domestic Product from 1990 to 2019. It shows the structure of the Lebanese economy that is skewed toward consumption of imports rather than striking a balance or favoring an export-oriented model (Table 1).

Lebanon's disastrous financial-economic policies come to life with Prime Minister Rafik Hariri in the early 1990s and receive the support of a wide segment of political elites thereafter. These political elites' adoption of neoliberal policies—commonly advocated by international financial institutions—privilege the financial and construction sectors at the expense of industry and agriculture, replace dependence on market movement for development, stress privatization, advance free trade instead of the protection of local industries and call for minimal state intervention especially in terms of distributive policies and subsidies (Traboulsi 2014: 26). Hariri's reconstruction project, dubbed "Horizon 2000," hinges on

[1] Author's Interview with Dan Azzi, Former CEO of Standard Chartered Bank in Lebanon, Former Harvard Fellow for Leadership and Financial Expert and Commentator on 14 November 2020.

Table 1 World Bank indicators

Year	*Agriculture, forestry and fishing, value added (% of GDP)*	*Manufacturing, value added (% of GDP)*	*Industry (including construction), value added (% of GDP)*	*Imports of goods and services (% of GDP)*	*Exports of goods and services (% of GDP)*	*Services, value added (% of GDP)*	*Gross domestic savings (% of GDP)*	*Foreign direct investment, net inflows (% of GDP)*
1990				99.91302198	17.99278605		−35.7272288	0.227233866
1991			23.63038218	79.9258363	12.52883571		−19.29665976	0.034751722
1992			22.82926066	71.99355128	10.85070516		−31.86950774	0.315902283
1993			22.62990249	67.17667586	9.174398857		−26.965881	0.09317852
1994	6.466640634	12.71716533	22.61620347	62.40150674	7.870608735	61.88864644	−17.87810462	0.235021371
1995	6.794856428	12.50239282	22.2342521	62.32893212	10.80645488	60.40601956	−15.0673936	0.298665506
1996	6.183078443	12.5429745	22.30642256	55.25029276	12.84632	61.23592578	−7.611474081	0.584358875
1997	6.705154639	12.52371134	22.27216495	47.55463918	14.53195876	60.61030928	−3.806185567	11.42467711
1998	6.088186929	12.42877357	23.68350606	40.95376496	13.76343264	58.32727829	1.212283453	6.58022463
1999	6.196850093	11.85600427	21.22945506	36.75018114	14.06398963	59.05884148	0.144910956	5.012211585
2000	6.237509608	11.41429669	19.96541122	35.9377402	14.17755573	61.52190623	−1.348962337	5.75581692
2001	5.863118728	11.32408765	20.43071372	39.95189236	15.604916	61.44247754	−1.533431052	8.237549526
2002	5.662926018	10.83056248	19.11540593	34.94389027	16.09864228	60.16209476	−0.096979773	6.975529146
2003	5.486374897	10.73162675	18.31213873	37.35425268	16.73658134	59.58051197	−1.456647399	14.24105903
2004	3.820910296	7.234334702	14.57516638	55.91789247	35.05490455	69.07399476	2.20462604	8.973516095
2005	3.579635058	7.51373215	14.76003823	56.13911756	36.62353871	71.14214226	3.45043211	12.20384959
2006	4.469734859	7.195471934	15.00815138	55.47973952	35.58965751	72.7943058	0.658009062	12.14443813
2007	4.892766618	7.235117057	15.50987735	60.82619096	37.141517	70.75109918	0.59741566	13.5978269
2008	4.52188618	8.303247575	16.21200383	68.08001874	38.54565444	72.43806776	−1.538043095	14.88051772
2009	3.950872086	7.916028888	15.43951679	57.20213716	33.1957017	69.18772882	2.914580701	13.56965891
2010	3.861959367	7.721533803	13.80937621	60.07353007	35.02312287	71.86177674	−0.113148696	11.13279363
2011	3.804609707	7.579314185	14.50621607	65.77678518	36.36309723	73.53999928	−2.5673222	7.856939895

Year	*Agriculture, forestry and fishing, value added (% of GDP)*	*Manufacturing, value added (% of GDP)*	*Industry (including construction), value added (% of GDP)*	*Imports of goods and services (% of GDP)*	*Exports of goods and services (% of GDP)*	*Services, value added (% of GDP)*	*Gross domestic savings (% of GDP)*	*Foreign direct investment, net inflows (% of GDP)*
2012	3.762977177	7.388618202	15.48435863	59.71025415	28.79665694	72.97264925	−6.137102381	7.065391006
2013	3.953061054	7.525200735	16.34163991	59.4904786	26.63014954	72.09209096	−5.26520697	5.672830621
2014	4.019752258	7.533680226	16.20695963	56.01525199	23.73678187	73.75549315	−7.331758895	5.946899255
2015	3.382572678	8.063245043	15.72948019	48.68958713	23.13501703	73.68578671	−3.33647604	4.323822553
2016	2.882851081	8.349852555	15.97507094	45.94552794	21.70162572	74.03369902	−1.543951658	5.016070552
2017	2.933449914	8.272872673	15.03945347	46.42959349	21.88032454	75.18025351	−3.174194813	4.746667966
2018	3.235901227	7.621206914	14.23786428	47.43262999	20.72220081	76.66918011	−5.918265607	4.82871161
2019	5.292462327		12.83131224	41.94829538	21.66337342	75.90712501	−1.932008261	4.166001685

nurturing the state's debt and rebuilding downtown Beirut as an international commercial and financial center (Butter 1995; Traboulsi 2014: 26). He gambled on politically conducive conditions for investment with a potential Arab–Israeli peace deal and therefore promised the Lebanese high living standards, the positioning of Lebanon at the head of the region's economies and annual growth of 9.4% in the period 1993–2003 (Traboulsi 2014: 26). His reconstruction plan and recapitalization of Lebanese banks—contributing to rising national debt—required the issuance of treasury bonds that paid a whopping interest rate of 20–42.5% in the 1990s (Traboulsi 2014: 27), which provided a pathway for convoluted political practices that manipulate markets, reap enormous profit for the select few and fossilize dooming public finance policies (see Salti 2019). While the Lebanese political, economic, and business elites sunk their teeth in these policies due to the magnanimous bank recapitalization and profits in the period 1992–2010, rising national debt reached $33.78 billion by the end of 2003 (today estimated at $90 billion) and thrust Lebanon into a debt trap where 50% of state expenditures service interest payments on this debt (Traboulsi 2014: 27). A coterie of select banks and Lebanese, Arab and foreign investors reaped huge profits through speculation on state debt. For instance, banks received $22 billion in interest payments on treasury bonds in 2002 when the national debt was estimated at $30 billion (ibid.).

These Lebanese government policies nudged the parties to the international donor conferences for Lebanon to demand a decrease in interest rates, cutting down on the size of the public sector, freezing the wages and fringe benefits of civil servants and the implementation of reforms. In the 2002 Paris II conference, participating states were asked to support Lebanon restructure its domestic debts at lower rates of interest (Nizameddine 2006). Saudi Arabia, France, and other states together provided $2.5 billion in soft loans (ibid.). In the most recent donor conference of "Conférence économique pour le développement, par les réformes et avec les entreprises"—dubbed CEDRE (French for Cedar but stressing reforms)—in 2018, international donors conditioned unlocking assistance to Lebanon only with a set of reforms and transparency measures which as yet face enormous political intransigence. International donors critique Lebanese government policies that spend generously on ailing infrastructure such as the country's cadaverous electric grid which continues to drain an annual amount of $1.5 billion from state coffers (see Stel and Naudé 2016; Verdeil 2018; Ratcliffe 2020).

However, the intimate linkages between Lebanon's political, economic-financial and religious elites help forge a pervasive political-economic system that withstands external and internal pressures and strains, while inspiring this lopsided economic model that privileges a select few at the expense of the majority. With 20% of the wealth concentrated in 1600 bank accounts (0.1% of all bank deposits), the top-14 Lebanese banks' profits equal to 4.5% of GDP, and the Lebanese state spending 3–5 of tax money to service its debt (Salti 2019), the rentier system forged by banks and elites exacerbates inequality in Lebanon. Lebanese elites build on the pretext of donors' conditionality to double down on the salaried strata of society often pitting industrialists and businesses in the non-financial sectors, whose fortunes remain quite sour due to the absence of incentives, protections, and other backing to rival their regional counterparts, against their employees whose purchasing power consistently dwindles (see Baroudi 1998, 2000, 2005). This wage freeze leads to rounds of social mobilization following 2010 until school teachers and other public servants receive a series of wage increases.[2] Then, a series of political events unpropitious for capital inflows starting with the Arab spring of 2011, the Syrian civil war, the participation of Hezbollah in several regional conflicts and consecutive institutional deadlocks in Lebanon not only sour relations with GCC states and the USA, but also drive a consistently negative balance of payments during this period.[3] As evinced in the following graph, as U.S. Dollar deposits dwindle, the Lebanese Central Bank looks for ways to attract some foreign capital into the country by offering up to 15% interest rates on dollar accounts and more on deposits in Lebanese pounds (Fig. 1).

During the same period, the Lebanese Central Bank issues Euro-bonds to attract foreign currency deposits into Lebanon. But with a majority of the bonds purchased by Lebanese individuals and institutions, the central

[2] Author's interview with Hanna Gharib, *Former Head of Public Secondary-School Teachers' Association and Current Secretary General of the Lebanese Communist Party* on 28 June 2018.

Author's interview with Rodolph Abboud, *Head of Private School Teachers' Syndicate* on 29 June 2018.

[3] The balance of payments (BOP) is a statement of all transactions made between entities in one country and the rest of the world over a defined period of time, such as a quarter or a year.

For more see Kenton, W. 2020. Balance of Payments definition. https://www.investopedia.com/terms/b/bop.asp.

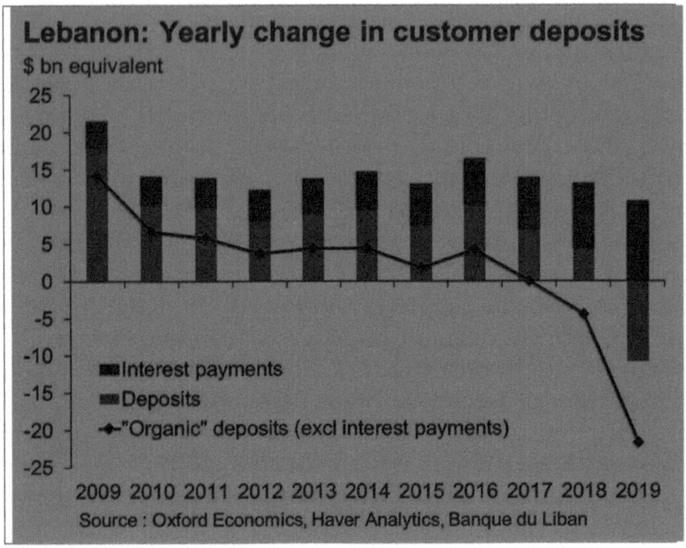

Fig. 1 Lebanon: yearly change in customer deposits ($ bn equivalent) (*Source* Oxford Economics, Haver analytics, Banque du Liban)

bank fails to attract fresh funds to shore up its balance of payments' deficit. Dan Azzi states that "For the past ten years, 95% of the Euro bonds, during their initial issuance, were being sold inside Lebanon. So it is like I lend you $1,000, then decide to give up Lebanese sovereignty when we disagree, and say we want to settle matters in the courts of New York… By the way, if we increased the interest rates to 9% and 10%, we would have been able to attract foreign money… Most of the Euro bonds we issued over the past ten years –including those issued in November 2019 which were 10% of the issuance– were issued inside Lebanon so from Lebanese-to-Lebanese. Therefore, ask yourself why we would give up Lebanese sovereignty of the legal resolution of this problem when we are borrowing internally?".[4]

Regarding the ownership patterns that tilt in favor of local bondholders, Dan Azzi states that "If I want to be outrageous in my

[4] Author's interview with Dan Azzi.

explanation, I could say that the issuance of bonds occurs to reap benefits because upon issuance different people and commercial banks collect commissions. But if I know that only Lebanese banks purchase these bonds, why do I pay commissions for commercial banks to issue these bonds. One explanation is the benefits reaped by a select few... And the other non-conspiritorial explanation –if we were to give them the benefit of the doubt– is that the issuance of Euro Bonds occurred because they wanted to sell them, retain them on the Lebanese market and then sell them on the secondary markets once things pick up."[5] This sounds the alarm bell in Lebanon's financial sector—subsequently dubbed Lebanon's Ponzi Scheme—as commercial banks, whose interests converge with the political elites—some directly owned by politicians, distribute lucrative sums to bond holders and depositors without offsetting them by actual financial gains. Since very little fresh funds flow into the country as a result of bond ownership patterns, banks distribute interest payments to Euro-Bondholders from the existing pool of foreign reserves, which foretells impending financial calamity.

In the Summer of 2019, foreign exchange offices start deviating from the official exchange rate of 1500 Lebanese Pounds to the U.S. Dollar as foreign currency reserves dwindle. Ensuing political decisions over a series of indirect taxes in the council of ministers sparks a national uprising on October 17, 2019 that forces bank closures for several weeks. During this closure, foreign exchange offices take the lead in currency trading, banks limit dollar withdrawals and a clear national crisis takes shape.

Parallel Currency Markets, Traders, and Informal Practices

Lebanon's sovereign debt default on March 7, 2020 is a grave Lebanese political elite blunder as the country's external debt payments constitute a tiny sliver of the overall sum owed.[6] This decision pulls the country into a financial abyss, blocks access to international credit markets, and obliges bank deposit holders to convert their now frozen accounts to cash or

[5] Ibid.

[6] Author's Interview with Dr. Fouad Zmokhol, President of the Global Lebanese Businessmen's Association (RDCL World), Professor of Business and Lebanese Businessman on 23 November 2020.

Author's Interview with Dan Azzi.

mobile assets. While some analysis to illuminate potential pathways out of this crisis is much needed, here we will explore how people's behavior on the market spawns informal practices in the conversion of their deposits to mobile assets. The analysis draws on information gathered through participant observation in the period between October 17, 2019 and November 2020. I was privy to numerous conversations and stories of individuals who rushed to convert their frozen deposits to more mobile assets. I also witnessed firsthand the traffic at foreign exchange offices whose hesitation to offer people Dollars or Lebanese Pounds at the parallel market rate seemed apparent lest it may invite state crackdowns. During this period, foreign exchange offices answered no phone calls, but selectively offered the market rates to individuals they believed were not government officers in disguise.

With approximately $110 billion in deposits and only $18 billion remaining in actual banknotes, the Lebanese economy is cash strapped. This forces the central bank to respond to the heavy demand on cash by printing Lebanese pounds. Such institutional behavior leads to an inflation in local currency driving up exchange rates to 9500 Lebanese Pounds to the U.S. Dollar by the time of this writing. However, at commercial banks, the official rate still stands at 1500 Lebanese Pounds to the Dollar, which reassures citizens who can pay small installments on a car loan in U.S. Dollars at this conversion rate. As for those citizens who possess bank deposits in Dollars, they may withdraw a limited monthly sum according to the rate of 3900 Lebanese Pounds to the Dollar. Moreover, the Lebanese Central Bank subsidizes a basket of goods at the rate of 3900 to minimize the inflationary impact on people's lives. Yet, with three visible exchange rates as a backdrop, persons whose Dollar deposits lay idle in bank accounts look for the best alternative to cash their deposits. This creates a high demand for jewelry in return for a Lollar bankers' cheque.[7]

Following the sovereign debt crisis and strict measures on bank withdrawals, the purchase of jewellery by means of a Lollars' cheque is an informal practice to convert one's frozen bank deposits to a trusted

[7] Lollars became the commonly known name for U.S. Dollars stuck in a Lebanese Bank Account during the sovereign Debt Crisis. Making reference to Lollar banker's cheques means that the U.S. Dollars are worth more than the official rate of 1500 to the Dollar but less than the market rate. Therefore, this creates a Lollar market in which Lollar bankers' cheques are valued according to a specific formula.

medium and value of exchange even if the client has to pay 2.5 times what the item's original cost would have been in cash U.S. Dollars. Since citizens remain unaware of the fate of their deposits in banks, The monetary logic undergirding such transactions focuses on salvaging part of a depositors' bank account (the sum spent typically follows every person's personal calculation of risk diversification). Illustrating this point is my key informant who states that "the price in Lollars is 2.5 times its original cash price in U.S. dollars. That would make a $30,000 diamond cost $75,000 if paid using a U.S. dollar cheque.[8] But I wouldn't explain that. I would say this diamond costs $75,000 if payable by U.S. dollar cheques. And you would be very happy to pay via banker's cheque."[9] This deal at 2.5 times the original price seems a very lucrative transaction if the seller is a real estate developer who owes commercial banks an amount in U.S. Dollars. In that case, this money can be used to settle their loans in U.S. Dollars. But in the case of jewellery, items need to be replaced by other imported gold or diamond products that are payable in U.S. Dollars so how does the jeweler cash the Lollar cheque?

The urgent need for cash inspires yet another market for cashing Lollar cheques. My key informant explains that "I would take this cheque and purchase Lebanese Liras in cash from someone who has a lot of them... I would buy the Liras at the Rate of 2,800/2,900 to the dollar... This rate is determined by the market... because, if you go to withdraw money from the bank, they will give you the money at the rate of 3,900 to the dollar. But if you don't cash them in all at once, what could you do? So this market was borne by individuals who possess Lebanese Liras but want to convert them to U.S. Dollar cheques because they fear devaluation or other market volatilities. So they sell Lebanese Liras for 2,800/2,900 for Lollars –having made these Liras through the 3,900 rate– they believe the profit is Lira 1,000. And the seller of the cheque makes 2,900 Liras

[8] Please note, this calculation is a function of the actual market rate of Lebanese Pounds. According to the informant, this operation stood when the Lebanese Pound exchanged at 7000 to the Dollar.

Of course, with Lebanese Pounds exchanging at 8500 to the Dollar, the price of goods such as diamonds in Lollar cheques would have to exceed threefold its original price in cash.

[9] Author's interview with Key Informant, speaking on condition of anonymity on 22 November 2020.

as opposed to the bank's official rate of 1,500 Liras to the dollar.[10] I take this money and buy U.S. dollars in cash." Of course, citizens whose money lay idle in Dollar accounts seek the best way to secure part of that sum in the form of cash or mobile assets that can act as a trusted medium of value. The fear of a potential haircut on bank deposits coupled with the absence of a viable government plan to restore an insolvent banking system lead to a noticeable increase in the purchases of jewelry, real estate, or cash dollars through the use of Lollar cheques.

Moreover, the desire to settle loans in U.S. dollars from individuals living abroad but who own real estate in Lebanon creates demand for Lollar cheques. These individuals would fly into the country with a specific cash dollar amount. They will buy a Lollar cheque from someone via a broker for almost three-fold the cash amount. This money will be used to settle their loans. Dan Azzi confirms this point by remarking that "Solider real estate development paid off $800 million in loans; their loans are now zero. Now, if you go to buy an apartment from Solider, they will not accept cheques but only cash dollars... From now to another year, all purchases in the real estate sector would be in cash dollars."[11] Eager for cash, some commercial banks briefly hopped on to the bandwagon by offering clients up to 2.5 or 2.7 times their cash Dollars in a Lollar bankers' cheque.[12] Now banks make this offer to clients only if fresh funds flow into their accounts from abroad. This bankers' cheque can then be used to settle loans in U.S. Dollars or purchase real estate.

However, with more than one licensed actor carrying out cash transactions at different rates as a backdrop, citizens' demand for such services contest both state institutions (here the central bank's role in fixing the exchange rate) and the unit of the Lebanese pound. Such Lollar cheque transactions oblige the seller to cash the cheque by purchasing Lebanese pounds, as explained above, and then use those Lebanese Pounds to

[10] This condition stood when the Lebanese Pound exchanged at 7000 to the U.S. Dollar.

Also only a limited amount of U.S. Dollars in Lebanese bank Accounts can be withdrawn according to the 3900 Lebanese Pound rate. Therefore, considerable sums exceeding 10,000 Dollars will take months to withdraw from accounts according to the 3900 rate else they would have to be withdrawn according to the official rate of 1500 Lebanese Pounds.

[11] Author's interview with Dan Azzi.

[12] Author's interview with key informant.

buy cash Dollars from foreign exchange offices. This business circuit requires the central bank to consistently print Lebanese Pounds to meet market demand, generates currency inflation, and excises a vast portion of business dealings from their otherwise formal banking channels. In other words, the more individuals decide to cash Lollar cheques for trusted media of value, the more the central bank will have to deal with currency inflation and informality on the Lebanese market that take place outside its immediate supervision. By carrying out these seemingly inconsequential transactions, citizens' consent to sale and purchase products through these means pose risks to state authority, challenge the central bank's management of the national currency, and provide a hotbed for informality. While the informality under scrutiny here do not pose enormous security risks, turning a considerable segment of business transactions away from formal channels toward the free market can constitute opportunities for the emergence of unforeseen informality.

While Ghani et al. (2005) require states to take the lead in public finances to assert their enforcement capacities, Hanau Santini et al. (2021) analyze governance spaces where states execute their functions alongside multiple other domestic and international actors. Such nuances in the conception of statehood help shed light on potential fluid avenues of incubation for informal practices. These informal practices occur alongside the state because both citizens and state institutions play an invasive role that influence markets. Keeping in mind that Lebanon's daily demand for U.S. Dollars is approximately 8/10 million, parties who possess Dollar sums—the central bank being a potent authority—can easily influence the exchange rate on the market. Private individuals who travel into Lebanon from abroad can also influence the market by merely exchanging a handsome sum of Dollars for Lebanese Pounds. On a cash strapped market, individuals, sub-state actors, and state institutions all play a role in determining the market exchange rate of the Lebanese Pound and, with that, the price of imported goods and other assets. But amidst a sovereign debt crisis that has spawned multiple exchange rates and a set of informal practices that wrestle with state regulations of the market, these informal practices—alongside many others—will remain omnipresent absent viable alternatives for cash on the market.

Conclusion: The Way Out

With life savings locked up in bank accounts, staggering unemployment, a 55% poverty rate, and vast percentage of the youth eager to hop onboard the first plane abroad (Reynolds 2020; UNESCWA 2020), Lebanon's prospects for an economic recovery seem grim at best. The emergent informality on Lebanese markets usher in an era where transactions will continue to operate outside the immediate reach of the state and through direct market channels—actor to actor—instead of conventional banking regulations. Yet, the reversal of these misfortunes can take place with a set of economic reforms that start with financial auditing of state accounts, transparency indicators, access to public information, clear auditing and budget practices, a clear bidding process for state contracts and anti-trust laws that end monopolistic markets (see Wood et al. 2020). The parties to the international donor conference of CEDRE in 2018 seem intent on overseeing the implementation of reforms in Lebanon as a conditionality for unlocking much-needed aid.[13] These demands have been echoed by the French President during his trips to Lebanon in 2020 and France's stewardship of an aid conference for the victims of the August 4, 2020 Beirut port blast.

As evinced in my analysis above, the intimate symbiotic ties between Lebanon's sectarian elites, economic interests, and the political process—notwithstanding some external parties—distort public finances and structure the Lebanese economy around unproductive sectors. This sectarian system must come to a standstill with the support of the international community. The International Monetary Fund and other state donors must condition aid to Lebanon on an economic plan that expands competition on the market, breaks monopolies and oligopolies, relies on domestic production in sectors where Lebanon holds a comparative advantage, builds necessary infrastructure such as the electric grid, strengthens institutional watchdogs and encourages foreign direct investment. They should supervise Lebanon's path to a positive balance of payments to ensure that any bailout package serves the purpose of foreign currency retention. These international actors should not fall prey to informal practices of Lebanon's sectarian elites lest they end up replicating those informal practices much as in Kosovo (Danielsson 2016a). International state donors and IMF conditionality must also stipulate a

[13] Author's interview with Dr. Fouad Zmokhol.

set of institutional reforms that strengthens judicial independence and fleshes out an economic plan to make depositors' frozen assets accessible in the long run. The reason for the proliferation of these informal practices examined in this chapter lie in the severe monetary dislocations in the currency market. While this follows no clear monetary logic, it reveals that certain economic shocks incentivize citizens to renounce part of their deposits in return for tangible assets.

Economics textbooks highlight the role of central banks as lenders of last resort, but fail to show what happens when that lender suffers a legitimacy deficit. In other words, when citizens and depositors lose their trust in the system, let alone the national currency, they frantically search for more reliable guardians of their assets, stable media of value, assets with a long lifespan and a high resale value. A sine qua non of market exchanges and formal institutional economic behavior is the ability to access, store and then convert capital—the fruit of labor—to various forms. In the case where such a possibility is out of reach, the prevailing market conditions, molded by economic institutional shifts, incentivize citizens to contest normal behavior through their adoption of informal practices.

Although palpable incentives may induce Lebanese citizens to revert business activity back to formal channels, the contestation of these channels through informality requires an overhaul of the entire political-economic structures undergirding sectarianism. Citizens whose money have been locked in bank accounts are now quite aware of the rationale behind their country's failed political-economic model. They question how politicians transfer their money to foreign bank accounts despite the central bank's restrictions on transfers, why the central bank spends its remaining foreign currency reserves on imports and—in perhaps simpler terms—why the Lebanese state did very little to nurture a positive balance of payments in the past 10 years (through the construction of domestic industries etc.). Their economic consciousness, fostered by a gaggle of famed economists speaking on television shows, youtube channels, podcasts, and other media, now points the blame at this political-economic model, which no longer services citizens. Even if citizens derived some benefit from the policy of a fixed exchange rate that hedges against the risk of severe currency fluctuations, citizens' trust in their elites, commercial banks, and the state's regulatory financial authorities have been shattered. The restoration of confidence will only come about with acute institutional reforms, meaning a metamorphosis of the financial-economic system undergirding sectarianism (Traboulsi 2014; Salloukh et al. 2015; Salloukh 2019). But what happens if these much-needed institutional reforms, the metamorphosis of sectarian financial-economic practices, and external support fail to deliver on citizens' expectations?

With the lack of severe institutional shifts that remold market behavior back to its formal channels, citizens' espousal of informality will most likely become the new normal well into the foreseeable future. This behavior will continue to strain an already cash-strapped economy by constantly calling for cash to execute day-to-day and business transactions. It will encourage citizens to avoid depositing their income in commercial banks but carefully contemplate other media of value such as land ownership, an apartment or summer house, gold, and other valuable items. It will oblige citizens' consideration of multiple asset portfolios for risk diversification as citizens learn that placing all their eggs in one basket is a dangerous bet in this distorted political-economic model. Such economic behavior will further erode the state and bank's capacities to provide credit as informality drives cash through extra-institutional channels. Citizens' trust, as the political elites will quickly learn in the forthcoming months and years, is vital for a healthy economy, but will only be restored when Lebanese elites reshape this failed economic model, else informality will contest state institutions until further notice.

Works Cited

Ayres, S., 2017. Assessing the Impact of Informal Governance on Political Innovation. *Public Management Review*, 19(1), pp. 90–107.

Baroudi, S. E., 1998. Economic Conflict in Postwar Lebanon: State–Labor Relations Between 1992 and 1997. *Middle East Journal*, Autumn, 52(4), pp. 531–550.

Baroudi, S. E., 2000. Business Associations and the Representation of Business Interests in Post-War Lebanon: The Case of the Association of Lebanese Industrialists. *Middle Eastern Studies*, July, 36(3), pp. 23–51.

Baroudi, S. E., 2005. Lebanon's Foreign Trade Relations in the Postwar Era: Scenarios for Integration (1990–Present). *Middle Eastern Studies*, March, 41(2), pp. 201–225.

Butter, D., 1995. How Hariri Pulled Lebanon from the Abyss. *Middle East Economic Digest*, October 27, p. 3.

Cammet, M., 2014. *Compassionate Communalism: Welfare and Sectarianism in Lebanon*. Ithaca: Cornell University Press.

Dabbagh, S., Deeb, G., El-khazen, F. & Kisirwani, M., 1997. The Lebanese Constitution. *Arab Law Quarterly*, 12(2), pp. 224–261.

Dabbous, D., 2013. *Regulating Lebanese Broadcasting: A Policy Analysis*. Germany: Lambert Academic Publishing.

Danielsson, A., 2016a. Reforming and Performing the Informal Economy: Constitutive Effects of the World Bank's Anti-Informality Practices in Kosovo. *Journal of Intervention and Statebuilding*, 10(2), pp. 241–260.

Danielsson, A., 2016b. Informality and Peacebuilding: Field Notes on the Intrinsic Reproductive Principles of Informal Economic Practices in Post-Conflict Kosovo. *International Peacekeeping*, 23(1), pp. 1–28.

Distler, W., 2018. 'And Everybody Did Whatever They Wanted to Do': Informal Practices of International Statebuilders in Kosovo. *Civil War*, 20(2), pp. 286–303.

El-Husseini, R., 2012. *Pax Syriana Elite Politics in Post-War Lebanon*. New York: Syracuse University Press.

El-Khazen, F., 2003. The Postwar Political Process: Authoritarianism by Diffusion. In: T. Hanf & N. Salam, eds. *Lebanon in Limbo: Postwar Society and State in an Uncertain Regional Environment*. Baden-Baden, Germany: Nomos.

Gaspard, T., 2003. *Competition in the Lebanese Economy* [Online]. Available at: http://www.economy.gov.lb/public/uploads/files/7982_8734_3466.pdf.

Ghani, A., Lockhart, C. & Carnahan, M., 2005. *Closing the Sovereignty Gap: An Approach to State-Building*. London: Overseas Development Institute.

Giddens, A., 1979. *Central Problems in Social Theory: Action, Structure and Contradiction in Social Analysis*. California: University of California Press.

Hanau Santini, R., Polese, A. & Kevlihan, R. eds., 2021. *Limited Statehood and Informal Governance in the Middle East and Africa*. London: Routledge.

Harris, W. A., 1997. *Faces of Lebanon: Sects, Wars, and Global Extensions*. New Jersey: Marcus Weiner Publishers.

Helou, J. P., 2015. Policy Overcomes Confessional Hurdles: A Policy Strategy Tackles Challenges in the Segmented Society and State of Lebanon. *Athens Journal of Mediterranean Studies*, 1(4), pp. 325–340.

Helou, J. P., 2020. *Activism, Change and Sectarianism in the Free Patriotic Movement in Lebanon*. London: Palgrave Macmillan.

Johnson, M., 1986. *Class and Client in Beirut: The Sunni Muslim Community and the Lebanese State 1840–1985*. London: Ithaca Press.

Khalaf, S., 1968. Primordial Ties and Politics in Lebanon. *Middle Eastern Studies*, April, 4(3), pp. 243–269.

Ledeneva, A., 2013. *Can Russia Modernise: Sistema, Power Networks and Informal Governance*. Cambridge: Cambridge University Press.

Leenders, R., 2012. *Spoils of Truce: Corruption and State-Building in Postwar Lebanon*. Ithaca: Cornell University Press.

Lijphart, A., 1984. *Democracies: Patterns of Majoritarian and Consensus Government in Twenty-One Countries*. Connecticut: Yale University Press.
Melki, J. P., 2007. *Television News and the State in Lebanon*. s.l.: University of Maryland.
Migdal, J. S., 1994. The State in Society: An Approach to Struggles for Domination. In: J. S. Migdal, A. Kohli & V. Shue, eds. *State Power and Social Forces: Domination and Transformation in the Third World*. Cambridge: Cambridge University Press.
Morris, J. & Polese, A., 2014. Introduction: Informality—Enduring Practices, Entwined Livelihoods. In: J. Morris & A. Polese, eds. *The Informal Post-Socialist Economy: Embedded Practices and Livelihoods*. London: Routledge.
Nizameddine, T., 2006. The Political Economy of Lebanon Under Rafiq Hariri: An Interpretation. *Middle East Journal*, Winter, 60(1), pp. 95–114.
Parry, J., & M. Bloch., 1989. *Money and the Morality of Exchange*. Cambridge: Cambridge University Press.
Picard, E., 1996. *Lebanon: A Shattered Country*. New York: Holmes & Meier.
Polese, A., 2010. The Formal and the Informal: Exploring 'Ukrainian' Education in Ukraine, Scenes from Odessa. *Comparative Education*, 46(1), pp. 47–62.
Polese, A., 2014. Informal Payments in Ukrainian Hospitals: On the Boundary Between Informal Payments, Gifts, and Bribes. *Anthropological Forum*, 24(4), pp. 381–395.
Polese, A., Kovács, B. & Jancsics, D., 2018. Informality 'In Spite Of' or 'Beyond' the State: Some Evidence from Hungary and Romania. *European Societies*, 20(2), pp. 207–235.
Polese, A., Morris, J. & Kovács, B., 2016. "States" of Informality in Post-Socialist Europe (and Beyond). *Journal of Contemporary Central and Eastern Europe*, 24(3), pp. 181–190.
Polese, A., Morris, J., Kovács, B. & Harboe, I., 2014. "Welfare State" in Central and Eastern Europe: Where Informality Fits In. *Journal of Contemporary European Studies*, 22(2), pp. 184–198.
Polese, A. & Rekhviashvili, L., 2017. Introduction: Informality and Power in the South Caucasus. *Caucasus Survey*, 5(1), pp. 1–10.
Pugh, M., 2004. Rubbing Salt into War Wounds: Shadow Economies and Peacebuilding in Bosnia and Kosovo. *Problems of Post-Communism*, 51(3), pp. 53–60.
Pugh, M., 2005a. Crime and Capitalism in Kosovo's Transformation. In: *46th International Studies Association Convention*, March 1–5.
Pugh, M., 2005b. The Political Economy of Peacebuilding: A Critical Theory Perspective. *International Journal of Peace Studies*, 10(2), pp. 23–42.
Pugh, M., Cooper, N. & Goodhand, J., 2004. *War Economies in a Regional Context: Challenges of Transformation*. Colorado: Lynne Rienner.

Ratcliffe, V., 2020. *Lebanon's Electricity Problems Are Poised to Become Much Worse* [Online]. Available at: https://www.aljazeera.com/economy/2020/7/29/lebanons-electricity-problems-are-poised-to-become-much-worse.

Reynolds, R., 2020. *Arab Youth Survey 2020: Young People Plan Exidous Amid Dismay at Region's Corruption and Economic Failure* [Online]. Available at: https://www.thenational.ae/arab-youth-survey-2020-young-people-plan-exodus-amid-dismay-at-region-s-corruption-and-economic-failure-1.1089153.

Salam, N., 2003. Taif Revisited. In: T. Hanf & N. Salam, eds. *Lebanon in Limbo: Postwar Society and State in an Uncertain Regional Environment*. Baden-Baden, Germany: Nomos.

Salamey, I., 2014. *The Government and Politics of Lebanon*. London: Routledge.

Salibi, K., 1988. *A House of Many Mansions: The History of Lebanon Reconsidered*. London: I.B. Tauris.

Salloukh, B. F., 2019. Taif and the Lebanese State: The Political Economy of a Very Sectarian Public Sector. *Nationalism and Ethnic Politics*, 25(1), pp. 43–60.

Salloukh, B. F. et al., 2015. *The Politics of Sectarianism in Postwar Lebanon*. London: Pluto Press.

Salti, N., 2019. *No Country for Poor Men: How Lebanon's Debt Has Exacerbated Inequality* [Online]. Available at: https://carnegie-mec.org/2019/09/17/no-country-for-poor-men-how-lebanon-s-debt-has-exacerbated-inequality-pub-79852 [Accessed 5 1 2021].

Steenberg, R., 2016. The Art of Not Seeing Like a State: On the Ideology of "Informality". *Journal of Contemporary Central and Eastern Europe*, 24(3), pp. 293–306.

Stel, N. & Naudé, W., 2016. 'Public–Private Entanglement': Entrepreneurship in Lebanon's Hybrid Political Order. *The Journal of Development Studies*, 52(2), pp. 254–268.

Taif, 1989. *Taif Accord* [Online]. Available at: http://peacemaker.un.org/lebanon-taifaccords89.

Traboulsi, F., 2012. *A History of Modern Lebanon*. Second ed. London: Pluto Press.

Traboulsi, F., 2014. Social Classes and Political Power in Lebanon. *Heinrich Böll Stiftung—Middle East Office*, pp. 1–132.

UNESCWA, 2020. *ESCWA Warns: More Than Half of Lebanon's Population Trapped in Poverty* [Online]. Available at: https://www.unescwa.org/news/Lebanon-poverty-2020 [Accessed 9 1 2021].

Van Tatenhove, J., Mak, J. & Liefferink, D., 2006. The Inter-Play Between Formal and Informal Practices. *Perspectives on European Politics and Society*, 7(1), pp. 8–24.

Verdeil, E., 2018. *Electricity Subsidies: Benefiting Some Regions More Than Others* [Online]. Available at: http://www.lcps-lebanon.org/featuredArticle.php?id=163 [Accessed 17 11 2018].

Wood, D., Boswall, J. & Minkara, Y., 2020. *Unfair Game: Lebanon's Rigged Markets Are Killing Competition* [Online]. Available at: https://www.thinktriangle.net/wp-content/uploads/2020/11/Unfair-Game-Lebanons-Rigged-Markets-Are-Killing-Competition.pdf [Accessed 28 12 2020].

Perceived Pull and Push Factors of Healthcare Professionals Intention for Mobility: The Case of Romania

Elena Druică and Rodica Ianole-Călin

Introduction

Understanding, measuring, and managing the global outflow of healthcare professionals is not a novel topic on the health agenda, but its increasing complexity continues to attract practical and academic interest from many social scientists. The multiple dimensions of the problem range from the general case of health professionals migration from poor to rich countries (Dovlo 2005; Chen et al. 2004) to the more subtle aspects of healthcare mobility paths observed mostly within the European Union (Jinks et al. 2000; Buchan et al. 2014; Glinos 2015), especially from Eastern and Southern Europe to the Western states.

A common explanatory paradigm cited very often is that doctors and nurses simply "follow the money" or, more formally expressed, the main determinant of a mobility decision is considered the level of wages (Vujicic

E. Druică · R. Ianole-Călin (✉)
Department of Applied Economics and Quantitative Analysis, University of Bucharest, Bucharest, Romania
e-mail: rodica.ianole@faa.unibuc.ro

A. Polese (ed.), *Informality, Labour Mobility and Precariousness*, International Political Economy Series,
https://doi.org/10.1007/978-3-030-82499-0_12

et al. 2004): low in source countries and high in destination countries. In this spirit, the last policy changes in the Romanian health sector have brought significant increases in the wages of doctors, as a reaction of the current deficit in supply (Stan and Erne 2016). The flaw of the measure is easily spotted in the fact that it was only addressed to the category of doctors, ignoring nurses and other healthcare support professionals. In consequence, the existing imbalance grew more acute and the measure failed to work as a standard regulation of the market through price adjustment. Having this perspective in mind, while the disparities at the level of healthcare expenditures are clearly shaping the framework for labor mobility, we believe that, at the level of the individual decision-making process, this is an overly simplified view, at least for the case of Romanian healthcare professionals.

This study focuses on two types of factors that may underline healthcare professionals' decision to work in a different country. On the one side, we considered factors that provide reasons to stay, measured through various dimensions of satisfaction at the workplace like working conditions, fair evaluation, opportunities for trainings and continuing medical education, and relations with others. On another side, we considered factors that provide reasons to leave. In this category, we included better opportunities for professional development, and a fairer and trustworthy working environment. Various authors refer to these types of motives as push and pull factors in healthcare migration (Eastwood et al. 2005; Clark et al. 2006). It is important to highlight the novel perspective brought by the working environment variable, connected to a specific form of informality. Namely, the variable refers to how informal payments are perceived by health professionals, specifically how they can negatively contribute to job dissatisfaction and the intention to leave the country. This is rather a counterintuitive view, opposing the popular perspective according to which healthcare professionals are simply benefiting from the informal payments, either from the position of actively requesting or passively accepting them. While the active position is easily condemnable, the passive one is not that easy to judge: being under constant public scrutiny is definitely a stress factor that may negatively impact social practices like gift giving and reciprocity, without ensuring that actual corruption is diminished.

The aim of the present chapter is to investigate if different dimensions of satisfaction at the workplace predict the temptation for mobility among the Romanian healthcare professionals. Additionally, we explore the extent to which expectations regarding better opportunities for professional development, and fairer and more trustful working environments drive the healthcare professionals into mobility. Unlike other approaches, where these factors were accounted for in objective terms, discussing the demand and supply on the source and destination countries, in our case they are measured as perceptions of healthcare professionals. Previous studies show that healthcare professionals from less developed countries are driven by economic motives, like stability, increased income, or improved prospects for one's children (Astor et al. 2005), while those from developed countries are mainly focused on personal motives (Gray and Johnson 2009). Our study focuses on Romania, positioned at the border of these two categories in terms of economic development (according to the World Bank classification it belongs to the upper-middle-income category). Many of the studies addressing the motives of the Romanian healthcare professionals to work abroad advocate patterns found in less developed countries (wages, improved working conditions, and better medical training for continuous development, to name only a few, see for example George and Reardon 2013). However, other factors like improved professional opportunities underline the official discourse and place the Romanian healthcare professional expectations among the factors that characterize more developed countries.

Thus, empirical proofs are needed to decide upon the most appropriate theoretical framework. Following the common perception and literature review stream focused on the healthcare systems weaknesses, we predict that a higher job satisfaction will be negatively associated with mobility temptation. In line with the literature that supports the idea that corruption and informal payments plague the Eastern European healthcare systems (Stepurko et al. 2015; Williams et al. 2016), being an important reason for the professionals to leave the country, we further expect that those who put a higher emphasize on a fairer and more trustable working environment, to be less satisfied than those who don't. Finally, considering the common opinion among officials and non-officials in Romania, that healthcare professionals leave the country in response to their low level of income, we also explore whether the temptation for mobility is predicted by one's satisfaction with one's wage.

Our first research model explores the determinants of the mobility intention and indicates three variables as having a significant, and positive, impact: contentedness with wage, professional motives, and environmental motives. The first determinant is the most interesting: a higher satisfaction with wage increases the intention to leave for an abroad job opportunity. The reverse correlation, by reference to both public and institutional point of views, stands as a first argument in the mobility versus migration debate, in the mobility corner. While economic reasons are still part of the equation, included in the professional motives latent variable, they are not the sole and dominant push factor. An evidence-based health public policy should further follow up this lead in entangling an even more precise understanding of the economic and non-economic forces behind the healthcare professionals' desire to leave the country.

The second model investigates the factors responsible for the overall satisfaction, among the same set of determinants. The "usual suspects" in assessing satisfaction prove to also be relevant in this case. Professional development, organizational support, working conditions, and evaluation are all positive influences on satisfaction, confirming the expectations and extant literature. However, with respect to the factors capturing reasons for leaving the country, there is one interesting finding: the environment, characterized by trust, less informality and more integrity, does not stand out as having any significant impact on satisfaction, as it has on the mobility intention.

Leaving aside the technicalities of our results, the crucial aspect is that the determinants of intention mobility and job satisfaction overlap only to a marginal extent, hinting that the case of Romanian healthcare professional is somewhat transitioning between the objective motivations encountered in developing countries and the subjective ones from developed countries.

Literature Review

The systemic view on the migration phenomena is often reflected in systemic reasons supporting a migration decision. The push–pull theory of migration classifies the various factors at play in two main categories, where the push factors belong to the source countries and the pull factors to the destination countries (Mej'ia et al. 1979; Schoorl et al. 2000). Depending on the context, different variables from the political, social, economic, legal, historical, cultural, and educational areas are discussed (Kline 2003), but the focus on individual determinants remains marginal. While this ratio of explanatory forces may stand relevant in the clear cases of migration, the balance needs to be readjusted when discussing about mobility intentions. Otrachshenko and Popova (2014) provide an interesting analysis connecting the dots between the individual and aggregate factors by illustrating how the effect of socioeconomic variables and macroeconomic conditions on the intention to migrate is an indirect one, mediated through life satisfaction. This opens up the discussion toward the importance of including more subjective indicators in the conceptualization of migration decision. Serving the purpose of our empirical study, the subjective approach is accounted by reviewing the literature focused on the determinants of job satisfaction for studying labor mobility (Freeman 1978) and behavioral intentions as primary antecedents for actual behaviors (Fishbein and Ajzen 1975).

Departing from a seminal definition, the mechanisms building up job satisfaction require a closer look into the realm of "cognitive and affective reactions to the differential perceptions of what an employee wants to receive compared with what he or she actually receives" (Cranny et al. 1992). Classical studies in personnel psychology indicate that satisfaction and commitment are in general negatively correlated with the intent to leave (Tett and Meyer 1993; Hellman 1997), with a fair share of variation concerning the magnitude of this relationship. The variation is quite natural if we take into account more recent conceptualization of job satisfaction as multidimensional, with a number of facets to consider: satisfaction with supervisor, satisfaction with variety, satisfaction with closure, satisfaction with compensation, satisfaction with co-workers, and satisfaction with management and human resources policies (Purani and Sahadev 2008). Moreover, situational specificity is an important aspect to consider and for the healthcare sector, this was mostly explored for the case of nurses.

In general, the literature is consistent and a low level of job satisfaction acts indeed in many cases as predictor of nurses' intention to leave (in the UK, Shields and Ward 2001, in Taiwan, Tzeng 2002, in Lebanon, El-Jardali et al. 2007, in Malaysia, Alam and Mohammad, 2010 etc.). However, Lum and her colleagues (1998) offer a novel interpretation of this relationship, showing only an indirect effect of nurses' job satisfaction on their intention to leave, the stronger and direct impact being hold by organizational commitment. This is a variable less investigated for the case of health professionals and its definition as "a global attitude serving as a stabilizing force that acts to maintain behavioral direction when expectancy/equity conditions are not met" (DeCotiis & Summers, 1987) offers a good explanatory paradigm of the somewhat counterintuitive results illustrated in the next section for the Romanian sample.

Data and Method

Participants

In our study there were 137 participants, age between 23 and 62 years (mean = 40.86, median = 42 and sd = 8.7), 83.2% females and 16.2% males. Of the entire sample, 38% are medical doctors, 37.2% are nurses, and the rest of 24.8% is administrative staff, with a number of years worked in healthcare ranging between 1 and 38 (mean = 16.17, median = 16, and sd = 9.72). The participants voluntarily completed an online survey. Data was analyzed in *R*, version 3.4.3.

Measurement

Dependent Variables

We measure two dependent variables, namely: (1) the healthcare professional intention to work in a different country, and (2) the overall satisfaction the participants report in relation to their workplaces. Both variables measured on a Likert scale 1–10. The mobility intention ranges from 1 to 10 (mean = 5.912, median = 6, sd = 3.27), and the overall satisfaction also ranges from 1 to 10 (mean = 7.175, median = 8, sd = 2.07).

Independent Variables

Satisfaction: To measure various dimensions of healthcare professional satisfaction at the workplace we used a shortened version of the Satisfaction of Employees in Health Care (SEHC) scale developed by Alpern et al. (2013). The original version of the SEHC comprises of 18 items, however, we chose only 14. The items are recorded in Table 1 and still cover three dimensions, as in the original study, which will be detailed later. All these dimensions have been documented before as relevant factors that predict the healthcare professionals intention to leave their country and work abroad (Clark et al. 2006).

Table 1 Factors measuring different dimensions of satisfaction at the workplace

Item	*Statements (Alpern et al. 2013)*	*Factor1*	*Factor2*	*Factor3*
Cronbach's Alpha by factor		*0.89*	*0.91*	*0.92*
SAT1	The management of this organization is supportive of me			0.429
SAT2	I receive the right amount of support and guidance from my direct supervisor			0.960
SAT3	I am provided with all useful trainings necessary to perform my job		0.786	
SAT4	I feel encouraged by my supervisor to offer suggestions and improvements			0.899
SAT5	I feel recognized and appreciated at my workplace		0.512	
SAT6	The organization rules make it easy for me to do a good job	0.646		
SAT7	I am satisfied with my chances for promotion		0.535	
SAT8	I am provided with adequate opportunities to develop my professional skills		0.968	
SAT9	I have an accurate written job description	0.866		
SAT10	The amount of work I am expected to finish each day is reasonable	0.843		
SAT11	My work assignments are always clearly explained to me	0.857		
SAT12	My work is evaluated based on a fair system of performance standards	0.450		
SAT13	The department provides all the equipment, supplies, and resources necessary for me to perform my duties	0.492		
SAT14	My co-workers and I work well together		0.467	

To derive the independent variables related to satisfaction, we conducted an exploratory factor analysis based on a maximum likelihood extraction method and a factor rotation "promax," and found that three dimensions describe satisfaction at the workplace, as presented in Table 1. The first factor refers to working conditions and fair evaluation, the second factor comprises items that capture professional recognition and professional development, while the third factor involves the quality of the relationships with the direct supervisor. The items involved in each factor capture one single dimension, as the Cronbach's Alpha values recorded on the second line of the table show. We will use the factor scores derived from this preliminary analysis as predictors in our final models. The cumulative variance explained by these three factors is 52.5%. Table 8 in Annex

Table 2 Factor loadings for what it is perceived as important in relationships for the healthcare professionals

Items	*Factor: relationships*
Relationships with co-workers	0.574
Relationships with superiors	0.846
Organizational recognition	0.555

presents the goodness of fit measurements for the confirmatory factor analysis that involves the extracted factors.

Importance of relationships: In the second stage, and also based on exploratory factor analysis, we derived the variable "relationships." The respondents were asked to rate on a scale from 1 to 10 how important are for them the relationships with their co-workers and with their superiors, and how important they consider the need for receiving recognition in their organization. With a Cronbach's Alpha of 0.69, we found one factor extracted using maximum likelihood, with the loadings presented in Table 2. Based on these results, we derived the score associated with this control variable. The cumulative variance explained by this factor is 45.1%.

To measure the intention of the professional to leave to work in a foreign country we have applied a short list of motives used previously in a preference elicitation study on Romanian medical students (Suciu et al. 2017). The first 5 listed reasons are an exact replication, while the last one regarding informal practices in the health system represents a novel addition, in line with the existing evidence pointing to the prevalence of informal payments for health services in Southeast Europe (Williams and Horodnic 2018).

Motives: The last exploratory factor analysis detected two dimensions related to the motives that would determine the healthcare professionals to make the decision to work in a foreign country. The cumulative variance explained by these two factors is 53.8%, and they capture aspects related to professional aspects (higher wages, better working conditions and better professional opportunities load in one of the factors), while the other one reflects reasons related to trust, informality and professional integrity. Table 3 presents these results, as well as illustrates through the Cronbach's Alpha values presented in the second line of the table that each subset of the scale has a good internal consistency. The goodness

Table 3 Factor loadings for what it is perceived as a motive for mobility

Items	*Factor 1 = Environment motives*	*Factor 2 = Professional motives*
Cronbach's Alpha by factor	*0.84*	*0.77*
Higher wages		0.749
Better working conditions		0.809
Better professional opportunities		0.502
Lack of trust in the Romanian healthcare	0.661	
A higher level of professional integrity	0.692	
Less informal practices	0.755	

of fit measures for the corresponding confirmatory factor analysis, are reported in Table 8 from Annex 1.

Control Variables

Consistent with previous research, we considered gender, age, and the time of employment in the healthcare system as control variables (Goštautaitė et al. 2018). Since respondents' age was correlated 85% with their experience in healthcare, we kept the later variable and discarded the former. To rule out other possible spurious correlations, we added the perceived importance assigned by the participants to various factors that may influence both their satisfaction at the workplace, and their mobility intention. In this respect, the participants were invited to rate on a scale from 1 to 10 how important are in their view: (a) the relationship with their co-workers, (b) the relationship with their superiors, and (c) the recognition they receive from their organization.

The last control variable is the self-assessed contentedness of the participants with their salaries. Other studies extensively discussed the role of the wage in healthcare professional intention to work abroad, and some authors focused on the differences between the salaries in the source country and the salaries in the destination country, although this difference was found inconclusive (Vujicic et al. 2004). In our study we focused on the self-reported satisfaction of the respondents regarding the salary they receive, and measured their answers on a scale from 1 to 10, where 1 means "extremely dissatisfied," while 10 stands for "extremely satisfied."

After controlling for the variables that have been considered above, we set the following research questions:

H1: overall satisfaction is positively associated with inclination to seek for fairer and trustworthy working environments
H2a: temptation for mobility is negatively associated with good working conditions and fair evaluation
H2b: temptation for mobility is negatively associated with perceived opportunities for professional development
H2c: temptation for mobility is negatively associated with better relationships with the supervisors
H3: temptation for mobility is negatively associated with contentedness with wage

Models and Estimation Method

To evaluate the effects of different dimensions of our latent variables accounting for satisfaction at the workplace and motive for mobility on the intention to work on a different country we estimated two different models. The purpose of the model that expresses the overall satisfaction is twofold: on the one side, it confirms that the three factors derived from the SEHC scale have an influence on the overall satisfaction of the Romanian healthcare professionals. On the other side, it aims to investigate whether people with higher inclination toward fairer and trustworthy working environments are less satisfied than those who value less this aspect. This model has the following functional form:

$$\begin{aligned} \mathrm{SAT}_i = {} & \beta_{10} + \beta_{11}^T X_i + \beta_{12}\mathrm{MOT\_PROF}_i + \beta_{13}\mathrm{MOT\_ENV}_i \\ & + \beta_{14}\mathrm{PROF\_DEV}_i + \beta_{15}\mathrm{ORG\_SUP}_i + \beta_{16}\mathrm{WORK\_COND}_i \end{aligned} \tag{1}$$

We denoted by SAT the dependent variable "overall satisfaction," X are the control variables, MOT_PROF stand for professional motives for mobility, MOT_ENV captures the motives related to fair and trustworthy environment, while professional development (PROF_DEV), organizational support (ORG_SUP) and working conditions and evaluation (WORK_COND) are the three dimensions derived from the SEHC scale. A similar model is developed to explain the temptation for mobility, as follows:

$$MOB_i = \text{ß}_{20} + \text{ß}_{21}^T X_i + \text{ß}_{22}MOT\_PROF_i + \text{ß}_{23}MOT\_ENV_i + \text{ß}_{24}PROF\_DEV_i + \text{ß}_{25}ORG\_SUP_i + \text{ß}_{26}WORK\_COND_i \quad (2)$$

The dependent variable MOB stands for the temptation for mobility, while the rest of the predictors have been described above. We estimated both models using the OLS robust regression.

Results

First, we investigate if there is any significant difference between the average score for mobility temptation across the three categories of healthcare professionals: medical doctors, nurses, and administrative staff. Table 4 presents the results. As the *p*-value is higher than the significance level 0.05, we conclude that there are no significant differences between these groups.

Given the recent pay raise received by medical doctors in the public sector, we conduct a similar analysis to assess if there is any difference between the average score for contentedness with one's salary across the three categories of healthcare professionals. The results presented in Table 5 shows that there are no significant differences between these groups, as the associated *p*-value is higher than the significance level 0.05.

Table 4 ANOVA for mobility temptation

	Df	*Sum Sq*	*Mean Sq*	*F value*	*Pr(>F)*
Professional category	2	9.1	4.531	0.42	0.658
Residuals	134	1443.9	10.775		

Table 5 ANOVA for contentedness with salary

	Df	*Sum Sq*	*Mean Sq*	*F value*	*Pr(>F)*
Professional category	2	9.1	4.531	0.42	0.658
Residuals	134	1443.9	10.775		

Table 6 ANOVA for overall satisfaction

	Df	*Sum Sq*	*Mean Sq*	*F value*	*Pr(>F)*
Professional category	2	40.3	20.168	4.991	0.00812**
Residuals	134	541.5	4.041		

**p < 0.01

Table 6 presents the results of the last ANOVA exploration, and proves that there are differences in reported satisfaction across categories. In particular, Fig. 1 shows that nurses report the lowest overall satisfaction.

Table 7 presents two models for each dependent variable. The first model in each case includes the control variables and the predictors derived from our factor analysis, while the second model is the final model obtained after applying a stepwise regression based on the Akaike Information Criterion to perform variable selection. This method is widely used in various types of studies, and provides an efficient list of final variables (see for example Zhang 2016).

The results show that in each case the adjusted R-squared of the second model increased compared with the corresponding first model, which confirms that by dropping some variables the quality of the information increased. Also, the variables that have been statistically significant in the

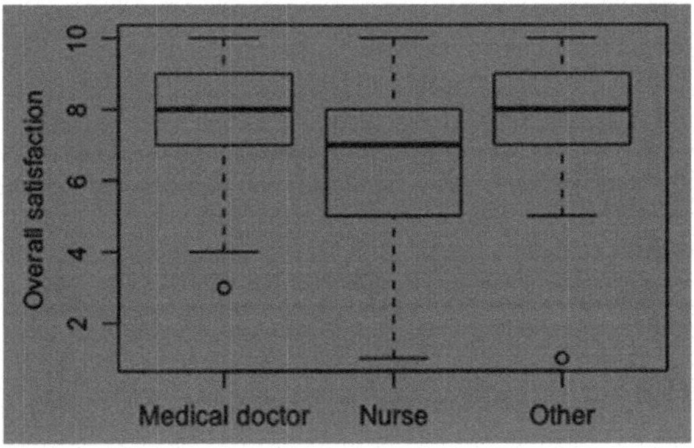

Fig. 1 Differences in overall satisfaction across categories of healthcare professionals

Table 7 Determinants of satisfaction and intention for mobility

Model	*Mobility temptation*		*Overall satisfaction*	
	Model 1	*Model 2*	*Model 1*	*Model 2*
Intercept	4.068*** (1.005)	4.259*** (0.970)	7.595*** (0.485)	7.536*** (0.220)
Gender	Reference	–	Reference	–
Female	0.022		−0.035	
Male	(0.638)		(0.308)	
Category	Reference	–	Reference	–
Medical	−0.179 (0.605)		−0.412	
doctor	−0.467 (0.625)		(0.289)	
Nurse			0.077 (0.299)	
Other				
Years in health	−0.034 (0.026)	−0.035 (0.024)	−0.028* (0.012)	−0.022 (0.012)
Contentedness with wage	0.333*** (0.121)	0.308* (0.116)	0.005 (0.058)	–
Relationships	−0.192 (0.308)	–	0.204 (0.149)	–
Professional motives	0.934* (0.415)	1.045* (0.394)	−0.329 (0.200)	−0.232* (0.109)
Environment motives	0.897** (0.308)	0.954** (0.285)	−0.175 (0.149)	–
Professional development	−0.166 (0.290)	–	1.465*** (0.140)	1.589*** (0.115)
Organizational support	0.009 (0.287)	–	1.276*** (0.139)	1.389*** (0.114)
Working conditions and evaluation	−0.222 (0.281)	–	1.344*** (0.136)	1.455*** (0.116)
R^2 / Adjusted R^2	37.84%/33.44%	36.5%/34.58%	63.78%/ 61.22%	62.93% / 61.52%
F Statistic	8.591 ($p < 0.0001$)	18.97 ($p < 0.0001$)	24.85 ($p < 0.0001$)	44.48 ($p < 0.0001$)

Robust standard errors in parentheses; $^*p < 0.05$. $^{**}p < 0.01$. $^{***}p < 0.001$

first place, remained statistically significant, or even gained significance. Therefore, we will discuss only the reduced versions of the models.

The first interesting result is that the more satisfied with the salary, the more tempted the healthcare professionals are to work in a different country. This result apparently contradicts the studies that show that migration intention is driven by economic reasons, and that lower wages support the quest for working abroad. Our findings suggest that higher wages act as signals for self-perceived higher competencies, self-confidence, and efficacy. Moreover, this is also an important argument to see the current outflow process more as mobility and not as migration. Deriving from here, precautionary measures should be enforced to a higher extent at an institutional level than the individual one.

"Professional motives" are statistically significant both in explaining the migration intention, but also the overall satisfaction. The respondents who report that higher wages, better working conditions, and better professional opportunities are the drivers for their intention to work abroad, are more tempted for mobility and less satisfied at the current workplace. This result positions our findings in the area of reasons specific to less developed countries. However, the "professional development" variable that captures aspects that complement professional motives is statistically significant only in predicting the level of satisfaction, but does not have any negative effect on professionals' intention for mobility. The result is surprising, given the nature of these two variables: "professional motives" accounts for what the professional expects to find in a different country, while "professional development" accounts for the healthcare professionals' satisfaction with the opportunities that they have now. Higher scores for "professional development" would be expected to decrease the temptation for mobility, as part of the expectations are already fulfilled.

In fact, the second interesting result is that none of the dimensions of the current satisfaction at the workplace is relevant in predicting the temptation for mobility. The third unexpected result is that, while seeking a fairer and trustworthy working environment is a good predictor for mobility intention, the variable is totally useless at predicting current satisfaction in the Romanian healthcare professionals.

Conclusions, Discussions, and Limitations of the Study

The chapter proposes a new angle of investigation for the timely and contemporary challenges brought by the acute outflow of healthcare professionals, around and the world, but in particular in Romania. At a first glance, the diagnosis of the situation would be that of an acute migration case but conceptually this goes against the understanding according to which "the movement of EU health professionals is not migration but mobility" (Glinos 2015). The two labels—mobility versus migration—enclose quite different connotations and choosing to use one or another may influencing the choice of a certain public policy path in addressing this issue.

As for theoretical implications, we align our research to the vast and heterogeneous thread on informal payments and corruption, the "popular flavor" and the most confusing category of informality, with reference to the healthcare section. The absence of a framework incorporating non-monetary transactions, as a defining legal aspect, seems to further expand into the absence of a framework looking carefully at the non-monetary determinants of the intention to leave the country, like our latent variable of environment motives, and its nature (mobility or migration). Moreover, the social perception of an under-developed institutional system characterizes very well the Romanian healthcare sector, along with "getting things done" mechanisms and lack of predictability.

This paves the way for evidence-based policy-making in two sensitive areas in Romanian society: public health (Druică et al. 2020) and informal economy (Ianole-Călin et al. 2017). The objective gap between what the state offers and what public health professionals receive is not only explained by institutional asymmetry (Williams and Horodnic 2017), but is also significantly amplified by divergent perceptions on the matter. Ignoring this dimension falls under an infrapolitics mechanism, in the sense of repeating almost identical actions in a state of incoordination and unawareness (Polese 2018), which shapes politics and creates further animosity in the relationship between informality and governance. While to a large extent this is relevant for other Eastern European countries, it may easily reach a critical stage in Romania given the prevalent perception of a weak state capacity (Vâlsan et al. 2020).

The overall results of this study suggest a disconnection between the healthcare professionals satisfaction at their current workplace, and what they would value if they were to choose working abroad. Of the research hypotheses that we set at the beginning of this chapter, we found that: H1, stating that overall satisfaction is positively associated with inclination to seek for fairer and trustworthy working environments, is rejected. Temptation for mobility proves to be unrelated to any of the dimensions accounting for professionals' satisfaction at their current workplace, therefore H2a, H2b and H2c are all rejected. The last hypothesis, H3, is also rejected, as temptation for mobility has been found as positively associated with contentedness with wage. Thus, our interpretation supports the meaning conveyed by mobility but the standard pull and push framework of migration determinations still proves expressive value in dealing with subjective factors of interest. More so, it provides a good background for further integrating aspects relating to the different features and perceptions associated with job satisfaction in the health sector.

From a practical point of view, we contribute to the literature by (1) exploring a small scale measuring intentions to leave on a Romanian sample, and (2) by attempting to understand the interplay between job satisfaction and the more traditional drivers of mobility for healthcare professionals. In line with the literature emphasizing a more complex model of the relationship between pay satisfaction and intentions to leave, our results reject the conventional perceptions and support the need for further exploration of the role of variables as professional motives, environmental motives (including trust, integrity and informality aspects) and personal development. Their interaction may even be conceptualized as organizational commitment, thus positioning the variable as having a potential mediating effect.

Acknowledgements This research was supported by a Marie Curie Research and Innovation Staff Exchange scheme within the H2020 Programme (grant acronym: SHADOW, no: 778118).

Annex 1

See Table 8.

Table 8 Goodness of fit measures in confirmatory factor analysis

Model	*Satisfaction*	*Motives*
Comparative Fit Index	0.932	0.972
Tucker-Lewis Index	0.914	0.948
RMSEA	0.112 (*P*-value RMSEA ≤ 0.05 is low)	0.101 (*P*-value RMSEA $\leq 0.05 = 0.070$)
SRMR	0.047	0.039

References

Alam, M. M., & Mohammad, J. F. (2010). Level of job satisfaction and intent to leave among Malaysian nurses. *Business Intelligence Journal, 3*(1), 123–137.

Alpern, R., Canavan, M. E., Thompson, J. T., McNatt, Z., Tatek, D., et al. (2013). Development of a brief instrument for assessing healthcare employee satisfaction in a low-income setting. *PLoS ONE, 8*(11): e79053. https://doi.org/10.1371/journal.pone.0079053.

Astor, A., Akhtar, T., Matallana, M. A., Muthuswamy, V., Olowu, F. A., Tallo, V., & Lie, R. K. (2005). Physician migration: Views from professionals in Colombia, Nigeria, India, Pakistan and the Philippines. *Social Science & Medicine, 61*(12), 2492–2500.

Buchan, J., Wismar, M., Glinos, I. A., & Bremner, J. (2014). Health professional mobility in a changing Europe. *New dynamics, mobile individuals and diverse responses*.

Chen, L., Evans, T., Anand, S., Boufford, J. I., Brown, H., Chowdhury, M., ... & Fee, E. (2004). Human resources for health: Overcoming the crisis. *The Lancet, 364*(9449), 1984–1990.

Clark, P. F., Stewart, B. J., Clark, A. D. (2006). The globalization of the labour market for health-care professionals. *International Labour Review;*145, 1/2; ABI/INFORM Global pg. 37.

Cranny, C. J., Smith, P. C., & Stone, E. F. (Eds.). (1992). *Job satisfaction: How people feel about their jobs and how it affects their performance*. Lexington Books.

DeCotiis, T. A., & Summers, T. P. (1987). A path analysis of a model of the antecedents and consequences of organizational commitment. *Human relations, 40*(7), 445–470.

Dovlo, D. (2005). Taking more than a fair share? The migration of health professionals from poor to rich countries. *PLoS Medicine, 2*(5), e109.

Druică, E., Mihăilă, V., Burcea, M., & Cepoi, V. (2020). Combining direct and indirect measurements to assess patients' satisfaction with the quality of public health services in Romania: Uncovering structural mechanisms and their implications. *International Journal of Environmental Research and Public Health, 17*(1), 152.

Eastwood, J. B., Conroy, R. E., Naicker, S., West, P. A., Tutt, R. C., & Plange-Rhule, J. (2005). Loss of health professionals from sub-Saharan Africa: The pivotal role of the UK. *Lancet, 365*(9474): 1893–1900.

El-Jardali, F., Jamal, D., Abdallah, A., & Kassak, K. (2007). Human resources for health planning and management in the Eastern Mediterranean region: Facts, gaps and forward thinking for research and policy. *Human Resources for Health, 5*(1), 9.

Fishbein, M., & Ajzen, I. (1975). *Belief, attitude, intention and behavior: An introduction to theory and research*. Reading, MA: Addison-Wesley.
Freeman, R. (1978). Job satisfaction as an economic variable. *American Economic Review, 68*(2), 135–141.
George, G., & Reardon, C. (2013). Preparing for export? Medical and nursing student migration intentions post-qualification in South Africa. *African Journal of Primary Health Care & Family Medicine* [S.l.], 5(1), 9 pages, May 2013. ISSN 2071-2936.
Glinos, I. A. (2015). Health professional mobility in the European Union: Exploring the equity and efficiency of free movement. *Health Policy, 119*(12), 1529–1536.
Gray J., & Johnson L. (2009) Intentions and motivations of nurses to migrate: A review of empirical studies. *International Journal of Migration, Health and Social Care,* 4(4), 41–48. https://doi.org/10.1108/17479894200800023.
Goštautaitė, B., Bučiūnienė, I., Milašauskienė, Ž., Bareikis, K., & Bertašiūtė, E., & Mikelionienė, G. (2018). Migration intentions of Lithuanian physicians, nurses, residents and medical students. *Health Policy.* https://doi.org/10.1016/j.healthpol.2018.07.001.
Hellman, C. M. (1997). Job satisfaction and intent to leave. *The Journal of Social Psychology, 137*(6), 677–689.
Ianole-Călin, R., Vâlsan, C., & Druică, E. (2017). The many shades of the grey economy in Romania. *The Informal Economy: Exploring Drivers and Practices*, 25–40.
Jinks, C., Ong, B. N., & Paton, C. (2000). Mobile medics? The mobility of doctors in the European Economic Area. *Health Policy, 54*(1), 45–64.
Kline, D. S. (2003). Push and pull factors in international nurse migration. *Journal of Nursing Scholarship, 35*(2), 107–111.
Lum, L., Kervin, J., Clark, K., Reid, F., & Sirola, W. (1998). Explaining nursing turnover intent: Job satisfaction, pay satisfaction, or organizational commitment? *Journal of Organizational Behavior: The International Journal of Industrial, Occupational and Organizational Psychology and Behavior, 19*(3), 305–320.
Mej'ia, A., Pizurki, H., & Royston, E. (1979). Physician and nurse migration: Analysis and policy implications, report on a WHO study.
Otrachshenko, V., & Popova, O. (2014). Life (dis) satisfaction and the intention to migrate: Evidence from Central and Eastern Europe. *The Journal of Socio-Economics, 48*, 40–49.
Polese, A. (2018). Informality and policy in the making: Four flavours to explain the essence of informality. *Interdisciplinary Political Studies, 4*(2), 7–20.

Purani, K., & Sahadev, S. (2008). The moderating role of industrial experience in the job satisfaction, intention to leave relationship: An empirical study among salesmen in India. *Journal of Business & Industrial Marketing, 23*(7), 475–485.

Schoorl, J., Heering, L., Esveldt, I., Groenewold, G., & Van der Erf, R. (2000). Push and pull factors of international migration: A comparative report.

Shields, M.A., & Ward, M. (2001). Improving nurse retention in the National Health Service in England: The impact of job satisfaction on intention to quit. *Journal of Health Economics, 20*, 677–701.

Stan, S., & Erne, R. (2016). Is migration from Central and Eastern Europe an opportunity for trade unions to demand higher wages? Evidence from the Romanian health sector. *European Journal of Industrial Relations, 22*(2), 167–183.

Stepurko, T., Pavlova, M., Gryga, I., & Groot, W. (2015). To pay or not to pay? A multicountry study on informal payments for health-care services and consumers' perceptions. *Health Expectations, 18*(6), 2978–2993.

Suciu, Ş. M., Popescu, C. A., Ciumageanu, M. D., & Buzoianu, A. D. (2017). Physician migration at its roots: A study on the emigration preferences and plans among medical students in Romania. *Human Resources for Health, 15*(1), 1–9.

Tett, R. P., & Meyer, J. P. (1993). Job satisfaction, organizational commitment, turnover intention, and turnover: Path analyses based on meta-analytic findings. *Personnel Psychology, 46*(2), 259–293.

Tzeng, H. M. (2002). The influence of nurses' working motivation and job satisfaction on intention to quit: An empirical investigation in Taiwan. *International Journal of Nursing Studies, 39*, 867–878.

Vâlsan, C., Druică, E., & Ianole-Călin, R. (2020). State capacity and tolerance towards tax evasion: First evidence from Romania. *Administrative Sciences, 10*(2), 33.

Vujicic, M., Zurn, P., Diallo, K., Adams, O., & Dal Poz, R. M. (2004). The role of wages in the migration of health care professionals from developing countries. *Human Resources for Health, 2*, 3. Available from: http://www.human-resources-health.com/content/2/1/3.

Williams, C. C., Horodnic, I. A., & Horodnic, A. V. (2016). Who is making informal payments for public healthcare in East-Central Europe? An evaluation of socio-economic and spatial variations. *Eastern Journal of European Studies, 7*(1), 49.

Williams, C. C., & Horodnic, I. A. (2017). Explaining the informal economy in post-communist societies: A study of the asymmetry between formal and informal institutions in Romania. In *The Informal Economy in Global Perspective* (pp. 117–140). Palgrave Macmillan, Cham.

Williams, C. C., & Horodnic, A. V. (2018). Evaluating the prevalence of informal payments for health services in Southeast Europe: An institutional approach. *Southeast European and Black Sea Studies*, 1–21.

Zhang, Z. (2016). Variable selection with stepwise and best subset approaches. *Annals of Translational Medicine, 4*(7), 136. https://doi.org/10.21037/atm.2016.03.35.

E-nformality: Smartphones as a New Regulatory Space for Informal Exchange of Formal Resources

Aksana Ismailbekova and Gulzat Baialieva

Introduction

Mobile phones and other technologies have transformed the nature and dynamics of informal social networks not only in Kyrgyzstan, but across the world. Some international organisations and government agencies[1] argue that the emergence of new ICT technologiesmay contribute to reduce the role of informality in public and societal life in that they have the potential to create transparent and efficient ways to access public services. This is usually done through implementing electronic queue systems, online payment platforms, registers, and other additions to public services, as well as creating transparency portals to provide access to

[1] Gosudarstvennaya registratsionnaya sluzhba (2015) Korporativnaya antikorruptsionnaya politika, https://grs.gov.kg/ru/systems/corporate-anti-corruption-policy/.

A. Ismailbekova (✉)
Leibniz-Zentrum Moderner Orient (ZMO), Berlin, Germany

G. Baialieva
Social Anthropology Department, University of Tuebingen, Tübingen, Germany

A. Polese (ed.), *Informality, Labour Mobility and Precariousness*, International Political Economy Series,
https://doi.org/10.1007/978-3-030-82499-0_13

government data, statistics, and state laws and regulations. By force of this, our work looks at e-nformality as the bodily and spatial relationship between people and the state and various services changes, available online, and now people from teachers to doctors can be contacted via WhatsApp if you have the right relations, when earlier it used to be phone calls or knocking doors. Besides, it illustrates how the above relationship becomes more visible for people and also for researchers as new semi-public and informal spaces emerge in the multiple WhatsApp groups which seem to surpass Diesel and other online platforms from old-school times (*odnoklassniki*). Furthermore, we trace here these informal networks in order to better understand how informal access to public resources works along inter-generational dynamics (elderly people, intergenerational solidarity, gendered division of responsibilities and support), and to better understand how it works through lateral support (exchange of services) networks, involving groupmates (*odnoklassniki*) and other social networks, especially in the context when the state institutions are weak. Informality here is not understood as corruption or an aggregate of illegal practices. It is an overarching framework taking into account economic, social, and political practices, far from being relegated to sweatshops and small-scale, ad hoc economic transactions' (Polese et al. 2017: 3) and taking into account their social significance and embeddedness with local sociocultural settings (Polese forthcoming). Indeed, as Ledeneva (2018: 1) puts it, informality concerns "the world's open secrets, unwritten rules and hidden practices… as 'ways of getting things done'. Informal practices may escape articulation in official discourse, but they capture the 'know-how' of what works in their vernacular representation".

Our paper builds on these notions and further develops insights into the informal system, which we argue should include considerations of new technologies and social media as well. In this respect, we see our paper at the intersection of different forms of informality before the "smartphone" revolutions (Sanghera 2020), informal favours, bribery in the context of migration, social welfare decline and the emergence of new forms of solidarity in times of digitalization (Urinboyev 2020; Ruget and Usmanalieva 2019). Indeed, mobile technology plays a crucial role in empowering migrants in many ways by staying in touch with family members (Thompson 2009) and controlling nannies in Paris by employers (Kim 2016). It also transforms migration and empowers marginalised migrants, enabling them to be engaged socially, politically, and economically (Ruget and Usmanalieva 2019). According to Ruget and Usmanalieva (2019: 175), "Kyrgyz migrants do not use technology to discuss important diaspora issues or the politics of their home country", we build on this and add that use of technology can also ease access, and creative use, of state-related resources. Our fieldwork is not representative of the whole country; rather, our statements are limited only to our own informants. In this way, we aim to go deep enough to explore the processes and dynamics of some of the practices instead of providing quantitative or representative picture (cf. Lonkila 1998). in the country that could be used to better understand the extent to which smartphones and new technologies are empowering different categories of people in different ways, especially in providing access public services in Kyrgyzstan through the logistics of informal networking. We observed that this is practiced across generations in dialectical ways both in Kyrgyzstan and in Russia. By using smartphones, young people and middle-aged people:

- Help each other in times of need (to expedite access to public resources).
- Share information about possible reliable networks (*svyazi*).
- Recommend reliable networks to each other.
- Prioritise these networks over others.

As many international organisations and government agencies are collaborating with national authorities to encourage formalisation of business practices, they tend to emphasise the role of digital technologies in the democratisation of access to services and the job market, they tend to rely on digital technologies. Yet, people in Kyrgyzstan are finding creative ways of bypassing this technology. This stems not necessarily from a desire to contradict the state or go against the institutions. It is rather the result of the fact that reforms are often slow and do not necessarily address what some citizens perceive as their real needs. The reaction prompted can be regarded as halfway between Scott's moral economy (1979), when people take care of themselves because nobody else is doing it, and Phelps innovativeness of informality (2021). Indeed, by using new technologies and gadgets in a creative way, Kyrgyz citizens contribute to the development and widening of the scope of informality and subsequently informality studies.

Generational aspect became unexpectedly important during the research because of the slight differences between the experiences and practices of informality between the younger and middle-aged generations. Both generations have similar contemporary experiences of insecurity, financial struggles, and difficult socio-economic conditions in the post-Soviet context, but these two groups have different experiences with the Soviet times themselves. Several young people pointed out that "informality, even corruption, for us is a normal thing". They perceive informality and bribery as a "normal" feature of daily life with two sides—hindering as well as enabling careers. The middle-aged generation, however, whose background is rooted in the Soviet period, see it in a negative light, even though they still practice informality and bribery through the use of technology. In this respect, the paper documents the shift from "in person" networking, performed by middle-aged generation to the digitalised networking more common amongst the younger generations, a thing pointing at the possible advantages of using digital ethnography for the study of informality in the future. The combination of more traditional ethnographic methods with those of digital ethnography (Kozinets 2016) ensured that we approached each group in a manner appropriate to them, and in a way which would facilitate efficient and suitable research. In this manner, web ethnography offers an opportunity to participate in the same settings and to have similar tools for articulation and interaction as the participants themselves. The use of web ethnography implies that studying virtual spaces is different from studying the "real" or offline social practices (Kozinets 2016). In her book on *Virtual ethnography*, a leading anthropologist of online methods (Hine 2000) writes about the distinction in terms of an emphasis on flow and connectivity, as opposed to ethnography's prior concentration on location and boundaries. Likewise, O'Reilly conceptualises virtual ethnography which challenges assumptions of "field site", in that "instead of thinking in terms of places or locations, our Internet ethnographer looks to connections between things" (O'Reilly 2009: 217). In this way, the mix of digital ethnography with traditional methods allows us to look at the patterns of how social actors rely on new technology.

The interesting digital generational gap became increasingly visible during our in-depth engagement with the field during our research on the role of informal social networks in expediting access to public services is based on extended research in the small town of Kant in the north of Kyrgyzstan and the Shamaldy-Sai community in the south of Kyrgyzstan, conducted over the last several years. We chose this region based on our previous commitment and established trust with informants. We started with participant observation in Shamaldy-Sai and an online group page on social media consisting of 312 people.

Thanks to this first step we identified 12 young people with whom we decided to conduct face-to-face interviews. Interviews usually lasted one or two hours. On some occasions, some follow-up questions were asked after the interview by phone or chat to integrate or further explain some processes and dynamics. In the process, we also identified potential members for focus two group discussions with 16 people aged 25–38 and 55–65 so that each group could freely express their opinions without being controlled either by young people or elders.

Field-penetration was a relatively smooth and quick process given our previous commitment with the region and the fact that several key informants and gatekeepers were already acquainted with our work. Eventually, this strategy gave us access to the younger generation aged from 25–38 years old; middle-aged generation, from 55 and onwards (in Kyrgyzstan they would belong to elderly generation).

The young people were recruited in various ways, for instance by means of snowball sampling, but mostly they were the ones who trusted us, and because of the nature of the context in which the research was being conducted. Some follow-up questions were asked by phone, Facebook, and WhatsApp, asking specific questions about details of their digital interactions.

The next section locates the paper within current debates and explains the situation in the region. In the following sections we will share the divergent experiences of younger and older people regarding informality through new technologies. The first section will focus on the experiences of the younger generation regarding access to medical services, kindergarten, school, information and networks and their use of technology to manage informal networks. Later, we will present the older generation's experiences in using technology for their own personal needs. In this way, we will identify similarities and differences in the use of the new technology in both generations to meet their particular needs.

The Context, the Field, the Literature

According to the State Communications Agency (SCA), 79% of the population has access to the internet and 50% had access to 3G mobile broadband coverage in 2014, (so half of the population has a smartphone). As broadband penetration varies significantly across oblasts (provinces or zones) and regions, mobile coverage is the most affordable means of internet access. Websites and locative applications are generated mainly from Russian-language content providers, with some limited options to change the language to Kyrgyz. Patterns of internet use in Kyrgyzstan generally are not different from the global phenomenon and range from recreational and social to professional use.

As a previous study of mobile phones among Uzbek migrants in Russia shows (Urinboyev 2017), migrants stay connected and informed via smartphones and use them to create a sense of community and solidarity. Urinboyev presents an interesting ethnographic account of the transnational lives of Central Asian migrants in Russia via smartphones. He describes that insecurities in the receiving country, along with poor informational and legal support, forces the migrants to cope together: "…the everyday life of Uzbek as well as Tajik and Kyrgyz migrants in Russia is characterized by a constant sense of insecurity, with the threat of exploitation, deportation, police corruption, racism, physical violence, and even death. This total lack of security compelled Central Asian migrants to create informal networks and structures for coping with the risks and uncertainties of their precarious livelihoods" (2017: 126).

The situation in Kyrgyzstan is slightly different and the two types of popular platforms available have, in some respects, a different function. Odnoklassniki (meaning "classmates") is a Russian social media platform roughly equivalent to Facebook. Shortened as OK, it has common social media features such as creating profiles and group pages, sending instant messages, publishing photos/videos, and launching events. OK users can access a wide variety of music and videos uploaded to the website, including full-length pirate versions of box-office movies. The users, who are predominantly over 25, visit the website to stay connected with classmates, make new friends and find people from their childhood. Most people who lived in the Soviet Union and later moved to different countries find OK appealing because they can search for their old friends, former neighbours, teachers, and classmates.

The second, WhatsApp, has the most users worldwide, followed by Facebook Messenger. Kyrgyzstan is not an exception to this trend (Telegram is more popular in Uzbekistan, IMO in Turkmenistan, Line in China). WhatsApp is the most widely used platform in the country both in urban and rural areas. Features like easy download, free audio and video calls, and real-time messages have elevated it above other instant communication technologies. Many businesses in Kyrgyz cities and villages advertise a special WhatsApp number to contact them. Moreover, nowadays many young people not only know the above-mentioned apps but also spend their time on Instagram, Telegram, Skype, Immo, and Zoom. These are also important apps that people actively use to share information (sharing photos, music, and videos), including some advice and support. WhatsApp thus plays a major role in connecting, supporting, and entertaining people as it is easy to understand and use, making it appealing across generations. The app also has many different functionalities that make it easy to use for different purposes, depending on your needs. It has significant effects on patterns of social relationships and sustains collective coordination. The use of this technology can therefore help to shed light on particular forms of expression and the negotiation of relationships. There exists a robust body of literature on the use of mobile phone technologies in Africa pointing at the crucial role of the transformation of the informal economy in developing countries (Wamala Larsson and Svensson 2018; Svensson and Larsson 2016). However, little has been written on inter-generational gaps and the way they influence the use of connections and the channels through which informality goes. We take our idea from a major work on the paradoxes of Soviet life from the perspective of the last Soviet generation is a work by Alexei Yurchak (2006). The main puzzle of his book is devoted to the contradictory experiences of the Soviet system as stable whilst the system itself gradually moved towards its unsurprising collapse. Our middle-aged generation in Kyrgyzstan belongs to the last Soviet generation of Yurchak, who are used to living outside the Soviet system *vne systemy* (2006: 127). In our research we focus on intergenerational dynamics, looking at both the Soviet generation and the younger generation. Our cases show that the last Soviet generation reappear in the younger one, thus pointing to a common denominator.

By following this argument further, closely aligning with Polese et al. (2017), Scott Radnitz (2010), Ledeneva (2018), and Ismailbekova (2018), we looked at informality as fundamental to the functioning of the formal system itself by proposing incentives to follow and obey the economic or political system's rules. However, the experiences of different generations of informality differ, mainly as a result of how the collapse of the Soviet Union in 1991 affected the lives of a whole generation and created a whole new proceeding generation with divergent views. In this respect, we refer to the "normalisation" of bribery and informality for the younger generation. This is related to the fact that the younger generation grew up after the collapse of the Soviet Union, under difficult socio-economic conditions (lack of jobs, deterioration in the level of education (schools, universities), social upheavals, lack of prospects for the future and distrust of the state), while at the same time being able to see the increased employment opportunities provided through informal networks and bribery. They therefore invested more of their energy in learning how to make contacts, deal with people and create connections. Nevertheless, this does not mean that the elderly generation do not also participate in digitalized networks in order to access state resources. It is also important to note that this was also the case in Soviet times, when connections were needed to access goods and services (Rivkin-Fish 2005; Polese 2014).

Yet, it is crucial to highlight that cooperation that developed there is still some degree of continuity between middle-aged generation and younger generation as evidence of both continuity and transformation (Hann et al. 2002). In this regard, also link to Ibanez-Tirado's (2015) idea about "How can I be post-Soviet if I was never Soviet?" in which she challenges the existing periodisation "Soviet/post-Soviet" and the different experiences of these two generations. There are "alternative temporalities" that shape people's perception in different ways. The way to synthesise the above debates and propose new direction is, thus, for us, the adoption of the concept of "e-nformality'" that allows to concentrate on the way new technologies, mass media, and smartphones create a new, largely self-regulating space for the informal exchange of formal resources. Informality is there for both younger and elder generations. The generational gap lays in the places where it can be found, its dynamics, and its digitalization mostly by those belonging to new generations.

Younger Generations Are Comfortable with New Technology

Health Facilities and Access to Medical Services

During a recent research project in Kyrgyzstan, we found that users of public health facilities often rely on the support of informal social networks (friends, acquaintances, relatives) due to a lack of trust in medical services (see Polese 2009; Rivkin-Fish 2005 for further detail). By relying on social acquaintances (*taanysh*), the informants smoothen the process by facilitating access to reliable medical treatment. If someone goes to a doctor without any *taanysh*, the medical personnel will ask them to wait and claim there was a shortage of medicine and a lack of free places (rooms) in the hospital. In the following, we present three case studies from our young informants to illustrate their experiences and the lack of options in such situations. These are: Azat's experience with doctors; Akai's experience in hospital; and Medina's solution to her mother-in-law's health problem.

Later we compare their experiences with those of the elder generation in order to see the similarities and contrasts. First is Azat, 33 years old, who works in the bank as a credit specialist and lives in the small town of Kant. Azat says that "for example, if I take my daughters to hospital, I always search for 'someone' whom I know well (i.e. I usually call or send a message to my elder brother's wife [zhenge]) and trust her by taking her medical prescriptions very seriously, otherwise unknown doctors keep prescribing 'Ibuprofen' even if you have a headache, knee, and other health related problems". Azat explained that the doctors do not try to investigate the reason of the disease but instead give other medicaments. The best way to recover from one's disease, one should rely only on "acquaintances" (taanysh).[2]

Another informant, Akai, 33 years old, works as a state clerk. His opinion differs from Azat's, but he too stressed the shortage of medicine and lack of free places in the ward. Akai claims that "in my case, when I first went to the hospital, the nurse told me there was no room in the hospital. I called my friend's father, who helped me find a place, and my friend's father asked me to come tomorrow morning. The next day I called him early, and he asked me to come". Akai's case also shows that medical staff usually do not give them the right information and they therefore must search for their own ways to access medical care.[3]

In a similar way, Medina, 31 years old, who works as a hairdresser and owns her own beauty salon, stressed the importance of relying on informal social networks in hospital. According to Medina: "Once my client offered me her help, since she worked in the hospital. She did it especially in the situation when my mother-in-law got sick and relatives were not aware of her exact health condition and diagnosis. My client not only helped us to provide an exact diagnosis, but she also brought

[2] Interview with the focus group members, 15 December 2019, Kant, Kyrgyzstan.

[3] Interview with the focus group members, 18 December 2019, Kant, Kyrgyzstan.

my mother-in-law to different further medical check-ups, therapy and treatment monitoring". In the eyes of Medina's relatives, her authority as a daughter-in-law had increased, as she says it. Medina also stressed the importance of having "important people" (svyazi) and acquaintances (znakomye) in the hospital in order to save time by not waiting in the queue, not running from one cabinet to another and saving some money. They keep in touch with her client on Facebook and WhatsApp.

For these young people, mistrust towards such health-related public institutions prevails. This can be quite frustrating, but young people find ways to deal with the system. Two important aspects are important to highlight here: First of all, a lack of trust towards public medical personnel, because they are not interested in diagnosing patients but are instead busy selling the foreign medicines for their own profit. Secondly, a lack of information available to patients, meaning that they cannot rely on medical personnel without asking for help from their extended informal social networks. They therefore try to rely on their own networks to have access to hospitals and to fully understand their medical treatment.

Kindergarten and School: Bypassing Digitalisation

When asked about having access to public kindergarten for children, many of our informants told us that it is difficult to get a place in the kindergarten through formal means. However, they have found different ways of getting a place there. The system has changed because of the electronic queue that was installed in every kindergarten. Nevertheless, people find creative ways to get around this rule and the required online registration. In this section, we describe in detail how our informants encountered difficulties in finding a kindergarten place. We discuss Azat's knowledge of the new programme "socially deprived people", Sultan's experience with online registration and Zhamily's friendship.

Azat, 33 years old, and who has three small children, shared his experiences of getting a place in the kindergarten: "There is an electronic queuing system to enrol a child. Every time I approach a kindergarten's administration, they ask me to wait. I remember recently we were in 80th place, but now we have been moved to the 125th place". Azat claims that the kindergarten administration uses the " special needs groups get the place first' (handicapped children, children only with one parent) argument but, in reality, his friend's healthy child got a place in the kindergarten".[4] Apparently, Azat's friend works for local municipal government.

Another informant, Sultan, 33 years old, who works as a consultant topographer in a construction company, says that "if you have informal networks, you are like a tank; without informal networks you are a loser".Sultan's family also used their "acquaintances" to get a place in kindergarten without waiting in line because his mother's best friend is a director there. There used to be no such queue; either a place exists or it doesn't, only the "selected" (svoi) people could get it by giving a bribe or using social networks.[5]

The situation of Zhamilya, 32 years old, a mother of one child, and employed as a hairstylist in a beauty salon is similar to Sultan. Zhamilya reported "one of my clients is the head of the kindergarten. I once told her that I could not enrol my child in kindergarten because I had to wait in the electronic queue, but she offered to let me come on Monday with all the necessary documents, pay a small contribution of 2000 soms (25 EUR) and found a way to enrol my child in the list under the auspices of 'specially needed ones' (i.e. children without one parent). I am happy now".[6]

Several important points can be taken from the above interviews. Firstly, they talk about how people find creative ways to bypass the electronic queue and online registration to get kindergarten places. One such way to bypass the electronic queue is to use quotas for "socially needed places", which allows kindergarten managers to use this "grey and uncontrolled" area for their own purposes, providing places for those who could either bribe or use their social networks. The informal network continues to play a crucial role. Furthermore, the notion of relying on informal social networks invokes not only personal relationships and reciprocal obligations at the moment, but also signifies power, security, and even happiness.

[4] Interview with the focus group members, 15 December 2019, Kant, Kyrgyzstan.

[5] Ibid.

[6] Interview with the focus group members, 18 December 2019, Kant, Kyrgyzstan.

WhatsApp: Access to Information and Networks

Our younger informants are well aware of the importance of new technologies. This technology provides them with access to information and communication at all times, giving them the ability and incentive to invest in and expand their informal networks and allowing them to be employed easily and quickly, highlighting their importance. The capacity of these technologies for immediate interaction saves time and helps the informants to extend and support existing networks.

Ainura, 32 years old, says that: “whenever I need some help, I always send a message to our WhatsApp group SOS: ‘who has someone in the hospital’ or ‘I need to bring my mother to the hospital’. Someone would respond to my request from my WhatsApp group friends”. In a similar way, Sultan says: “personally I have several WhatsApp groups like ‘relatives’, consisting of 25 people, ‘classmates’ (‘klasstashtar’), ‘friends’ (‘My schastlivye’), 17 people, ‘football team ‘Laliga’ ‘ consisting of 25 people’, relatives from my mother’s side ‘Bololor’; 25 people, and Brigada ‘Abdysh-Ata 2011’ consisting of 17 people.. I also have friends among state registers, bank officers, tax officers, doctors, and those who work in social funds. It is very hard to get an appointment with a state register, for example, because they have always a long queue of elderly people, sometimes up to 60–70 people per day. They can afford waiting because they have enough time. Therefore, I just send a WhatsApp message to our state register friend by sending photos of registration certificate, receipt, and identification code in advance, he would prepare all the necessary documents in before I come”.[7]

According to Arman, father of four children, 34 years old: “we have a State Civil Register Office (ZAGS) in Kant and we get all our documents from ZAGS, such as birth certificates, marriage certificates, death certificates, personal identification numbers, etc. Those without acquaintances, they usually come at 6.45 in the morning and wait until 3.00 p.m. Of course, one can be stubborn and wait for a long time. Everyday more than 70 people go there. It happens that some people cannot enter the room on the first day and they have to continue waiting again on the second day. I got the birth certificates of my two daughters without any

[7] In-depth interview with Sultan, 7 January 2020, Kant, Kyrgyzstan.

problem by sending a WhatsApp message to my friend in advance: my friend came out and took the documents from outside".[8]

Young people face considerable challenges in accessing public services—including bureaucracy—which they have described as "they're clearly blowing us off" (*futbolit*). Problems include a lack of time, inability to get permission from employers, long queues, the inability of public services to accommodate needs and the non-transparent information available to young people, which also includes some forms of discrimination. Those who are in Russia cannot physically be in Kyrgyzstan and thus do not have immediate access to public resources (see more below).

Saving time is important for the younger generation and investment in informal social networks is crucial, but both are mutually exclusive. Since investing in a social network in a traditional, face-to-face way takes more time and thus makes it difficult to save time, they actively employ new technologies in this respect to make both investing in social networks and saving time mutually constitutive. Notably, all of the young people we interviewed are aware of the difficult structural conditions which contextualise their struggles and negotiations. At the same time, they were raised under such systems. They pointed to different hindrances from public state sectors, and, as a result, they cultivate their social networks by being actively involved in various kinds of circles. For many young people, stories of parents with well-paid jobs and support from government remain a fantasy.

Older Generations Are Less Comfortable with Technology

Our informants from the eldest generations did not feel comfortable with new technologies and social media, especially when it came to asking personal questions, doing favours, and solving delicate personal problems. Instead, our older informants are now using technology to entertain themselves, surf YouTube, watch concerts, read the news, and communicate with children and other family members abroad. Technology for them is an area of the home domain or even private life, not business. They exchange e-cards, stories, anecdotes, songs, and even movies amongst themselves.

[8] Interview with the focus group members, 15 December 2019, Kant, Kyrgyzstan.

In the interviews, the older generation emphasised more personal relationships, direct contacts, and expressing gratitude as the means of accelerating access to public services. They preferred to give back to what they had personally benefited from, through direct gifts such as food, an invitation to a foreign café, and gifts, with an emphasis on family and regional identity. In other words, the older generation is actively dependent on informal social networks but prefers to give them moral connotations such as sense of duty, loyalty, reciprocity, and the exchange of gifts (food, gratitude) as well as some bribes. They prefer to do this physically, without using new technologies or social media. But in reality, when their children ask, they still have to use it. They can sometimes ask for some favours over the phone, but these favours must be returned or repaid.

Munara, 60 years old, told us how she once used her informal network to get a place in the kindergarten. Her experience highlights that, in the beginning, she usually gets a refusal, especially when she had applied for kindergarten. Knowing that there is a place in kindergarten, she usually employs her kin and other networks. For example, Munara says: "I figured out that the head of kindergarten was apparently from Kochkor, where my husband came from. I told her that she had come from my husband's small village and she became my young sister-in-law". By using this kin network, she got a place in kindergarten.[9]

In addition, Munara compares the current situation with Soviet times and says that, during the Soviet Union, people used to reply on informal networks but they were hidden. There was still law functioning during the Soviet times, and there were enforced rules. Nowadays if you go to court, if you have a case or not, the outcomes might be paradoxical. If you are guilty but if you have someone in the court, you come out of court as not guilty. If you are not guilty but if you do not have anyone in the court, you could end up being found guilty. For example, yesterday there was an accident in Tokmok city. A young man was killed, and his car was burnt. They found the main suspect, but he was released on the basis of him being insane. Apparently, this man was not psychologically ill or irresponsible, but he could get a doctor's medical certificate from the hospital because of his strong "roof" (*krysha)* contact, who was working in the White House.

[9] Focus group discussion 22 December, 2019 Munara.

Another story highlights access to higher education and its relation to informal networks and corruption. Klara, 58, told us how she recently relied on her informal network to solve a case of corruption. Her son is studying at the National University and the lecturer asked her son to buy a mobile phone, whereupon he refused and was punished for it with a bad grade. Klara found out about this situation late, and she decided to take academic leave for her son due to illness. In order to do this, she had approached the director of the hospital R.M., who had helped her get the medical certificate without any problems. In return, Klara invited her to eat at the café and gave her some gifts (scarf). Now she used her relative who works in the government to influence the university management to let them take a year's leave from the university.[10] Klara says that she used smartphones to manage all these tasks.

Almas, 62 years old, shares his experience of relying on informal networks in career opportunities and promotion. His colleague Ak-Jol planned to take a gift (i.e. sheep) to Issyk-Kul to his boss Satybaldiev. Apparently, Satybaldiev was on holiday in one of the holiday houses in Issyk-Kul. He asked Almas if it was okay to take him to Issyk-Kul, since they were colleagues, and so Almas agreed to drive the car. There were two big boxes: one was filled with meat and the other with vegetables and fruit. He asked Almas to stay in the car and wait near Satybaldiev's house. He took both boxes with him and came back in less than 30 minutes. He asked me to drive home. Apparently, he came all the way from Osh to Issyk-Kul (12 hours) for less than 30 minutes, saw his boss briefly and left two boxes for him. After his return to Osh, Ak-Jol was appointed head of the agency, which dealt mainly with foreign investment in road construction in southern Kyrgyzstan.[11]

The case studies above demonstrate various aspects of informal newtwoks: subversive and supportive roles, functional and dysfunctional tasks, and sociability and instrumentality (Ledeneva 2018: 9). Following Ledeneva, we argue that in the interviews mentioned above, an aspect of tension predominates. There is a double standard: the tension between the question of why it is acceptable to ask in face-to-face contacts and the question of why it is unacceptable to demand personal services using technology; because of the lack of opportunities to deal with informality

[10] Ibid.

[11] Focus group discussion 22 December 2019, Abas.

and various opportunities that they might otherwise take advantage of; because of the lack of trust in their own personal networks and distrust of government agencies. The core of the informality of the older generation lies in this duality. In all these cases the ideals of how things should be according to the rules and regulations prevail, but the results are different, which is even more visible when comparing the attitudes of different generations.

Informal Markets in the Context of migration: Obtaining Fake Qualifications

One day there was a call from Tahir,[12] Gulzat Baialieva's classmate—a Kyrgyz labour migrant in Russia. After a short chat, he threw in his question in a very routine and straightforward manner. "How much would the cheapest university diploma cost? I need it urgently by summer".

He explained that he needs to get a university degree in order to find a better job in Moscow. It was hard to react on the spot, although the informal markets of selling university diplomas, transcripts, and certificates of school graduation are nothing new. He got the unwanted response that he'd better apply for the university and that his idea implies corruption. He has never written or called back, but after two years we met him in Kyrgyzstan and took the opportunity to chat with on that matter. Tahir did buy the university degree from one of the provincial universities in Kyrgyzstan and paid 700 USD for it. He sounded routine and confident when he was explaining the details of how he accessed the formally allocated resources (university degree with transcripts) informally.

After his first failed inquiry to get him a university diploma, Tahir phoned our retired schoolteacher Batma eje, whose son worked as a university teacher in one of the Kyrgyz cities. He knew his son as well and talked to him over the phone about the situation, about the safety of the deal, the amount to pay and how to pay. As he was in Russia, he was urgently searching for "black" ways to get a formal "white" paper that could grant him the status of a graduate specialist—in whatever specialisation. He got one in economics. We didn't show our discomfort and further developed the conversation on this very intimate topic

[12] All the names are fictive to guarantee anonymity to informants.

over a shashlik[13] lunch. Tahir explained why he decided to call us and why it is quite "normal" to buy degree without studying at the university: "Everyone in the education system is engaged with it because they need to survive, and the salary is too low to afford life. They have documents to sell but no money, we have money but no documents. The needs and supply you know....You came to my mind since I saw you active in social media Odnoklassniki, which is a very friendly space for migrants. I thought you could understand my situation and help but you live abroad and don't need money I realised. Otherwise, you would also help me, wouldn't you?".

He continued this conversation, telling us that he was forced to drop out of school, although he dreamed of getting proof of higher education. Now due to loyal and strong connections he possessed the desired and required documents: a school certificate and university degree.

> Now I have everything. Every barber knows it (Po sekretu vsemu svetu, ne govoryat, no eto vse znayut) that with good connections you can buy everything. I have a school certificate as well. No one suffers and no one loses anything. I dreamed of studying at the university, but it was difficult for my mother when my dad died. She took me with her to work in Russia.

Tahir was a good student but didn't finish school because of family hardship and migrated to Russia to help his mother to make money. He worked in different sectors, including as a waiter, construction worker, and taxi driver in different Russian cities, Saratov, Yekaterinburg and now in Moscow. His mother died from a heart attack. Tahir stays connected with his home community, invests money in repairing his parents' house there and hopes to return forever one day. Sometimes his family stays in the town for winter and summer seasons as their children are still in pre-school.

Based on Tahir's open and confident narratives it can be concluded that such informal practices in the educational system are not bizarre: "every barber knows it". Likewise, "the needs and supply" regulate the informal market in the context of migration and new technologies. Yet, it is not an everyday disclosed talk and is shared only among trusted groups.

[13] Local food similar to barbeque, common to have as a fast food in tea-houses in the southern part of Kyrgyzstan.

Although Tahir was thousands of kilometers away from these educational institutions and was not eligible for the documents he obtained, his connections, mediated via smartphone, connected him to the desired informal services.

The signature, where needed, was signed on his behalf with the help of notarised consent (also informally legalised), and the personal information was provided as a photocopy via WhatsApp. Emails or fax are almost never used by most labour migrants, who would need extra skills to access them. They are also considered insecure, whereas the video and audio functions of WhatsApp and instant messaging give a sense of security and are perceived as a safer way to interact and exchange private matters.

A former school vice-director, Batma eje, with her authority and diplomacy, justified the system and appreciated that new technologies can serve the migrants. She recognises that it is risky and undesirable to produce the formal documentation informally. However, she doesn't see it as a destructive practice. Unlike other mediators engaged in providing illegal service to migrants at a distance, Batma eje doesn't make money and just connects "people who know" (bilgen kishiler). She said that it has been tightened now and only those who have higher positions and loyal assistants can deal with the informal services. It takes more time, more trust and more money.

We met Batma eje couple of times over lunch and asked different moral questions. She criticised the collapse of the economy, political corruption and the so-called lost generation that had to migrate. The teacher concluded that, at a time when everyone has to do all they can just to survive, to find bread to eat, and to achieve stability, such an exchange of formal resources through informal means doesn't destroy an already ruined system.

Conclusion

New technology and mobile phones connect a wide range of networks (family members, friends, relatives, team members, acquaintances) in Kyrgyzstan, and help significantly, especially when it comes to access to public services. But this experience is not equal to different generations of our informants because they have contrasting practices and views in relations with new technology and new social media.

Younger members of the network exchange information via mobile phones and provide each other with advice, especially to those who have to go to hospital, police, state registers, etc. By this way they economise their time but at the same time do not lose their informal networks. The new technology thus facilitates two mutually exclusive aspects as mutually constitutive. In other ways, they facilitate access to information. It does not "go under the table" (pod stolom), as our elderly informants, rather it goes everywhere via the internet. This makes easier to facilitate the dynamics of informal practices in relation to access to public services, mainly to easily make appointments, to find important people, and to obtain and send diplomas, and other important documents.

In contrast, we see the different dynamics of informal practices that members of middle-aged generation utilise. They prefer that their informal practices remain tangible, personal and moral. They highly stress obligation, loyalty, reciprocity, and gift exchange. They want to use this network for socially acceptable needs, despite the fact bribes and favours are involved. Some middle-aged generations use telephones for logistical tasks and communication with friends and family members. Yet, it does not mean that they do not go beyond these merely logistical tasks because they also have children and learn from their children to use smartphones for their own needs.

The normalisation of e-nformality can be observed through the prism of new technology and social media, more so by the younger generation than the older generation. This is mainly because, as Ibanez-Tirado argues, there are alternative temporalities that shape in our case, people's perception of informality. This is mainly because the younger people did not experience the "Soviet times", their experience was linked to the difficult socio-economic situation in which the creative use of informality can be a way to succeed, thus contributing to the normalisation of informality. The older generation, on the other hand, has a different understanding of informality in the sense of duality (to live according to rules and regulations, but when things don't work, they are not so surprised and use other means). Yet, elder informants are also quickly adapting to the new realities of technology. We also showed how features from the older generation reappear in the younger one, thus pointing to a common denominator as the story of Klara above highlights.

New technology and social media are supporting the dynamics of informal networks and changing their nature and dynamics along with creating new optics of entering public services. Technology goes beyond bureaucracy and government rules; rather, it is an effective way to share information and knowledge about the easiest ways to inform each other about who to contact. This simplifies the communication of those who are involved in informal networks. They have created social groups to help each other, and which serve as an effective way of coping with state irregularities. At the same time, they contribute to the functioning of the state by getting closer to people, giving money, and strengthening connections with state authorities.

References

Hann, Chris, Caroline Humphrey, and Katherine Verdery. 2002. Introduction: Postsocialism as a Topic of Anthropological Investigation. In *Postsocialism: Ideals, Ideologies and Practices in Eurasia*, edited by Chris Hann, 1–28. London: Routledge.

Hine, Christine. 2000. *Virtual Ethnography*. London: Sage Publications.

Ibañez-Tirado, Diana. 2015. 'How Can I Be Post-Soviet if I Was Never SOVIET?' Rethinking Categories of Time and Social Change—A Perspective from Kulob, Southern Tajikistan. *Central Asian Survey* 34 (2): 190–203.

Ismailbekova, Aksana. 2018. Mapping Lineage Leadership in Kyrgyzstan: Lineage Associations and Informal Governance. *Zeitschrift für Ethnologie, V* 143 (2): 195–221.

Kim, Y. 2016. Mobile Phone for Empowerment? Global Nannies in Paris. *Media, Culture & Society* 38 (4): 525–539. https://doi.org/10.1177/0163443715613638.

Kozinets, Robert V. 2016 Netnography. In *The Blackwell Encyclopedia of Sociology*, edited by George Ritzer, 1–2. Oxford, UK: John Wiley & Sons, Ltd. https://doi.org/10.1002/9781405165518.wbeos0782.

Ledeneva, A. 2018. Introduction: The Informal View of the World-Key Challenges and Main Findings of the Global Informality Project. In *The Global Encyclopaedia of Informality. Towards Understanding of Social and Cultural Complexity Volume 1*, edited by A Ledeneva. London: UCL Press. http://discovery.ucl.ac.uk/10041167/1/The-Global-Encyclopaedia-of-Informality-Volume.1.pdf.

Lonkila, M. 1998. The Social Meaning of Work: Aspects of the Teaching Profession in Post-Soviet Russia. *Europe—Asia Studies* 50 (4): 699–712.

O'Reilly, K. 2009. *Key Concepts in Ethnography*. London: Sage. http://www.uk.sagepub.com/booksProdDesc.nav?prodId=Book229834.

Phelps, Nicholas A. 2021. The Inventiveness of Informality: An Introduction. *International Development Planning Review*, 43 (1): 1–12.

Polese, A. 2009. 'If I Receive It, It Is a Gift; if I Demand It, Then It Is a Bribe' on the Local Meaning of Economic Transactions in Post-soviet Ukraine.

Anthropology in Action 15 (3): 47–60. http://www.ingentaconnect.com/content/berghahn/antiac/2008/00000015/00000003/art00005.
Polese, Abel. 2014. Informal Payments in Ukrainian Hospitals: On the Boundary between Informal Payments, Gifts and Bribes. Anthropological Forum.
Polese, A. forthcoming. What Is Informality? About "The Art of Bypassing the State" and Its Theorizations in Post-Socialist Spaces (and beyond). Eurasian Geography and Economics.
Polese, A., C. Williams, I. Horodnic, P. Bejakovic. 2017. Informal Economies as Varieties of Governance. In *The Informal Economy in Global Perspective: Varieties of Governance*. London: Palgrave.
Radnitz, Scott. 2010. *Weapons of the Wealthy: Predatory Regimes and Elite-Led Protests in Central Asia*. Cornell University Press.
Rivkin-Fish, Michele. 2005. *Women's Health in Post-Soviet Russia: The Politics of Intervention*. Bloomington: Indiana University Press.
Ruget, Vanessa, and Burul Usmanalieva. 2019. Can Smartphones Empower Labour Migrants? The Case of Kyrgyzstani Migrants in Russia. *Central Asian Survey* 38 (2): 165–180.
Sanghera, Balihar. 2020. Justice, Power and Informal Settlements: Understanding the Juridical View of Property Rights in Central Asia. *International Sociology* 35 (1): 22–44.
Scott, James. 1979. *The Moral Economy of the Peasant: Rebellion and Subsistence in Southeast Asia*. Herausgeber: Yale University Press.
Svensson, Jakob, and Caroline Wamala Larsson. 2016 Situated Empowerment. Mobile Phones Practices among Market Women in Kampala. *Mobile Media and Communication* 4 (2): 205–220.
Thompson, E. 2009. Mobile Phones, Communities, and Social Networks Among Foreign Workers in Singapore. *Global Networks* 9 (3): 359–380.
Urinboyev, R. 2017. "Establishing an Uzbek Mahalla" via Smartphones and Social Media: Everyday Transnational Lives of Uzbek Labour Migrants in Russia." In *Constructing the Uzbek State: Narratives of Post-Soviet Years*, edited by M. Laruelle, 119–148. Boulder: Lexington Books.
Urinboyev, R. 2020. Migration and Hybrid Political Regimes Navigating the Legal Landscape in Russia. California: University of California Press.
Yurchak, Alexei. 2006. *Everything Was Forever, Until It Was No More: The Last Soviet Generation*. Princeton, NJ: Princeton University Press.
Wamala Larsson, Caroline, and Jakob Svensson. 2018. Mobile Phones in the Transformation of the Informal Economy: Stories from Market Women in Kampala, Uganda. *Journal of Eastern African Studies* 12 (3): 533–551.

Work, Subsistence and Distress of the Homeless in Moldova

Petru Negură

Introduction

'They do not want to work; they really don't want to work …'—I have come across this sentence formulated in different ways many times in discussions with various domiciled people while conducting my research on the homeless in Moldova. It is an assumption that everyone, including professionals in the field—social workers, police officers, medical assistants—seem to share in a quasi-general consensus. The homeless people have a more nuanced opinion on that. Yet several homeless people interviewed within this research project had a rather disdainful attitude towards their peers (with whom they did not identify fully) claiming that they were 'lazy' since they 'refused to work' and would lead a 'parasitic'

P. Negură (✉)
Leibniz-Institut Für Ost- Und Südosteuropaforschung, Regensburg, Germany

Free University of Moldova, Chisinau, Republic of Moldova

The Institute for Legal, Political, and Sociological Research, Chisinau, Republic of Moldova

A. Polese (ed.), *Informality, Labour Mobility and Precariousness*, International Political Economy Series,
https://doi.org/10.1007/978-3-030-82499-0_14

lifestyle. This chapter will try to check this assumption and to understand the relationship that homeless people in Moldova have with the labour market. Sociobiographical records and interviews conducted with homeless and professionals in the area of social protection provide factual and contextual information on the relations of the homeless with labour and various work activities. I will try to fit this data into the broader context of the developments of the labour market in Moldova during the last three decades. This paper will contribute to investigating the employment situation of homeless people in Moldova, with an analysis focused on the case of the capital city in the context of the post-Soviet transformation. It maintains that most homeless people who are fit for work are systematically excluded from the formal sector of the labour market. Yet they are engaged in various subsistence economic activities which help them survive and include them into informal support networks among peers. Nevertheless, the informal labour activities further perpetuate the homeless people's social and economic vulnerability.

The majority of homeless in Moldova have no identity documents and many of those who have identity documents have no residence visa. For these people getting a job without a formal contract between employer and employee is the only possible way to occupy a place in the labour market. According to official estimates, during the 2000s in Moldova, more than 30% of the total employed population has worked in an informal job. The 'informal sector' of the labour market offers thus the most realistic opportunities to be employed to the homeless people and other categories of vulnerable persons. Unlike the restricted access to the 'formal sector', the informal employment provides the homeless with a source of subsistence, which is refused to them elsewhere.

Drawing on neo-Marxist theoretical perspectives and Axel Honneth's social recognition theory, this chapter suggests that, beyond the institutional status of employment (formal, informal, or semi-formal), the degree of the workers' alienation or emancipation depends on the relationship of subordination and autonomy to the employer and the level of control and involvement in the labour process (Honneth 2010). From this perspective, informal work in which homeless people are involved can give employers more freedom, but generally reduces the ability of employees to control the labour process and their autonomy to the employer.

A significant number of studies focus exclusively on individual factors that prevent homeless people from exercising a formal occupation (see literature reviews in this regard in Gwadz et al. 2009; Lei 2013). Several recent studies have examined the systemic factors that lead to the exclusion of socially vulnerable people, including the homeless, from the labour market in Western Europe and North America. These factors are related to the radical transformation of the economic system and the decline of the public health and social protection systems in the years 1980–1990 (see Wacquant 2009; Kus 2006; Hong and Wernet 2007). This transformation has led to an outbreak in the number of the poor, unemployed, and detainees in the western countries (Wacquant 2009). In the case of the Republic of Moldova and other post-Soviet states, the liberalization and privatization of the market in the 1990s went hand in hand with the dismantling of the state industrial system and the drastic reduction of social spending.

The *neo-Marxist* theoretical model analyses informal work as a form of precarious wage labour understood as a palliative solution for workers without employment alternatives in the formal sector. The expansion of the informal sector would have the adverse effect of capital accumulation and the formation of a 'reserve army' of the capital-controlled economy. According to this perspective, there would be three main categories of informal work: subsistence work, subordinate work, and autonomous employment (Portes and Schauffler 1993; Biles 2009). This theoretical perspective is consistent with the analysis model synthesized by Honneth (1995, 2010) according to which the *alienation* and *emancipation* of the worker depend on the degree of control on the labour process and autonomy to the employer. Also, the emancipation of the worker in the work process depends on the existence (or not) of institutional and social structures of social recognition for the work performed, including through decent salary, as a contribution to the common good and social integration.

The difficulty of homeless people in getting stable employment is seen by most social service providers in Moldova as a consequence of a cumulus of individual causes. Similarly, social policies to address this issue are developed and applied most often from an individual perspective (see also Gaetz and O'Grady 2002; Shier et al. 2012). Social welfare services and shelters are only designed to provide temporary shelter and meet basic needs, often ignoring the employment needs of users. Even when these needs are considered, only unstable and poorly paid jobs that homeless people are supposed to have access to are sought after. Certain aspects of the labour market (inadequate work, inappropriate remuneration, and bad relationships with employers) contribute to maintaining the housing vulnerability of employed homeless people (Gaetz and O'Grady 2002; Shier et al. 2012). The disparity between labour income and the cost of renting and maintaining a home lowers the chances of reintegration and stabilization of the homeless (see also Balkin 1992). Therefore, a strategy that aims to reduce the homelessness would be to provide social housing and to raise employment income, in addition to providing support to ensure stability. Other strategies would be to assist homeless people to increase their employability on their own and participate in certain entrepreneurial activities (Balkin 1992). However, as this paper shows, professional assistance strategies would not be sufficient and sustainable if systemic issues faced by homeless people are not addressed: structural discrimination, low pay, job instability in the informal sector and risks of abuses by employers. In addition to these difficulties and labour protection deficiencies, in Moldova, like in Russia and other post-Soviet societies, the socio-professional vulnerability of homeless people is aggravated by the lack of identity papers for many of them (Stephenson 2006; Höjdestrand 2009, pp. 47–76; Ryabchuk 2007, pp. 72–76).

In contrast to the mainstream literature on homeless privileging individual approaches, this study addresses the employment status of the homeless person from a systemic perspective. In a labour market controlled by state institutions and private companies, the homeless people resort to informal labour activity as an individual strategy aiming to avoid systemic exclusion factors. This study joins other contributions to this book in trying to understand the complexity of informal labour by disentangling various social logics intertwined in this process. From the perspective of the direct and indirect harmfulness model (Polese et al. 2017; Ledeneva 2013; see also the Introduction to this book), homeless persons are involved in informal work that allows the economic system to function and helps these vulnerable people to survive, although harming them in the long run. It is thus a strategy going 'in spite' and, at some point, 'beyond the state' and its rigidity to regulate the vulnerable people's exclusion from the formal labour market (Polese et al. 2017). Yet this strategy does not contribute to reducing the mechanisms of exclusion of homeless people from the labour market, but only bypasses them.

This analysis builds on sociobiographical data of all 810 users of the Centre for Homeless in Chisinau (the Shelter), from 2004 to February 2017. They contain general information about the user's birth date, birthplace, gender, marital status, education, professional status, work experience, last residence(s), health, relations with family and relatives, short-term and long-term objectives, work plans, and other related categories. These biographic data have been anonymized and submitted to a descriptive analysis with SPSS software.

To compare data from the analysed admission forms to the Shelter with data provided by the National Bureau of Statistics (NBS) for economically active population, I took into account the specific relationship of each group with the labour market. The sample of persons who filled out the admission form to the Shelter is made up of the following subgroups (defined by the persons' relationship with the labour market): (1) active urban population (made up of economically active population and unemployed—persons aged 15 and over) and (2) inactive urban population aged 15 and over. From the second group, I removed the share of some statistically unrepresented or insignificantly represented groups in the group of Shelter users such as migrant people, students (only 0.59% indicated such an occupation), and persons on parental leave or looking after another person.

The author also conducted in-depth interviews, from September 2013 to February 2017, with 90 people (of whom 65 homeless people, five people at risk of homelessness and 20 professionals). From the homeless interviewees, 50 were men, and 16 were women. 38 respondents were interviewed in the Shelter and 27 on the streets or in squatter camps. A sustained observation also accompanied the interviews. Thematic analysis according to 'grounded theory' model (observing the distribution of thematic categories in the corpus) has been applied to narrative sources with the aid of NVivo software (Corbin and Strauss 1990; Bazeley and Jackson 2013). This study complies with internationally accepted ethical guidelines in sociological and social work research. All respondents' names mentioned in this paper have been changed. All participants gave written consent to participate in the research.

Post-Soviet Transformation of the Labour Market, the Disintegration of the Social Welfare System, and Massive Pauperization

After the collapse of the Soviet Union and its independence (in 1991), Moldova has faced systemic economic difficulties, endemic poverty, and low quality of governance. The massive pauperization of the population in the 1990s took place in the background of the brutal transformation of the labour market in the 1990s, especially due to the closure of a large number of state-owned enterprises and the emergence of private enterprises, together with the weakening and shrinking of the social welfare system (Negură 2016). The fall of a massive population below the limit of absolute poverty (up to 70% of the population in 1999) has generated a series of economic survival strategies and practices, including (self) employment in various informal economic activities (Burrawoy and Verdery 1999; Morris and Polese 2015).

According to official data, in the early 2000s, almost 40% of the workforce in Moldova was employed in an informal job. The proportion of informally employed people has fallen proportionately by the end of the 2000s to below 30% to further increase since 2010. By 2018, the proportion of informally employed people has increased to 38% (*Statistical Yearbook* 2009, 2019). These people were employed informally in formal and informal enterprises and individual households. The main areas of informal employment in 2013 were agriculture (69%), trade, hotels and restaurants (12%), construction (11%), and transport and communications (4%). (*Statistical Yearbook* 2009, 2019; Popa et al. 2015, p. 16). During the 2000s, informal employment in agriculture contracted, but it increased in the field of construction by 27% from 2003 to 2013 (Popa et al. 2015, p. 16). Men are overrepresented in the informal labour sector (34% of men compared to 27% of women in 2013) but women are disadvantaged, being employed mainly in households. Most of the people employed informally are unskilled workers (66%), 59% have a primary and general education and 26% hold vocational education (Popa et al. 2015, pp. 17–19).

The social welfare system in the post-socialist states, including Moldova, was affected in the post-1990 years by under-financing and lower institutional capacity (Cerami 2009; Polese et al. 2014). The public system of social apartments and workers' dormitories was almost entirely privatized during the 1990s (Negură 2016). Today, the social housing system in Moldova is far from being able to satisfy the demands of this service. To deal with an increasing number of socially vulnerable people becoming homeless in the late 1990s and the early 2000s, the Chisinau municipality opened in 2004 the first service to homeless persons, the Centre for Hosting and Orientation for Homeless People (hereafter, the Shelter).

The Shelter provides night-time accommodation (from 20:00 to 8:00), social assistance, basic medical care, and food (breakfast and dinner) for up to 100 users. Many users stay in the Shelter for more than three months as stipulated in the internal regulation. Also, most users return to the Shelter after several months. This explains the relatively small number of unique users from 2004 to early 2017 (810 people). In the following years, city authorities opened shelters in two other cities. In 2006, an international Catholic Association opened a shelter for 20 users in Chisinau (with a total of 600 users until April 2019). This shelter was, however, closed down in December 2018 after an interruption in funding by international donors and refusal of central and local governments to take over this project. Some non-governmental organizations provide homeless people with soup kitchens, showers, laundry, and set up measures to prevent tuberculosis amongst homeless people in Chisinau. However, these services are insufficient, given the high number of homeless people, estimated at 4000–5000 only in Chisinau city.

The Shelter has become a 'catch-all institution' for a range of categories of impoverished and vulnerable persons, not adapted to socioeconomic conditions of post-Soviet transformation. The Shelter users are made up of several subgroups: elderly (33%); people with various disabilities (16.8%); chronically ill (7.6%); persons at various stages of alcoholism (50%); and an over-proportionate number of people who are considered to be products of public institutions, such as the justice and social protection systems, specifically prisons (23%) and state boarding schools (11%). In a sense, the Shelter users represent the non-assimilated surplus of social protection institutions of Soviet origin (nursing homes for elderly and disabled and orphanages) which, since the mid-2000s, have been undergoing a process of deinstitutionalization, assisted by international non-governmental organisations, and of repressive institutions, especially the prisons (Negură 2018).

Men are overrepresented in the homeless population, making up 75% of the Shelter users. The numerical predominance of men among the homeless is characteristic of other societies as well. In the case studied here, the high number of homeless men could be related to some factors that contribute to the residential vulnerability of men in particular, namely the experience of detention. The number of former detainees is significantly higher among homeless men than homeless women in the sample (27.9% from men compared to 9.7% of women). This gender proportion is specific to the population of detainees in general, in Moldova and other countries (Negură 2018). Similarly, the results of the qualitative study show that the proportion of people suffering from alcoholism is higher among homeless men than among women (30 from 48 men and 6 from 16 women interviewed). As some studies suggest for other contexts, the high proportion of men among the homeless is also related to the social expectations of main breadwinners attributed to men as husbands and fathers. Alcohol abuse, manifestations of violence, self-aggression, and other anomic behaviours can be seen as psychosocial responses of men who could not meet these social expectations (Liu et al. 2009). In this sense, the social expectation to work is all the more acutely felt by the male participants in this research.

Work Experience of Homeless People: A Quantitative Estimate

The descriptive analysis of biographic records of 810 Shelter users in 2004–2017 provides an estimate of the professional profile of the homeless people, based on their work records. Comparing these data with information on active urban population (aged 15 and over) available at the national level, though keeping the proportion and adjusting the terminologies accordingly to each group, the following section draws the socioprofessional profile of homeless people in the past and at the moment of research and examines their plans.

According to these data (Table 1), 1.32% of the homeless users said they had worked in agriculture (it is a predominantly urban population); 23.58% had worked as skilled workers (all categories); 5.18% were employed as technicians or foremen; 10.67% worked in an 'intellectual' occupation (by an extended definition of the term, including administrative activities, according to the Classifier of Occupations of the Republic of Moldova); 4.08% worked as sellers; 13.44% have been unskilled workers (porters, street sweepers, and gatekeepers); 1.05% were housewives; 4.48% were unemployed; 4.08% said they had not worked; finally, 32.15% declared they had *another* occupation than those mentioned above, without specifying which one.

The evidence that stands out upon considering these data is the relatively low share of 'intellectual' occupations exercised in the past by Shelter users—10.67%, compared to 23% among the specific active and inactive urban population (without migrants, students and caregivers of children or adults) and compared to 34.66% among the general active

Table 1 Distribution of answers in the Shelter admission form under the heading 'Occupational area you have worked in'

		Frequency	*Per cent*	*Valid Per cent*
Valid	Farmer	10	1.2	1.3
	Qualified worker	179	22.1	23.6
	Technician/foreman	39	4.8	5.1
	Intellectual	81	10.0	10.7
	Porter, sweeper, guard	102	12.6	13.4
	Housewife	8	1.0	1.1
	Unemployed	34	4.2	4.5
	Seller	31	3.8	4.1
	Other	244	30.1	32.1
	Did not work	31	3.8	4.1
	Total	759	93.7	100.0
Missing	Data	51	6.3	
Total		810	100.0	

urban population. Another detail that may draw the reader's attention is the high share of people who have been employed in unskilled jobs, at least 14% if one includes in this category porters, street sweepers, gatekeepers, domestic workers, and others, compared to 6.6% for the same category within the general specific active urban and inactive population and compared to 9.27% among urban active population. Finally, a high share (32.15%) stated being employed in 'another' occupational area than those listed without specifying it. This percentage might conceal some occupation that respondents preferred not to declare or the lack of any professional occupation in the strict sense of it.

The share of unskilled workers among the homeless users is even more significant if one considers the data from the second answer choice for the category 'other occupations' (besides those mentioned in the first answer choice—see Table 2). Thus, among the 274 individuals who answered this question (33.8% from the sample), 216 persons (78.8% from who answered) stated to having performed an unskilled job whereas 52% (or 19%) have had a qualified occupation.

The share of people who worked in an unskilled job (26.7% of all respondents at the first question about users' profession and 78.83% of those who answered the second question) is much higher than the

Table 2 Distribution of answers in the Shelter admission questionnaire under the heading 'Other occupational areas you have worked in'

		Frequency	*Per cent*	*Valid Per cent*
Valid	Unskilled worker	216	26.7	78.8
	Qualified worker	44	5.4	16.1
	Technician	6	0.7	2.2
	Functionary	2	0.2	0.7
	Farmer	5	0.6	1.8
	Student	1	0.1	0.4
	Total	274	33.8	100.0
Missing	Data	536	66.2	
Total		810	100.0	

proportion of individuals with no education (1.9%), with primary education (2.6%) and with lower (eight-grade) secondary education (9.5%) of the sample, in total 14%. Also, the share of persons who worked in medium-skilled occupations (skilled workers, foremen, technicians, and sellers), 33% of the users, is more than two times lower than the share of persons with general secondary education, specialized secondary education, high school and secondary vocational education altogether (78% in total).

This piece of findings suggests that a significant number of homeless people have gone through a process of downward social mobility, being compelled by circumstances to perform occupations that are below their professional qualifications. Moreover, at least 41% of the Shelter users did not work recently in the generally accepted sense of the term. This percentage is higher than the share of 31.35% of the inactive urban population (without the categories of inactive persons unrepresented in the sample of homeless people: migrants, students, and caregivers of children or adults). The share of inactive users is lower than the aggregate proportion of aged (33%), people with disabilities (16.8%), and chronically ill (7.6%) in the sample.

The form filled out upon admission in the Shelter and updated during the stay in the institution provides information on the Shelter users' current labour status and their employment plans. One of the questions asks the applicants about their plans with two choices of answer. For the first choice of answer, 42.91% said they want to get their identity documents and 16.6% wanted to find a job (see Table 3).

Table 3 Distribution of answers in the Shelter admission form to the Shelter to the question 'What are your plans?'

		Frequency	*Per cent*	*Valid Per cent*
Valid	Completing papers	336	41.5	42.9
	Getting a job	130	16.0	16.6
	Placed in a home for the elderly	49	6.0	6.3
	Multiple aims	83	10.2	10.6
	Family reintegration	113	14.0	14.4
	Social integration	72	8.9	9.2
	Total	783	96.7	100.0
Missing	Data	27	3.3	
Total		810	100.0	

For the second answer (out of 86.8% of those who gave a second answer, see Table 4), 45.23% expressed their wish to find a job. Results of the two answer choices suggest that the priority plans of the Shelter users were, first, to get identity documents (for 50% of all respondents who did not have them) and, secondly, finding a job (for working-age individuals—about 66% of the sample). The fact that 16.6% of individuals opted, in the first choice, for 'finding a job' as a plan and 45.23% gave this answer in the second choice ('other plans') suggest that at least 60% of the Shelter users were not working at the time of filling in the admission form or were not satisfied by the work they did and were willing to find a job. The share of persons looking for a job among the Shelter users

Table 4 Distribution of answers in the Shelter admission form to the question 'What other plans do you have?'

		Frequency	*Per cent*	*Valid Per cent*
Valid	Get a job	318	39.3	45.2
	Get papers	51	6.3	7.3
	Placed in a home for the elderly	104	12.8	14.8
	Reintegration into family	160	19.8	22.8
	Emigration	70	8.6	10.0
	Total	703	86.8	100.0
Missing	Data	107	13.2	
Total		810	100.0	

Table 5 Distribution of results of the users' stay at the Shelter assessed by the Shelter administration

		Frequency	*Per cent*	*Valid Per cent*
Valid	Got employment	53	6.5	6.6
	Accommodation outside the Shelter	130	16.0	16.1
	Reintegration family	196	24.2	24.2
	Returning Centre	55	6.8	6.8
	Left unannounced	311	38.4	38.4
	Deceased	41	5.1	5.1
	Placed in a home for elderly	5	0.6	0.6
	Remained in the Shelter	18	2.2	2.2
	Total	809	99.9	100.0
Missing	Data	1	0.1	
Total		810	100.0	

exceeds by at least ten times the share of the unemployed in the general population (between 3 and 6% during 2010–2019).

Yet according to the data provided by the Shelter administration, very few users, only 53 persons out of 820 (or 6.6%), managed to find a job during their stay at the Shelter, with or without the help of the Shelter social workers (see Table 5). What are the reasons for such a low 'success rate' of homeless users to find a job? In the following part of this paper, I will try to answer this question drawing on interviews with 90 people (65 homeless people, five people at risk of homelessness and 20 professionals). To get a more balanced picture of the relation of the homeless with the labour market and their work experience in Moldova, I will contextualize this information locally and in a regional context, from an immediate and a historical perspective.

The Lack of Identity Papers and *Propiska*

At least half (51.24%) of the Shelter users included in the quantitative research did not have identity documents. The lack of identity documents is one of the most serious problems for the homeless people in Moldova and other countries of the region. The research reported such a problem about the homeless in Russia (Stephenson 2006, p. 78; Kovalenko 2009, p. 529). Formally, it is impossible to be hired without identity papers, since concluding an employment contract involves the identification of parties through an original and valid identity document. The National Agency for Employment (ANOFM) itself does not provide formal advisory services to applicants with no identity documents, although the employees of the ANOFM sometimes agree to offer such services informally upon request of applicants without identity documents. Furthermore, one of the main requirements of eligibility for obtaining the unemployment status and allowance is holding a valid identity document.

Recovery procedure of documents is extremely troublesome and lengthy. This procedure becomes de-motivating for anyone, much more so for people lacking resources and social skills, as is often the case of homeless people with no identity documents. Even when holding valid identity documents, the homeless persons usually lack residence visa (permanent address) which is printed on the ID card. Employers tend to reject the applications of people with no residence visa by the effect of the inertia of former legal status, especially in the Soviet times when residence visa, the so-called *propiska*, was a prerequisite for benefitting from all civil rights and a way of state control over citizen's status and mobility (Pipko and Pucciarelli 1985). Residence visa is a central element of exclusion of the homeless persons from the labour market in Russia and other post-Soviet countries (Höjdestrand 2009, pp. 5–6, 20–26; Stephenson 2006, pp. 6, 197). Rejection of job applicants without *propiska* would be also, according to a social worker from the Shelter, a tacit measure for employees to protect themselves against potential undisciplined employees.

The unequal access to identity documents and residence visa continues to be a way of control over the mobility and a form of exclusion, exercised by the state and economic agents over the citizens. Citizens who do not have a permanent place of residence on the reverse of their identity document cannot enjoy the same civil rights, social and medical protection, and employment opportunities as the domiciled citizens. In this way, homeless people without valid documents and a residence visa thicken the ranks of the 'labour reserve army' of undocumented workers.

The 'Informal Sector': A Labour Market Accessible for the Vulnerable Workforce

In many cases, informal work satisfies the interests of all stakeholders. Employees find, for better or worse, a job and an income source, while employers employ a workforce for whom they do not pay contributions to the public social security fund and income taxes (see Pfau-Effinger 2017; Chen 2012). Certain homeless persons interviewed told me with some pride that they work in an enterprise and have a fairly good income. The fact that they work in an 'informal' workplace does not worry much either the employees, as long as they receive an income agreed upon beforehand, or the employers. Some informal employees confessed that employers had an understanding attitude to their lack of identity documents and a stable domicile. Moreover, according to one of them, the enterprise owner (an enterprise dealing with plastic processing) provided temporary housing to him, even though not in a place intended for a living.

Informal employment (in the formal or informal sector, with the full range of 'hybrid' variations between the two "sectors" of labour) implies a series of risks and hardships, both immediate and long-term, especially for employees (on the variations of the informal labour in the Eastern Europe and ex-Soviet countries, including Moldova, see Pfau-Effinger 2017). Among long-term risks, inherent in informal employment because of lack of contributions to the social insurance system, one can mention the following: limited access to medical services, inability to receive unemployment benefits if one loses his or her job and, finally, low amount of retirement payment one would benefit after reaching retirement age (Popa et al. 2015). Employment in an informal job and informal employment in a formal enterprise implies, particularly for the homeless, an advanced risk of vulnerability to situations of abuse by employers. Lack of identity papers, social skills, and legal knowledge worsen this vulnerability that many employers take advantage of and abuse in different situations.

Some homeless people interviewed for this research have told me stories of abuses they were subjected to by employers, the little recognition (both in symbolic and economic terms) granted to them for their work, and the lack of social rights and protection. Maxim's testimony is emblematic in this respect:

> The work of a man should be respected. It is not respected here. I was hoeing in a brigade. I worked for 9 months in Tiraspol on a large farm. I was promised a salary of 3500 [Moldovan Lei, i.e., 265USD in January 2014] a month. After 9 months of hard work, I was given only 1000. Go home, I say! I was hoping to be given 9000, at least 1000 per month. They just said: go wherever you want. 1000 and that is it. Here I had another experience. I worked in a kolkhoz [collective farm]. There is one

> in the region. I worked even Saturday and Sunday with no days off and I was given 1200 lei. Life is very hard. (…) Most often gipsies employ you for 1000. They would not give more. All this for the hardest work: as porters, harvesting apples with no days off… You are expecting something in the end… (…) I worked at ponds. At two ponds. It was a seasonal job. I was told: Max, you will work a season and then I'll pay you. I worked for four months and went to take my money. He told me to work a bit longer and he would pay me more. I thought that it was OK … It was New Year already. I went to him and said: Andrei, stop telling me lies. I need money. He said: Max, I need a worker but not one I have to pay. I went to another man. He also made a pond and was growing fish. I worked there for a year and a half. He was the mayor. He had enough fish but would not give me money. I worked on a construction site. They had to pay me 25.000. I was told that I would be paid when we finished the construction. We agreed. However, he did not give me money for the ticket from Briceni to Drochia. I had to ask people for money to be able to come back. Who to complain to? I had to sign a work contract but I do not have identity papers… We reached an agreement. I have done my part of work. A lawyer told me I had to sign a document with him. But why did he hire me then? (Man, 55–60 years old, January 2014, hereafter M55-60-Jan2014)

Andrei, another interviewee, told me frankly that because of lack of trust in employers in the construction area where he used to work, he does not even try to find a job:

> I do not work because everybody deceives me. Everybody tells me lies. Everybody. I start plastering, I start working. They give me an advance, a small one but when I finish the work and they have to pay me: "Hey, you. Take your belongings and beat it!" – and this did not happen for the first time. This time I do not want to work here. I have identity papers, I have everything. I carry everything on me. I have no [insurance] policy. I have an identity card, driving license, everything. I even have a certificate confirming when I came out of prison, for the pension fund to have retirement payment. (M40-45-Aug2015)

As one can see in these two testimonies and others of this kind, the main form of abuse suffered by the homeless employed informally is the remuneration that is much lower than the amounts agreed or even the lack of any payment. As appears from these interviews, employers do not hesitate to make use of their status of power to abuse and exploit powerless workers. This balance of power is even more unequal to the detriment of the employee as the latter feels helpless and deprived of any protection by the state while its justice system is perceived as corrupt and unfair. Abuses in the workplace of homelessness in the informal sector in Russia and Ukraine are also common practice (see Ryabchuk 2014; Höjdestrand 2009).

Some professionals who deal with the homeless said that employers also feel unsafe with employees recruited from the homeless, perceived as untrustworthy, particularly the homeless with no identity documents that cannot be controlled in any way. An official from the Shelter administration told me that several times he had helped some Shelter users to be hired on a construction site. They agreed with the employer but the next day some of them did not come to work on time. On the other hand, according to a reportage broadcast on a local TV channel and also some testimonies, some Shelter users were employed informally at a furniture factory belonging to the Shelter's Director. Therefore, the town hall decided to fire him. One way or another, the Shelter used to be a pool of cheap labour for informal jobs in constructions and the agricultural market.

Due to the lack of formal regulation and control, both sides—employers and employees—are watching each other with suspicion (see also Ryabchuk 2014, p. 17; Höjdestrand 2009, p. 60). To protect themselves and ensure the correctness of the other, according to the interviews with homeless persons and experts, employers and employees prefer to establish employment relations based on verbal short-term agreements, usually on daily work called *kalym*, the payment is made at the end of the working day, depending on the previous agreement and the work is done (about 'kalym' arrangements, see Morris 2015). Areas in which the homeless people from Chisinau can find work as day labourers are quite physically demanding and hazardous as a rule: porters at the market, unskilled construction workers, or seasonal agricultural workers. If the relationship of trust between the employer and employee is established, the work agreement—verbal or written—can be extended and even stabilized. A controversial bill 'on activities carried out by unskilled occasional day labourers', issued by the Ministry of Labour, Social Protection and Family and approved by the Government, provides for legalization of a part of the informal labour sector. Yet some experts estimate that this project further affected the workers' rights and would prejudice the tax system (Cozonac 2016).

The Economy of the Homeless and Refuse Economics

When, for some reasons (e.g. disability), homeless persons are not able to work even informally as day labourers or seasonal workers or when they are constantly refused by employers, for instance, because of their appearance (as visible homeless people), they withdraw in an 'economy of the homeless' (Gaetz and O'Grady 2002; Lei 2013) by engaging in various subsistence practices which Höjdestrand analyses within the concept of 'refuse economics' (Höjdestrand 2009, p. 47). These informal economic practices include activities that ordinary citizens refuse to do because they consider them too 'dirty' literally and symbolically. Waste collection and recycling is one of these jobs and subsistence strategies to which usually resort the neediest homeless, deprived of any source of income.

Collection and exchange of bottles, one of the basic occupations of the most deprived homeless people, requires persistent and laborious involvement in exchange for a minute 'remuneration' and hardly acceptable even by the poorest citizens. At private bottle collection points, one bottle is bought for 5 bani, the most widespread, to 50 bani apiece. Therefore, for an amount equivalent to one Euro a person must collect no fewer than 200 bottles of 10 bani apiece (or 40 bottles of 50 bani apiece). Other recyclable materials—plastic or cardboard—are even less rewarding, without a functional system of waste recycling.

During a discussion I had with several homeless persons in a squat at the edge of a park in Chisinau I was told by them repeatedly that they work to make money for buying some bread and are not just staying without doing anything:

- *We do work.*
- *Yes, we work, collect scrap metal, cardboard, give a hand to someone. Maybe somebody needs help with some work. We are paid about 100 lei.*
- *Recently we removed a huge pile of trash for 50 lei. We do not steal.*
- *If we have nothing to eat, we go to the garbage bin and find something. Sometimes people give us some food. They know us and when they see us, they give us something every time.*
- *I'm a cripple, but I go and work.*
- *Yesterday I went and took some garbage from* [a furniture factory]. *A woman let us in and we took out the trash. She gave us 50 lei. We went*

away and had a meal. And that's it. It's difficult. There are problems, annoyances ... But we do not give up. We support each other because we are like a family here and we are in the same boat (group interview, August 2015).

This statement seeking to provide self-justification ('We work, we don't steal') can be heard in various forms also in the case of other subsistence practices, otherwise considered blameworthy, such as begging: 'What could I do? Do I have to steal? We have to survive somehow'. Denial of unemployment in these testimonies is in agreement with the 'work ethic', built up in Soviet times through cultural and legal means, and strengthened during the post-Soviet times through various forms of coercion on those who evade 'production process' (Negura 2007; Heintz 2005). In this type of statement, the homeless person, placed in the deontological space delimited by two normative constraints, a prescriptive one, that of the need to work, and a prohibitive one, that of not to steal, actually is forced to choose a middle way in the situation where he or she is denied the right to work formally. From this moral perspective, any occupation that does not directly violate the legal code, although it might not comply with certain social norms, becomes acceptable from the outset and is positively valued within the group as an occupation among many others.

The subsistence practices, such as the collection of bottles, although legal, are at the limit of the illicitness from the admitted social norms (see the taxonomy of informal activities between (il)legal and (il)licit in the introduction to this book by Polese, adapted from van Schendel and Abraham 2005). In this context, the insistence of these precarious workers to justify their work as a legitimate occupation can be seen as the expression of a symbolic struggle, with minimal chances to win, to gain a minimum of symbolic recognition for a precarious informal work considered illegitimate, and though legal (see Honneth 1995, 2010). At the same time, some forms of precarious informal work can create and consolidate communities that offer 'countercultures of respect' in compensation for the lack of other forms of social recognition (Sennett and Cobb 1972; Honneth 1995, p. 124; see also Karabanow et al. 2011).

Compared to long-term informal work, in construction, in the market trade or in fishing, in which several interviewees were employed, the subsistence activities in which vulnerable people are (self) employed, such as collecting materials for recycling, involve a higher autonomy status and lower risks of abuse by third parties. In this respect, it can be said that informal work activities in which homeless people and other vulnerable people are employed can be defined according to the relationship with the employer, the degree of autonomy of the employee and the control over the activity performed (Portes and Schauffler 1993; Biles 2009; see also Honneth 2010). Any work performed in an informal relationship with an employer usually involves, in the case of homeless employees, risks of abuse, exploitation, and oppression. This risk decreases sharply in the case of self-employment of homeless people in precarious informal work with a lower income, but also with lower risks (see also Balkin 1992). It is therefore not surprising that many homeless people take refuge in such precarious economic activity, instead of informally engaging in enterprises for the promise of higher pay.

The normative constraint to work emerges as a leitmotiv in the discourse of people who are marginalized in the labour market. This ethical constraint is perceived, apparently, more intensely by the men of working age since society has the highest expectations about them in this regard. These expectations are associated the most with the role of breadwinner in a family that is traditionally considered to be the responsibility of men (see Liu et al. 2009; Thébaud 2010; Riabchuk 2012). Hence the insistence of people belonging to this category in interviews on the activity of 'work' they carry out and the qualifier of 'worker' they apply to themselves, rejecting any criminal activity ('we do not steal') and eventually the label of 'drunkard' ('We are not drunkards, we are workers!' [Group, Aug2015]). Even when some individuals we spoke to were noticeably affected by some chronic condition or disability, they insisted on displaying involvement in an activity described as 'work' or at least they expressed their wish to work despite all obvious impediments.

> I am a cripple but I work. (Vanya, M40-46-Aug2015)
> Give me some work. Give me work because I come from the countryside and I know how to work. I have taken the cane to be able to walk. Still, I can work. I need to work too. It does not matter a lot or a little. I would like to have an occupation at least. (Dan, M40-46-Mar2014)

We do not know exactly where the (self) legitimizing discourse ends and where the biographical reality begins. This discourse shows at least the reality of a strong moral and ideological pressure to work, internalized and expressed as such, to the detriment of illegal occupations and incomes (such as theft) and those considered deviant (such as alcoholism).

During a conversation that was not recorded, the former director of the Shelter expressed his viewpoint that is shared by other professionals in this area (social workers, civil servants, current and former police officers) to a large extent. According to him, citizens must have rights as well as obligations. In a democracy, work is considered a right. It should, however, be an obligation, so that those who refuse to work, the 'tramp' in the first place, be forced to do so by legal coercion. According to his viewpoint, the social protection system developed for vulnerable people encourages laziness. This position, in harmony with the neoliberal workfare policies which forces the social benefits recipients to work in generally unskilled jobs, at meagre wages and with low social status, is strikingly similar to the policies applied by the USSR which obliged its citizens to work, while 'voluntary and malicious evasion' from work was punishable by up to three years in prison especially after the anti-Parasite Law of 1961 (on American 'workfare reform', see Wacquant 2009, pp. 103–109; on the Soviet anti-parasite law and its social impact, see Fitzpatrick 2013; Lastovka 2011; Stephenson 2006, pp. 71–112).

A Fragmented Professional Career

The life stories of many persons I discussed with are evidence of a fragmented professional trajectory in the aftermath of crises experienced by Moldovan economy after 1990 (manifested, inter alia, by closing down former Soviet enterprises, massive layoffs of employees, and failure to pay regular salaries in public institutions). Precariousness and loss of work security were experienced by many people who in the late 1990s fell below absolute and extreme poverty thresholds (Negură 2019, p. 555). A significant part of the homeless in Moldova emerged from this mass of impoverished people, losers of post-socialist transition. Most impoverished people during transition years were not necessarily unemployed, quite many of them were employed and held identity documents. A notable difference that stands out between the domiciled poor and the homeless is that the former usually have a home while the latter have lost even this basic resource. The discontinuous professional career further weakened the social and professional status of many of these people looking for a source of subsistence and doing occasional, unsafe, and unskilled work. The abilities to adapt to a market economy, which came unexpectedly for the unprepared population, were even lower in the case of people who, in the 1990s, were turning 50 years old: about ¼ of the population and who represented 33.5% of the users at the time of admission to the Shelter between 2004 and 2017. A personal or family issue, a chronic disease, or an injury precipitated the process of sliding down to the lower layers of poverty.

Also, 23% of our sample went through imprisonment experience that brutally interrupted or altered their socio-professional path. The decrease in professional skills during detention is aggravated in their case by the stigma of former offenders. The share of former detainees among the Shelter users (and its increase from 21% in 2015 to 23% in 2017) is in line with the repressive approach exercised by the state to the vulnerable people over the last decades. Thus, the detainees' population in Moldova continued to grow with 27% from 2010 to 2017, although the number of persons committing crimes decreased over the same period by 24%. Also, the proportion of the unemployed among detainees increased from 23% in 2002 to 60% in 2018 (Negură 2018).

In the situation where all sorts of precarious work are not sufficient for improving the economic condition and increase the worker's social status, many people that found themselves in this situation are tempted to make use of smaller effort of various coping strategies. Alcohol use and abuse is such a strategy. Almost half of the people I discussed with were suffering from alcoholism problems of different levels. This finding is consistent with another research (Doltu et al. 2016, p. 32). Some studies on homeless people show that in 50% of cases these persons suffered from alcoholism before becoming homeless. In other 50% of cases, alcohol addiction developed after they became chronic homeless (Johnson and Chamberlain 2008). Alcohol addiction becomes an obstacle to employment and social integration and perpetuates the vicious circle of precariousness and exclusion.

The situation is not entirely hopeless for people who find sufficient resources, enough motivation and strong support to overcome this existential deadlock while employees of public institutions and social welfare associations could provide some assistance in this regard. However, most of the interviewees, although endowed with some resources, confess their inability to overcome, through constant employment efforts, the condition of precariousness and exclusion.

Discussion of Findings and Some Conclusions

Despite their image of 'lazy' and 'not working' people, 'assisted' by the state social protection system, most homeless in Moldova do work or want to work while only a small part of them do not work at all. On the contrary, the state does little to facilitate the access of these people to formal employment and does not protect the rights of those employed informally. Informal jobs are the only way for most of them to get hired. The professional profile of the homeless in conjunction with the education level, compared to the profile of the general urban population, suggest that most of the homeless people in Moldova experienced a drastic decrease in their social and professional status. In the aftermath of the massive pauperization of the 1990s and the context of weakening of the social protection system and labour rights, many people went through an abrupt process of downward social mobility. A part of them ended up in residential precariousness doubled by fragmentation of their professional career and persistent unemployment.

A high number of Shelter users were looking for a job when filling out the admission form (at least 60%), either because they were not working or because they were not satisfied with the work they were doing at the time. One of the most significant reasons why at least half of the homeless did not work at the time of the research is the lack of identity documents. Employment in informal jobs offers most homeless people the most realistic job opportunities.

Many of those employed in informal jobs were subject to various forms of abuse, mainly refusal to pay or lower payment compared to the amount agreed. Being informally employed in physically demanding occupations (as porters or in constructions), these persons are also exposed to greater risks of work accidents since they have no medical insurance policy. Because of their vulnerability in the labour market, many homeless persons withdraw in the 'economy of the homeless', engaging in occupations that most ordinary citizens reject. This precarious and low-paid work perpetuates the precarious and vulnerable status of the homeless.

The findings of the quantitative research and qualitative analysis of interviews with homeless people in Chisinau, contextualized with data on the labour market developments in the decades after 1990 and corroborated with other studies, suggest that employment in a formal or informal job is insufficient for most homeless people to rent a home and integrate into the mainstream society. Due to the inadequacy of economic, social, and cultural capitals, these people have only access to the 'secondary sector' of the labour market, *i.e.* to precarious and poorly paid jobs. The jobs in the 'secondary sector' only perpetuate their vulnerability (see also Gaetz and O'Grady 2002; Hong and Wernet 2007). Informal work, with the inherent risks of labour insecurity, becomes the only space for professional activity accessible to homeless people, given the systemic exclusion from employment and the economic and ethical pressure to work. Homeless men feel all the more acutely the pressure to work. Employment in low-paid jobs and in the absence of any form of labour protection provides them with a minimal source of survival and social recognition, but limits their chances of social recovery and economic stabilization. Therefore, most precarious work the homeless people are performing to survive is *social reproductive labour* preventing these people to upgrade their social position and overcome their marginality (Bhattacharya 2017).

The precariousness of employees in informal jobs, and the most vulnerable among them—the homeless, was aggravated in the context of a 'deregulated' labour market during the 1990s and the disintegration of the former system of labour and social protection while legal protection mechanisms for employees and the unemployed have weakened to extinction. Since the state and civil society have suspended any form of control over the observance of employees' rights, especially in the informal labour sector, the working poor is thickening the ranks of the 'reserve army' of the new market economy.

No workfare policy will contribute to overcoming the precarious situation if it does not address structural labour market conditions, namely if it does not establish a sufficient minimum income and ensure equitable access to public health services and a decent old-age pension. For the homeless, one of the most determining factors to get employment is the possession of a home. A social service that would help the homeless to find a precarious job would not be enough to overcome the condition of homelessness and deprivation. In the case of Moldova (and other countries in the region), a coherent support program for homeless people should start with the implementation of a simplified procedure for recovering identity documents. Besides, the printing of the residence visa on the identity card, which leads to employment discrimination, should be waived. State institutions, trade unions, and non-governmental organizations should create an effective legal framework and mechanisms for implementing it to prevent the risks of abuse and exploitation of employees in the informal labour sector. In the absence of such a coherent and comprehensive program, homeless people will continue to be excluded and discriminated against, and their numbers will continue to grow.

References

Balkin, S. (1992) 'Entrepreneurial Activities of Homeless Men', *The Journal of Sociology & Social Welfare*, 19(4), pp. 129–150.

Bazeley, P. and Jackson, K. (2013) *Qualitative Analysis with NVivo*. Los Angeles, CA: Sage.

Bhattacharya, T. (2017) 'How Not to Skip Class: Social Reproduction of Labor and the Global Working Class', in *Social Reproduction Theory: Remapping Class, Recentering Oppression*. London: Pluto Press, pp. 68–93.

Biles, J. J. (2009) 'Informal Work in Latin America: Competing Perspectives and Recent Debates', *Geography Compass*, 3(1), pp. 214–236. https://doi.org/10.1111/j.1749-8198.2008.00188.x.

Burrawoy, M. and Verdery, K., eds. (1999) *Uncertain Transition: Ethnography's of Change in the Post-Socialist World*. Oxford: Rowman & Littlefield.

Cerami, A. (2009) 'Mechanisms of Institutional Change in Central and Eastern European Welfare State Restructuring', in Cerami, A. and Vanhuysse, P. (eds) *Post-Communist Welfare Pathways: Theorizing Social Policy Transformations in Central and Eastern Europe*. Basingstoke: Palgrave Macmillan, pp. 35–52.

Chen, M. A. (2012) *The Informal Economy: Definitions, Theories and Policies*. Cambridge, MA: Women in Informal Employment Globalizing and Organizing.

Corbin, J. and Strauss, A. (1990) 'Grounded Theory Research: Procedures, Canons and Evaluative Criteria', *Qualitative Sociology*, 13(1), pp. 3–21.

Cozonac, C. (2016) 'Munca cu ziua – sclavie modernă. Legi şi realităţi [Day Labour—Modern Slavery: Laws and Realities]', *Platzforma. Revistă de critică socială*, 21.10.2016. Available at: https://www.platzforma.md/arhive/35789 (accessed: 07.09.2020).

Doltu, S., et al. (2016) *Studiu operaţional privind prevalenţa şi profilul comorbidităţilor la persoanele fără adăpost, cu tuberculoză activă şi fără tuberculoză, în municipiul Chişinău (Republica Moldova)*. Chişinău.

Fitzpatrick, S. (2013) 'Social Parasites: How Tramps, Idle Youth, and Busy Entrepreneurs Impeded the Soviet March to Communism', *Cahiers du Monde Russe*, 47(1), pp. 377–408. Available at: http://www.jstor.org/stable/20175002.

Gaetz, S. and O'Grady, B. (2002) 'Making Money: Exploring the Economy of Young Homeless Workers', *Work, Employment and Society*, 16(3), pp. 433–456. https://doi.org/10.1177/095001702762217425.

Gwadz, M. V., et al. (2009) 'The Initiation of Homeless Youth into the Street Economy', *Journal of Adolescence*, 32(2), pp. 357–377. https://doi.org/10.1016/j.adolescence.2008.01.004.

Heintz, M. (2005) *Etica muncii la românii de azi* [Work Ethic of Today's Romanians]. Bucharest: Curtea Veche.

Höjdestrand, T. (2009) *Needed by Nobody: Homelessness and Humanness in Post-Socialist Russia*. Ithaca and London: Cornell University Press.

Hong, P. Y. P. and Wernet, S. P. (2007) 'Structural Reinterpretation of Poverty by Examining Working Poverty: Implications for Community and Policy Practice', *Families in Society*, 88(3), pp. 361–373. https://doi.org/10.1606/1044-3894.3645.

Honneth, A. (1995). *The Struggle for Recognition: The Moral Grammar of Social Conflicts*. The MIT Press.

Honneth, A. (2010) 'Work and Recognition: A Redefinition', in Schmidt am Busch, H.-C. and Zurn, C. F. (eds) *The Philosophy of Recognition: Historical and Contemporary Perspectives*. Lanham, Boulder, and New York: Lexington Books, pp. 223–239.

Johnson, G. and Chamberlain, Ch. (2008) 'Homelessness and Alcohol Abuse: Which Comes First?', *Australian Social Work*, 61(4), pp. 342–356. https://doi.org/10.1080/03124070802428191.

Karabanow, J., Hughes, J., Ticknor, J., Kidd, S. and Patterson, D. (2011) 'The Economics of Being Young and Poor: How Homeless Youth Survive in Neo-Liberal Times', *Journal of Sociology and Social Welfare*, 37(4), pp. 39–63.

Kovalenko, E. A. (2009) 'Voronka bezdomnosti i ogranichenie potentsyal'nykh vozmozhnostei', *Zhurnal issledovanii sotsial'noi politiki*, 8(4), pp. 519–536.

Kus, B. (2006) 'Neoliberalism, Institutional Change and the Welfare State: The Case of Britain and France', *International Journal of Comparative Sociology*, 47(6), pp. 488–525. https://doi.org/10.1177/0020715206070268.

Lastovka, T. (2011). Tunejadstvo v SSSR (1961–1991): iuridicheskaia teoriia i sotsial 'naia praktika', *Antropologicheskij Forum*, 14, pp. 212–229.

Ledeneva, A. (2013) *Can Russia Modernise?: Sistema, Power Networks and Informal Governance*. Cambridge: Cambridge University Press. https://doi.org/10.1017/CBO9780511978494.

Lei, L. (2013) 'Employment, Day Labor, and Shadow Work Among Homeless Assistance Clients in the United States', *Journal of Poverty*, 17(3), pp. 253–272. https://doi.org/10.1080/10875549.2013.804479.

Liu, W. M., Stinson, R., Hernandez, J., Shepard, S., and Haag, S. (2009) 'A Qualitative Examination of Masculinity, Homelessness, and Social Class among Men in a Transitional Shelter', *Psychology of Men and Masculinity*, 10(2), pp. 131–148. https://doi.org/10.1037/a0014999.

Morris, J. (2015) 'Kalym (Russia)', in Ledeneva, A. V. (ed.) *The Global Encyclopaedia of Informality: Understanding Social and Cultural Complexity* (Volume 1). London: UCL Press, pp. 157–160.

Morris, J. and Polese, A., eds. (2015) *Informal Economies in Post-Socialist Spaces: Practices, Institutions and Networks*. Houndmills: Palgrave.

Negură, L. (2007) *Le travail après le communisme. L'émergence d'une nouvelle représentation sociale dans l'espace postsoviétique*. Québec: Presses de l'Université Laval.

Negură, P. (2016) 'Republic of Moldova's Transition: From a Failed Communism to an Un-Commenced Capitalism', *Studia Politica. Romanian Political Science Review*, 26(4), pp. 541–568.

Negură, P. (2018) 'Penitenciarele din Republica Moldova: un sistem de penalizare a săracilor (I)', *Revista de Filosofie, Sociologie și, Științe Politice*, 3(178), pp. 155–66.

Negură, P. (2019) 'The State Policy Towards the Homeless in Moldova Between the "Left Hand" and the "Right Hand": The Case of Chisinau Shelter', *Südosteuropa. Journal of Politics and Society*, 67(2), pp. 175–195.

Pfau-Effinger, B. (2017) 'Informal Employment in the Poor European Periphery', *International Journal of Sociology and Social Policy*, 37(7–8), pp. 387–399. https://doi.org/10.1108/IJSSP-07-2016-0080.

Pipko, S. and Pucciarelli, A. (1985) 'The Soviet Internal Passport System', *International Lawyer*, 19(3), pp. 915–919.

Polese, A., Kovács, B. and Jancsics, D. (2017) 'Informality "In Spite Of" or "Beyond" the State: Some Evidence from Hungary and Romania', *European Societies*, 20(2), 207–235. https://doi.org/10.1080/14616696.2017.1354390.

Polese, A., Morris, J., Kovacs, B. and Harboe, I. (2014) 'Welfare States' and Social Policies in Eastern Europe and the Former USSR: Where Informality Fits In?', *Journal of Contemporary European Studies*, 22(2), pp. 184–198.

Popa, A., Lupusor, A., Rosca, I., Hirose, K. and Hettes, M. (2015) *Economia informală în Republica Moldova: O analiză cuprinzătoare*. Chişinău.

Portes, A. and Schauffler, R. (1993) 'Competing Perspectives on the Latin American Informal Sector', *Population and Development Review*, 19(1), pp. 33–60.

Riabchuk, A. (2012) 'Homeless Men and the Crisis of Masculinity in Contemporary Ukraine', in Hankivsky, O. and Salnykova, A. (eds) *Gender, Politics, and Society in Ukraine*. Totonto: University of Toronto Press, pp. 204–221.

Ryabchuk, A. (2007) 'Work "in the Shadow": Experiences of Homelessness Among Casual Workers in Construction Industry', *Naukovi Zapysky. Sociodobichni Nauky*, 70, pp. 72–76.

Ryabchuk, A. (2014) 'Homelessness in Ukraine: Structural Causes and Moral Evaluation', *Debatte*, 22(3), pp. 288–306. https://doi.org/10.1080/0965156X.2014.988964.

Sennett, R. and Cobb, J. (1972) *The Hidden Injuries of Class*. New York: Knopf.

Shier, M. L., Jones, M. E. and Graham, J. R. (2012) 'Employment Difficulties Experienced by Employed Homeless People: Labor Market Factors That Contribute to and Maintain Homelessness', *Journal of Poverty*, 16(1), pp. 27–47. https://doi.org/10.1080/10875549.2012.640522.

Statistical Yearbook of the Republic of Moldova. (2009). Chişinău: Statistica Moldovei.

Statistical Yearbook of the Republic of Moldova. (2019). Chişinău: Statistica Moldovei.

Stephenson, S. (2006) *Crossing the Line: Vagrancy, Homelessness and Social Displacement in Russia*. London: Ashgate.

Thébaud, S. (2010) 'Masculinity, Bargaining, and Breadwinning: Understanding Men's Housework in the Cultural Context of Paid Work', *Gender and Society*, 24(3), pp. 330–354. https://doi.org/10.1177/0891243210369105.

van Schendel, W. and Abraham, I. (2005) 'Introduction', in van Schendel, W. and Abraham, I. (eds) *Illicit Flows and Criminal Things: States, Borders, and the Other Side of Globalization*. Bloomington, IN: Indiana University Press, pp. 1–37.

Wacquant, L. (2009) *Punishing the Poor: The Neoliberal Government of Social Insecurity*. Durham and London: Duke University Press.

Walker, I. and Pettigrew, Th. F. (1984) 'Relative Deprivation Theory: An Overview and Conceptual Critique', *British Journal of Social Psychology*, 23(4), pp. 301–310. https://doi.org/10.1111/j.2044-8309.1984.tb00645.x.

Index

A. Polese (ed.), *Informality, Labour Mobility and Precariousness*, International Political Economy Series,
https://doi.org/10.1007/978-3-030-82499-0

The manufacturer's authorised representative in the EU is Springer
Nature Customer Service Centre GmbH, Europaplatz 3, 69115 Heidelberg,
Germany. If you have any concerns regarding our products, please
contact ProductSafety@springernature.com

Printed and bound by CPI Group (UK) Ltd, Croydon, CR0 4YY
30/09/2026
02225456-0012